SCIENCE UNDER FIRE

SCIENCE UNDER FIRE

*Challenges to Scientific Authority
in Modern America*

ANDREW JEWETT

Harvard University Press

*Cambridge, Massachusetts
London, England*
2020

First printing

Publication of this book has been supported through the generous provisions of the
Maurice and Lula Bradley Smith Memorial Fund.

Library of Congress Cataloging-in-Publication Data
Names: Jewett, Andrew, 1970– author.
Title: Science under fire : challenges to scientific authority in modern
America / Andrew Jewett.
Description: First. | Cambridge, Massachusetts : Harvard University Press, 2020. |
Includes bibliographical references and index.
Identifiers: LCCN 2019044010 | ISBN 9780674987913 (cloth)
Subjects: LCSH: Science—Social aspects—United States—History—20th century. |
Science—Social aspects—United States—History—21st century. | Science—
Political aspects—United States—History—20th century. | Science—Political aspects—
United States—History—21st century. | Science—Moral and ethical aspects—
United States—History—20th century. | Science—Moral and ethical aspects—
United States—History—21st century. | Political culture—United States. |
Truthfulness and falsehood—Political aspects—United States.
Classification: LCC Q175.52.U5 J49 2020 | DDC 303.48/30973—dc23
LC record available at https://lccn.loc.gov/2019044010

CONTENTS

SCIENCE UNDER FIRE

INTRODUCTION

Science as a Cultural Threat

TODAY, SCIENCE IS UNDER FIRE as never before in the United States. Even amid the COVID-19 pandemic, Donald Trump and his Republican allies dismiss the findings of health experts as casually as they do those of climate scientists. Most top Republicans also challenge Darwinism, and many of their followers reject the Big Bang theory as well. Indeed, conservatives sometimes portray scientists as agents of a liberal conspiracy against American institutions and values. Since the 1990s, GOP leaders have worked to limit the influence of scientists in areas ranging from global warming to contraception to high school biology curricula. By 2016, Republican skepticism toward the scientific establishment had grown so strong that the Democratic presidential candidate Hillary Clinton drew wild applause for simply stating, "I believe in science." Since then, Trump's administration has challenged climate researchers, epidemiologists, and other experts more forcefully than ever.

But it is not just conservatives who question scientific authority in the United States. Alarm at many applications of biological research, for example, crosses party lines. This impulse usually targets genetic engineering and biotechnology, but it also fosters skepticism toward vaccination and other medical practices. Across the political spectrum, in fact, citizens tend to pick and choose among scientific theories and applications based on preexisting commitments. They are frequently suspicious of basic research procedures as well. Many believe that peer review and other internal policing mechanisms fail to remove powerful biases. Conservatives often charge that peer review enforces liberal groupthink, while some progressives say it leaves conventional social norms unexamined.

Even as individuals, scientists face growing skepticism. Fifty-five percent of Republicans and 36 percent of Democrats consider scientists just as biased as nonscientists. Fewer than 20 percent believe that scientists consistently report conflicts of interest arising from sources of funding. Concern about scientific misconduct is widespread, and most Americans doubt that the perpetrators face serious repercussions. Significant numbers trust the experts who apply knowledge more than those who produce it. And such suspicions are especially strong among Black and Latinx Americans—largely Democratic constituencies—as well as Republicans. Viewing these patterns, many scientists fear that they now live in a "post-truth" world where much of the citizenry has turned against them. The March for Science movement launched in 2017 represents an unprecedented mobilization of rank-and-file researchers against perceived cultural and political threats to the scientific enterprise as a whole.[1]

As the 2020s dawn, it is crucial to understand the sources and contours of this skepticism toward science and scientists. Today, new data, theories, and devices reshape our lives daily. We stand on the brink of revolutions in fields from biotechnology to robotics to computing, even as global warming accelerates. As a result, arguments over science underlie some of our most divisive and consequential policy debates. Whether the issue is climate change, fracking, abortion, genetically modified foods, or even immigration, contemporary political battles generate disputes over the legitimacy of scientific theories, methodologies, institutions, concepts, and even facts. In this context, scholars, citizens, and policymakers must think carefully about science and its cultural and political ramifications. The prevailing views on these matters will significantly determine our future—and perhaps even our survival as a species.

To understand why science is so widely distrusted in the United States, one must obviously start with the political influence of theologically conservative Christians in recent decades. Since Ronald Reagan's election in 1980, a fraught but durable coalition of free-market advocates and Christian conservatives has anchored the Republican Party. The Christian Right has targeted myriad scientific theories and innovations as part of its "culture war" against modern liberalism. Today, its power is such that Republican leaders routinely speak out against "secularism," in such varied guises as abortion rights, strict church-state separation, and Darwinism in the schools. Theological conservatives also tend to reject climate science, viewing environmentalism as a dangerous, socialistic religion.

Yet the rise of the Christian Right cannot fully explain phenomena such as the breadth of antivaccination sentiment and concerns about genetic engineering. A second narrative, common among working scientists and scholarly interpreters, holds that a broad-gauged revolt against science took place in the wake of the 1960s. That period brought not only the conservative backlash but also a host of countercultural impulses, including New Age spirituality and belief in UFOs, astrology, and the paranormal. The era's political movements also fueled opposition, as a new generation of critics identified science as an ideological tool of the establishment. Plummeting levels of trust in institutions, especially after Watergate, implicated science as well. At the level of research funding, meanwhile, the 1970s brought tighter budgets, new layers of bureaucratic procedure, and intense pressure to generate immediate, practical outcomes.

Thus, we have a number of ready explanations for science's contemporary travails. Since the 1970s, Christian conservatives have elevated the Bible over science while various left-leaning groups have harbored a countercultural distaste for rationality, a populist distrust of experts, or both. But as this book shows, there is much more to the story than these familiar impulses. Skepticism toward science was hardly new in the 1970s, despite its changed forms and heightened impact on research funding. Going back to the 1920s, in fact, prominent groups of Americans have also challenged scientific authority in a different way, decrying its moral implications and ascribing to it a host of negative social effects. Ever since World War I, many critics who accepted Darwinian evolution have nevertheless identified science as a dangerous cultural presence that causes profound moral harm. They have argued that science advances a faulty view of human persons and human relations, injecting a pernicious social philosophy into the cultural bloodstream. To fully explain today's distrust of science, we must account for the long-standing fear that it authorizes false and damaging understandings of who we are and how we behave. Often this response has focused on broad philosophical frameworks associated with science, but the methods and findings of the social sciences have also drawn considerable criticism, as have extrapolations from biology to human behavior.[2]

By now, the charge that scientists' pernicious conceptions of human behavior have warped our collective self-understanding has reverberated through American public culture for a century. Flip through old magazines, or eavesdrop on past public debates, and you will notice this assertion again and again—seldom the dominant note, but always vigorously sounded. Metaphors of pollution and contagion abound: Science seeps, creeps, infects. Indeed, this charge of cultural and moral corruption still resonates today, even as its contours have shifted over time. For example, the campaign against Darwinism—which began, not coincidentally, in the 1920s—has always reflected deep fears about changing conceptions of human behavior as well as a commitment to the literal truth of Genesis. "Why does the public care so passionately about a theory of biology?" asks the evangelical writer Nancy Pearcey. "Because people sense intuitively that there's much more at stake than a scientific theory. They know that when naturalistic evolution is taught in the science classroom, then a naturalistic view of ethics will be taught down the hallway in the history classroom, the sociology classroom, the family life classroom, and in all areas of the curriculum." In short, science authorizes a misguided, dangerous view of humanity. It delivers material progress but also sows moral degradation.[3]

This belief has hardly been confined to theological conservatives. Since the 1920s, many other critics have argued that science poisons the wells of culture, although these groups have typically traced the offending moral framework to the social sciences or naturalistic philosophies associated with science rather than Darwin's theory. This style of argumentation spread especially widely after World War II, reorienting public images of science as it did. In the 1950s and early 1960s, a remarkably broad array of mainline Protestants, humanities scholars, conservative political commentators, and even establishment liberals joined theological conservatives in arguing that science represented a moral, and even existential, threat to civilization. They often employed classic tropes from nineteenth-century romanticism, contrasting the vital force of living, organic, subjective beings with the dead hand of cold, rational, reductive machines. Many argued that scientists had concocted a speculative and damaging view of human behavior by illegitimately extending the reductive, mechanistic, materialistic approach of science to the study of human beings. Scholarly critics dubbed this philosophical error "scientism" in the 1940s, and the term came into wide usage by the late 1950s.

It is no coincidence that such arguments proliferated just as science's influence reached new heights. The postwar period, which we now remember as the "golden age" of American science, brought a society-wide reckoning with the place of science in modern culture. Critics of varied political and religious persuasions argued that even the horrors of atomic warfare paled in comparison to science's capacity to unravel the social fabric itself. Science, they contended, replaced the familiar view of human beings as moral actors with a new conception that ignored their capacity for moral choice and reduced them to the status of animals or machines. Such arguments helped to pave the way for that upheavals of the late 1960s and 1970s, even as the radical theorists of that era altered critiques of science's cultural effects to fit their own purposes. A tendency to trace social ills to the cultural sway of an ideologically infected science carried through that transformative period and up to our own day, even as the details of the indictment have changed. Across the decades, critics of numerous stripes have argued that scientists—knowingly or not—had corrupted modern culture by entrenching faulty conceptions of human behavior.

SCIENCE AS A CULTURAL CATEGORY

Claims about science's deleterious cultural impact have resonated broadly in American public life over the past century. In this tradition of argumentation, the term "science" holds distinctive and often conflicting meanings. When philosophers and working researchers define science, they typically invoke specific methods and subjects of inquiry: Science is the empirical study of natural phenomena, eschewing supernatural explanations. It deals with phenomena that are concrete, observable, manipulable, and ideally quantifiable. It involves experimentation under carefully controlled circumstances. It requires peer review. But the tradition of criticism described in this book reflects the widespread tendency to define science in terms more relevant to nonscientists: What is science to *us*, as ordinary citizens and individuals? How does it shape our institutions, our workplaces, our families? What are its promises and dangers? What can we expect from it in the future? This book explores conceptions of science as a social and cultural force, rather than a method or a subject.

Images of science as a social and cultural force tend to bring a range of real-world phenomena under its aegis. Listen to the Columbia University historian Jacques Barzun, writing in 1964. Since World War II, he noted, "the articulate have unceasingly cried out against the tyranny of scientific thought, the oppression of machinery, the hegemony of things, the dehumanization brought about by the sway of number and quantity." Indeed, Barzun continued, Americans, despite being the most modern of peoples, had cheered the loudest for "the wounded poets, the philosophers of myth, the devotees of agrarianism, the novelists of the life of instinct, and the painters and dramatists who scorn civilization—to say nothing of the day-to-day critics of mass culture." Science's social impact was on everyone's minds in postwar America, Barzun noted, and the judgment was often harsh.[4]

Note how easily Barzun slid from "the tyranny of scientific thought" to rule by machinery and things, as well as the oppressive character of mass culture and even civilization. On a strict definition, hardly any of these would qualify as "scientific." But Barzun and his counterparts were not operating with strict definitions, or the precise distinctions that philosophers prize. Rather, their interpretations of the contemporary scene—of science's impact on their worlds of experience—reflected the powerful associative logic of everyday language. For Barzun, as for the other characters in this book, the label "scientific" applied to much more than the findings and methods of working researchers. It also adhered to broader ways of thinking, to cultural patterns and social institutions, and even to certain modes of politics.

This has always been true of the terms "science" and especially "scientific," ever since their first use and right up to the present. Looking at various cultural settings, one finds all kinds of practices, innovations, institutions, actors, beliefs, and values coded as scientific. Today, for example, many Americans would identify in vitro fertilization, the Paris Agreement on climate change, the computer industry, and perhaps even technocratic, expert-centered modes of governance as scientific in nature. Sometimes, this attribution depends on a causal relationship, as when developments in genetics make possible new forms of bioengineering. Conceptual leaps from science to technology have been especially common, due in part to the long-standing attempts of many leading scientists to portray technology as merely applied science. At other times, applications of the adjective "scientific" reflect the use of empirical arguments to justify social patterns such as racism—or antiracism. A third kind of case involves more diffuse forms of inspiration or analogical likeness that seem to link science to abstract, impersonal practices or ideals, such as bureaucracy. These understandings of how science shapes our world rest on assumptions about what science does, not just definitions of what it is.[5]

The resonances of science as a cultural category vary dramatically over time and from group to group, reflecting their particular concerns. Barzun, for example, identified the bureaucratized genre of the grant application and the fluoridation of drinking water as two of science's most odious manifestations. Back in the 1930s, by contrast, an educational theorist had captured sentiments common among social scientists when he argued that individuals with a "scientific attitude" would oppose bigotry, warmongering, wasteful consumption, self-seeking, political corruption, and much else. Innumerable other meanings of "scientific" have prevailed at other times and places. It has been particularly common to link science to technological advances, engineering feats, and medical practices. But large-scale social and cultural tendencies have often been considered scientific as well. Such links between science and dangerous or controversial phenomena have created deep wells of anxiety about science's potential for misapplication, even among groups that value science highly in the abstract.[6]

The history of science as a broad cultural phenomenon, constituted by webs of potent associations, is distinct from the history of science as a research and teaching enterprise, despite significant areas of overlap. Science's cultural meanings stand in complex and multidirectional relationships to the techniques and claims of working scientists. Technical developments such as mastering the process of nuclear fission or learning to manipulate DNA strands certainly influence those meanings. But widely shared meanings also shape how individuals and groups respond to new findings and theories—and even impact the direction of the research enterprise itself. They do so by helping to determine what readers and listeners hear when writers and speakers use the word "science." Such meanings shape not only the definition of that term but also the concrete, real-world examples it brings immediately to mind—the cascading chains of mental associations set off by any mention of science.[7]

Writing the history of science as a cultural category thus requires scholars to range far beyond the technical pursuits of particular researchers or fields. The critics in this book were not typically responding to seminal publications in the disciplines, or sometimes even the leading theoretical frameworks. Often, in fact, they ignored such phenomena and fastened onto the public expressions that reinforced their preexisting fears about science's cultural implications. They offered compendia of striking examples that captured the real meaning of science as they understood it—often an inner logic or inexorable drift that operated largely independently from the motives of practitioners. On such grounds, a fictional character such as Dr. Strangelove, from Stanley Kubrick's 1964 film, seemed to reveal the thoroughly deviant personality and motives of the modern expert, despite the protestations of working researchers. But scholarly publications could also provide grist for the mill, insofar as they reinforced perceptions about science's real-world implications.[8]

Historians should also widen their scope beyond scientists' own public representations of their work. We would never look back at a particular moment and assume that Catholics alone determined the cultural meanings of Catholicism, or baseball players alone the meanings of their game. This is particularly true of practices and institutions that take on positions of critical national importance, as science did after World War II. Scientists can influence the cultural meanings of their vocation, but they hold no monopoly in that domain. (This is doubly true of the social sciences, which are perpetually embattled and intertwined with widely shared values.) As a cultural category, science is made and remade in private, personal exchanges as well as public settings. Everywhere there is culture—which is to say, everywhere—one finds cultural meanings of science and actors crafting such meanings, even as the existing meanings also influence them. Virtually every social group has contributed to defining science as a cultural object—or rather, many different cultural objects. As these various constructions circulate through the media, education, and other domains, science comes to bear multiple, contradictory meanings.[9]

A CRITIQUE EMERGES

This book traces how a particular set of constructions of science took shape and then changed over time, constituting a robust tradition of criticism that has traced a remarkable range of social problems to science's defective understanding of humanity. The narrative proceeds in a roughly chronological fashion, although it lingers at two points to highlight the range of perspectives that flourished at key historical moments. Thus, Chapter 1 explores a host of intellectual and social changes that convinced some skeptics a scientific culture had arrived in the 1920s, while Chapter 2 details how those critics understood science's character and influence during the same years. After Chapter 3 examines the growing association of science with New Deal liberalism in the 1930s and 1940s, the next five chapters adopt a variety of perspectives on the crucial period from the late 1940s to the early 1960s, when this tradition of criticism spread widely and many of its familiar themes emerged. These chapters survey the postwar developments that led new groups of critics to conclude that science had taken the cultural reins (Chapter 4); explore the responses of religious leaders (Chapter 5), humanities scholars and internal critics of the social sciences (Chapter 6), and political conservatives (Chapter 7); and identify a series of sites at which these networks of critics and lines of argumentation converged (Chapter 8). The final three chapters follow critiques of science's cultural influence through the seismic shifts of the 1960s and 1970s and into the culture wars of the late twentieth century.

Given the ubiquity and influence of such critical accounts, it may seem surprising that few Americans ascribed pernicious social effects to science before the 1920s. After all, a populist suspicion of elites and experts runs deep in American political culture. Since the founding era, Americans have distrusted centralized power—especially political power, but often cultural authority as well, whether that of ministers or professors. Yet that populist sentiment rarely targeted scientists before the 1920s. Many Americans viewed science as a kind of "people's knowledge," a practical, common-sense mode of reasoning that stood against all forms of elite authority. The political ascendance of Progressivism after 1890 made science increasingly central to governance, but the habitual identification of science with a populist rejection of authority largely persisted. Meanwhile, hardly any Americans believed that science had given their culture its distinctive character. Even those religious leaders who equated Darwinism with materialism thought that it threatened American culture in the future, not that it had already remade that culture. Up through World War I, the vast majority of Americans assumed that they lived in a Christian country, for better or worse. Indeed, the early twentieth century brought some of the loftiest expectations to date that the United States, and indeed the world, would be Christianized in every aspect.

Such hopes survived the 1920s in many circles. But small groups of cultural critics began to trace social changes that alarmed them to the cultural influence of science. Some lamented the mobilization of science by city and state governments: in classrooms, where biology lessons and sex education courses violated conventional norms, and in mandatory vaccination programs, which involved state agencies intervening directly in citizens' bodies. Other critics worried about the growing federal bureaucracy, which continued to gain regulatory authority despite the rightward shift in electoral politics after 1920. Still others thought a climate of utilitarianism and industrialism had corrupted politics and learning alike. The hedonistic tenor of the 1920s consumer culture and the violations of sexual propriety by Jazz Age youth also signaled to some critics a widespread loss of moral guideposts.[10]

Above all else, however, loomed the popular vogue of psychology, with its emphasis on cultural conditioning, childhood traumas, and other nonmoral, nonrational causes of behavior. The post–World War I years witnessed an explosion of popular interest in all of the natural and social sciences. But psychology became a veritable craze, with millions of readers devouring popular treatments and applying the new interpretive categories to themselves and others. Small cadres of literary scholars, southern writers, and mainline Protestants, along with larger groups of Catholic leaders and conservative Protestants, connected the vogue of psychology to wider social and cultural changes. They identified science as the source of a dangerously amoral worldview that had captured the public mind and eroded society's cultural foundations. These critics of the 1920s levied a charge that would become increasingly common in subsequent decades: modern science had dissolved conventional understandings of the human person and led the entire culture astray.

Over time, the specific contours of this argument shifted with cultural and political changes. In the 1930s, for example, the emergence of a moderate welfare state under Franklin D. Roosevelt reshaped perceptions of

science's cultural impact, and additional groups came on board. The bureaucratic innovations of the New Deal fed into the powerful associative
logic of commonsense reasoning, leading a number of Americans to equate
science with the technocratic, managerial liberalism of Roosevelt and his
allies. Over the next few decades, this association would take firm hold,
leading many of the New Deal's challengers to question the authority of
science and turning some critics of the social sciences against the welfare
state. Meanwhile, many other skeptics argued in the 1930s and 1940s that
the secularization of modern societies at the hands of scientists and their
allies had created a moral vacuum that was filled by the totalitarian state.
"When civilization is built on technologies and machines, ignores the dignity of man, and rejects responsibility to God, then Nazism is the logical
result," one group wrote in 1941. By then, a growing number of Protestants
and some Jews were sounding the alarm, as the rise of neoorthodox theology and the emigration from Central Europe sharpened and magnified
this critique of secularism.[11]

The association of science with a secular form of welfare liberalism deepened in the 1950s and early 1960s, but the details of postwar critiques
also reflected new conditions. Overlaid on concerns about nuclear destruction after World War II was what many saw as the imminent threat of manipulative, implicitly totalitarian programs of control by experts. As the
Cold War took hold, the New Deal state became the "national security
state" and birthed the military-industrial complex. With science growing
ever more central to American governance, both instrumentally and ideologically, all manner of critics concluded that a spiritually deadening, technocratic outlook was forcing American society into science's inhuman mold.
A scientific understanding of humanity, in this view, permeated the culture
at large, having radiated outward from the universities to shape public
opinion and policy formation. Reality itself, many thought, was changing
to fit the narrow, reductive interpretation of the scientists and planners:
People were treating one another like machines, and behaving more and
more like machines themselves. Such fears often centered on the alarming
prospect of social engineering—the possibility that social scientists could reshape personalities and social practices in keeping with predetermined ends.
The power-hungry social engineer and the mindless technocrat became
stock figures in American cultural criticism after 1945.[12]

The threat here lay in science's apparent denial of the moral freedom of
the individual, which many critics believed had turned American liberalism
into a near-copy of Soviet ideology. ("Communism is based upon a scientific and value-free methodology," a letter to the Catholic journal *America*
declared.) Such critics identified science as a materialistic and deterministic

mode of thought that reduced all phenomena to unchanging patterns of cause and effect, ruling out the existence of minds, ideals, values, and other nonmaterial entities. Applying this model to human behavior destroyed human autonomy and dignity, they argued. One observer called the resulting condition "self-tyranny," featuring a comprehensive "incapacity to think for oneself" that was rooted in a disavowal of "personal initiative and decision." Innumerable postwar critics deplored the emergence of what the political scientist Andrew Hacker called "Predictable Man." All around them, they saw "machine men" with "machine values," "faceless ciphers" lacking any "consciousness or aims." These critics sought to save the public from the social chaos created by science's application to human behavior by subordinating empirical knowledge to the normative resources of religious, literary, or political traditions.[13]

Such concerns became deeply entrenched in American culture by the 1950s. The postwar years produced not only the institutions and funding structures that still shape scientific research today but also many of our foundational assumptions about science's contours and cultural meanings. In recent years, our histories of that crucial period have foregrounded science's growing authority. Citing lavish research budgets, the prestige of physics, the ascension of psychological expertise, the cultural sway of white-coated experts, and the technocratic character of Cold War–era politics, they portray the postwar United States as the scene of naïve and almost universal trust in science. Yet there is another side to this story. The postwar era also brought potent fears that science had spread into intrinsically moral realms and cast its pall on the culture at large. Some of the critics have disappeared from our histories; many others, like Barzun, are remembered as standard-issue liberals. But these commentators were numerous, prominent, and influential. Even as science and scientists took on important new roles in American society, the expansion of their authority also inspired a national referendum on the social, cultural, and political meanings of science that featured deep undercurrents of fear and mistrust alongside assertions of beneficence. A wide variety of critics argued that science, rather than big business, the welfare state, the military, or the churches, set the tone for American public life.[14]

Few American commentators embraced the frank irrationalism and illiberalism that had spread in some European circles since the mid-nineteenth century. Most aimed to reorient modern institutions toward religious or literary values, not to destroy or supersede those institutions. Indeed, American critics often described Friedrich Nietzsche and other European antimodernists as products of science's nihilistic influence. They preferred Europe's moderate Christian critics—Arnold Toynbee, Jacques Maritain,

Christopher Dawson, Barbara Ward, Pierre Lecomte du Noüy—to the stark pessimism of, say, Oswald Spengler. In this view, the "modern crisis" could be solved largely by restoring a robust tradition of moral reasoning that science had marginalized or discredited. Rather than indicting whole social systems, these figures tended to criticize specific policies or cultural commitments, especially those associated with education. American critics of modernity generally accommodated the prevailing economic and political institutions, which made their arguments especially influential over the decades.

Above all, mid-twentieth-century critics emphasized science's impact on philosophical anthropology: theories of the human person. "It is here, on the nature of man, between those who would respect him as an autonomous person and those who would degrade him to a living instrument, that the issue is joined," the political journalist Walter Lippmann wrote. "From these opposing conceptions are bred radically different attitudes towards the whole of human experience, in all the realms of action and feeling, from the greatest to the smallest." Most critics sought to bring a seemingly amoral, nihilistic science under the sway of some version of "humanism" that emphasized the moral freedom of the individual. Varieties of humanism proliferated in response to the apparent threat from science: there were Christian, conservative, Marxist, classical, and literary humanisms, and many hybrids as well. Each portrayed the human person in immaterial, voluntaristic terms, ignoring the body and identifying exercises of individual subjectivity—valuing, preferring, choosing—as the truly human modes of behavior.[15]

THEMES AND CONSTITUENCIES

The postwar equation of science with a cold, depersonalized stance contradicted the alternative conceptions of nature and science that had flourished among many working scientists and naturalistic philosophers during the interwar years. By the early twentieth century, most biologists and some philosophers had made room for human values, ideals, and purposes within their understandings of nature. Such views also permeated the social sciences in the 1920s and 1930s. Although physical scientists and engineers tended to view morality in fairly traditional terms, as the product of Christian faith, by the 1920s most biologists, social scientists, and allied philosophers had concluded that the moral freedom of the individual figured centrally in a naturalistic, evolutionary understanding of life on earth. But theorists of this variety struggled in vain against skeptics who insisted that applying the scientific method to the human world entailed squeezing out or imagining away its moral content.[16]

For these commentators, studying any subject scientifically meant applying a particular conceptual framework—a reductive, materialistic, mechanistic, and often quantitative lens—to that phenomenon. Science was the study of matter in motion, guided by strict causal relations that could be discerned through sensory evidence and expressed in quantitative terms—ideally, rigorous, mathematical formulas like Newton's laws. On this view, science embodied the mechanistic viewpoint of classical, nineteenth-century physics; it was isolated from normative claims and confined solely to spatiotemporal phenomena. By definition, such a science was strictly neutral with regard to morality—and thus, the critics declared, utterly impotent as a guide in human affairs, except insofar as one's goals were purely technical and instrumental. Although the scientific method fit the physical world, human dynamics stood outside the "nature" that scientists could explore. From this perspective, studying human beings scientifically meant assuming that they acted like physical objects. After World War II, a growing number of critics considered this kind of "scientism" not only a faulty philosophy but also a dangerous cultural force that permeated the modern age—increasingly seen as an "age of science"—and had produced its characteristic problems. They traced social norms, cultural practices, government policies, and even wars to science's amoral outlook.

In the years after World War II, such arguments appeared among critics from across the political spectrum and religious believers of virtually all theological persuasions. Christian and Jewish leaders, for example, argued again and again that moderns had been cut loose from basic understandings of right and wrong, creating a dangerous moral vacuum. When the American Jewish Committee launched its journal *Commentary* in 1945 to explore the causes and implications of the Holocaust, founding editor Elliot E. Cohen argued that the forces of modernity were obliterating "the inviolability of the individual" through invisible but powerful cultural means. "It is not so much that this ideal has been crushed by tyrannical rulers," Cohen wrote, "but that it is dying in the hearts and minds of men." Indeed, such concerns strengthened an ongoing turn, dating back to the 1930s, against the theological liberalism that permeated American thought and culture during the early twentieth century. Not all postwar religious leaders worried that their predecessors had ceded too much authority to scientists and left moderns adrift, lacking the moral guidance that only religion could provide. But many concluded that genuine commitment to religion required active resistance to the project of understanding human beings in purely scientific terms. Even the rather genteel forms of Christianity and Judaism that predominated during the postwar surge of public religiosity nurtured tough critiques of scientific thinking and the modern condition.[17]

Few of these religious critics questioned Darwinism or other aspects of the natural sciences. Instead, they sought to limit science to the study of physical and biological phenomena, giving religion full jurisdiction over human subjectivity and its products. Many argued that religious traditions offered genuine, if nonempirical, knowledge of the human condition. From the standpoint of science-religion dynamics, in fact, the biggest story of the twentieth century was not the running battle over Darwinism in the schools but rather the much broader and more variegated patterns of re- sistance to the extension of scientific methods into the morally charged domain of social relations. Twentieth-century understandings of science were profoundly shaped by controversies around the social sciences and philosophy, not just the natural sciences. Those controversies produced a complex set of conflicts and alliances that have structured much of Amer- ican intellectual life, and a good deal of American politics, since World War II. As the chemist James B. Conant observed in 1951, from his post as president of Harvard University, the "theologian and agnostic now quarrel most violently in their interpretations of the views of psychologists, an- thropologists, and sociologists."[18]

Theologians were hardly the only disputants, however. Hosts of critics outside the churches and synagogues—including Conant himself—also challenged science's cultural influence in the middle decades of the twen- tieth century. Many scholars in the humanities, for example, agreed that the modern era labored under a science-worship that was inimical to human meanings and purposes. Yet they looked to Western art and literature, not religion, for the needed ethical complement to science's technical, value- neutral findings. Postwar conservatives sounded similar themes, joining cen- trists such as Lippmann in lamenting the cultural sway of "positivists who hold that the only world which has reality is the physical world." A wide range of conservatives associated their politics with "a humanist reverence for the dignity of the individual soul" and argued that modern science's "purely mechanistic view of man" produced "an ant-heap age" that "guil- lotines whoever is individual, superior, or just different." (They added a charge that radical commentators would also adopt in the 1960s and 1970s, arguing that scientism was not just a philosophical mistake and a corro- sive cultural force but also a potent social ideology that buttressed the power of a dominant elite.) The era's theistic, humanistic, and conservative cri- tiques of scientism often reinforced one another, despite the many differ- ences. Indeed, the portrait of a science-obsessed culture also appealed to other groups as well, including Conant and other natural scientists who distrusted the social sciences, and even dissidents within the social sciences themselves. These otherwise disparate critics agreed that modern science was producing a new type of person, "stripped of all qualities not accessible to the scientific method."[19]

Such views of science and modern culture carried through to the 1960s, shaping that decade's multiple, overlapping revolts. The left-wing humanism of many student radicals and activist professors often hewed surprisingly closely to the views of liberal, centrist, and even conservative commentators from the 1950s. They, too, portrayed a society relentlessly driven by technical imperatives to trample on human values at every turn. But the stream of humanistic criticism also flowed into new channels as it was caught up in the political earthquakes of the 1960s and 1970s. Left critics now joined conservatives in arguing that scientism represented the characteristic ideology of a ruling elite. Yet these radicals identified science as a bulwark of traditional social norms, not a corrosive threat to such norms. They increasingly argued that science's cultural influence buttressed social inequalities, keeping favored groups in power and others down. In the 1970s, left-leaning critics also ascribed pernicious effects to biology as well as the social sciences. Here, the target of criticism was not science's morally relativistic character but rather its entanglement with assertions of innate group differences. This kind of argument circulated among critical scholars with growing frequency in the late twentieth century and shaped debates over biotechnology and other controversial issues.

Meanwhile, many free-market thinkers of the 1970s continued to link modern science to socialism as well as moral relativism. Over time, some eventually warmed to the anti-Darwinism of the burgeoning Christian Right, which would later deepen the convergence by adopting the economic conservatives' climate denialism in the 1990s. Conservatives deplored the regulatory initiatives launched by Richard Nixon and congressional Democrats in the early 1970s, which yoked scientific research to federal power at new bureaus such as the Environmental Protection Agency and the Occupational Safety and Health Administration. A shared dislike of expert-driven policy initiatives helped broker the alliance between Christian conservatives dismayed by the secularity of the American state and economic conservatives alarmed by its size and scope.

Since the 1970s, claims about science's baleful cultural influence have anchored important strands of radicalism and conservatism, even as they have largely disappeared from the rhetorical arsenals of liberals and centrists. From both ends of the political spectrum, one hears sweeping challenges to modernity, defined as an age of enthrallment to scientific rationality. Today's critics often trace modern culture back to Descartes, Bacon, and Newton, not the shifts of the nineteenth and twentieth centuries. Recent critics have also linked science to capitalism and state power, while adopting a more pluralistic tenor than their postwar counterparts, who usually proposed a universal framework of values. Even religious traditionalists often adopt a pluralistic approach today, arguing that science must share the stage with an array of religious views. Despite these profound changes, however, the underlying assertion persists: science causes serious social and political problems by enforcing faulty understandings of humanity. That mode of analysis is much less common among mainstream commentators today than it was in the 1950s and early 1960s, but it remains influential in the universities and among theological conservatives.

SOCIAL CONTEXTS

Exploring this tradition of criticism sheds new light on some of the most important dynamics of the twentieth century. Science's cultural influence has preoccupied a remarkably broad and varied range of commentators. Viewing American history from this angle highlights lines of connection, intersection, and influence between individuals, groups, and movements that historians tend to portray as fundamentally different from one another—not least, the postwar New Right and the 1960s New Left. Although a concern with science's corrupting cultural effects has never been the dominant strain in American thinking about science, it has been persistent, influential, and consequential for nearly a century—above all, in the post–World War II "golden age." The chapters that follow track the efforts of prominent figures, including scholars, journalists, novelists, educators, and politicians, to alert their contemporaries to the cultural dangers that they associated with science.

The narrative also includes ordinary citizens, insofar as they made themselves heard through letters to journals, opinion polls, or pressure on congressional representatives. Not surprisingly, however, intellectuals have been especially likely to see political, social, and cultural phenomena as manifestations of faulty ideas drawn from science. Viewing everyday practices as instantiations of broader principles suggests to intellectuals that they can change their societies by making arguments—that is, by mobilizing their characteristic skills and proclivities, without abandoning their posts and becoming activists, lobbyists, or politicians. (Indeed, this understanding of social causation often leads them to take up intellectual work in the first place.) Critics of scientism have tended to favor idealist conceptions of historical causation, in which immaterial phenomena—ideas, thoughts, concepts, mentalités, discourses, sensibilities, imaginaries, "isms"—spin off the features of the experienced world as by-products of their interactions. Even thinkers who devoted their careers to puncturing the illusion of rationality in human affairs, such as the theologian Reinhold Niebuhr and the international relations theorist Hans Morgenthau, tended to describe social and political realities as outgrowths of philosophies. Whereas other groups have tended to pick and choose, criticizing only certain scientific expressions and products, intellectuals have often challenged science as a generalized, abstract, philosophical phenomenon.[20]

Equally unsurprising, those intellectuals whose vocations committed them to a cultural practice associated with forthright value judgments—a religious tradition, moral philosophy, Western political ideals, Marxist thought, an artistic or literary canon—were the most likely to decry the extension of scientific methods to the study of human beings. It is no mere accident of methodology that this book's pages teem with theologians, practitioners of the humanities, and political theorists. These groups have been especially likely to trace social and political problems back to foundational ideas. Thus, when totalitarianism loomed in the 1930s, many critics attributed it to a faulty philosophy of knowledge. After the war, they continued urging citizens to throw off the yoke of scientific rationality and to recapture a robust understanding of social morality—an authoritative set of values drawn from religion, or literature and the arts, or the centuries-long experience of the West, or everyday common sense, or some combination of these sources. Like science's advocates, the critics described in this book were not simply heroic resistance fighters speaking truth to power, just as they were not simply unhinged crackpots either. Their arguments frequently harmonized with their professional interests, as well as the personal proclivities that influenced their choices of vocation in the first place. This critical discourse on science's cultural effects cannot be reduced to mere special pleading. But likewise, it cannot be understood without acknowledging the influence of professional imperatives. Time and time again, critics identified their own religious perspective, or their version of the humanities, or their school of art, or their interpretivist methodology, as an effective antidote to science's cultural depredations.[21]

Social interests often operated in a different manner among marginalized groups. The critical discourse described in this book was largely the province of white men until the 1970s, when radical theorists increasingly connected racial and gender discrimination to the cultural influence of the sciences. This is not solely because American intellectual life was exclusive, although it certainly was. Rather, the most outspoken members of marginalized groups often focused on gaining access to scientific institutions and challenging particular scientific theories and applications rather than deploring science's influence as a broad cultural force. Moreover, scientific thinkers—social scientists from the 1920s forward, and most biologists by the 1930s—offered some of the strongest challenges to the concept of fixed racial differences, although most were slower to analyze sex and gender in this manner. To be sure, the biological racism and eugenics movement of the late nineteenth and early twentieth centuries cast a long shadow among Black Americans; many were deeply suspicious of Darwinism in the 1920s, and often for decades afterward. The theologically conservative piety of many Black churches led even mathematicians such as Howard University's Kelly Miller to decry the "Pagan and Godless" character of modern thought and education. Black women also joined immigrant and working-class mothers in resisting the advice of childrearing experts, who frequently denigrated their practices. At the same time, however, important cadres of women and Black scholars worked to claim the mantle of intelligence and expertise for themselves by finding places in the research community and turning science's unique forms of authority—and its close association with social progress—against injustice.[22]

Overall, debates about science's cultural impact have reflected the jousting of many different sets of aspirants to cultural authority, as well as ordinary citizens, journalists, and politicians favoring a wide range of cultural programs. The goal here is not to debunk the claims under study, although there is much wisdom in the rebuttal that scientists have rarely, if ever, enjoyed such influence over the surrounding culture. Rather, it is to assess the differences that such charges of cultural corruption have made. These arguments constitute a distinctive mode of social thought that has shaped the course of history—both American and global—in ways that call for detailed excavation and careful analysis. They have profoundly affected public understandings of science and its cultural influence, of course, but also broader currents in American life.

Ironically, this mode of criticism gained some of its influence by piggybacking on the authority of science itself. It mirrored the thoroughly detached, value-neutral model of science that prominent scientists also embraced in the mid-twentieth century. As the political climate shifted in the 1940s, leading scientific thinkers insisted that their enterprise was not only politically and theologically neutral but also morally neutral. Social scientists, whose left-leaning politics raised suspicion, were especially likely to claim value-neutrality amid the surge of domestic anticommunism. They identified the physicists' rigorous experimentation, systematization, and quantification as the hallmarks of science. In disavowing partisanship, however, these social scientists opened themselves to the charge that their work undermined morality. Celebrating value-neutral social science—even as a uniquely powerful tool for implementing public values—reinforced the arguments of those who viewed science as a uniquely powerful *solvent* of public values. Such critics seized on the starkest and most uncompromising statements of the value-neutral ideal, especially those by social scientists. They also collected and quoted predictions about science's capacity to radically alter social practices and norms, giving wide circulation to statements that reinforced their understandings of science's character and influence. Indeed, the scientists' claims sometimes reached broad audiences by way of critics. To these commentators, the most outrageous and frightening statements by particular scientists revealed the inner meaning of science itself.[23]

In an earlier book, I showed that a rigidly value-neutral ideal was comparatively novel in the American social sciences and argued that it represented in part an attempt to deflect conservative charges that scientists leaned toward communism and other subversive views. In the long run, however, pleading ethical neutrality and comparing social science to physics may have proven counterproductive from the standpoint of public perceptions. This technical, reductive approach seemed to prove that thinking scientifically about human beings meant reducing them to something subhuman: machines, or animals, or tables of statistics at best. It reinforced the critics' tendency to blame the bomb and other alarming technological developments on the cultural enervation caused by social-scientific thinking, not on the highly valued physical sciences. Indeed, many of these critics drew a line of scientific validity at the border between the natural and social sciences and inveighed strongly against incursions across that line. Their contributions to what sociologists call "boundary work"—defining science through contrasts with alternatives such as religion, philosophy, politics, and art—sometimes found broader audiences than the equivalent efforts of scientists themselves. In the postwar years, when science's cultural meanings were everywhere under discussion, the critics provided a particularly influential source of popular understandings.[24]

Indeed, one of the main effects of the critical tradition described in this book was to help establish what remains the prevailing understanding of science in the United States: that it is rigorously value-neutral and aims to reduce complex phenomena to materialistic, mechanistic causes and effects. A motley coalition of cultural commentators enforced the value-neutral model, often by taking it for granted and then lamenting its implications for American public life. As the Jewish writer Will Herberg put it, "The world of science is indeed very much as it is pictured by modern positivism—but the world of science is very far from being the world of reality!" Moreover, such critics gained additional influence by embedding their abstract, methodological portraits of science in concrete, morally charged narratives about how science was reshaping modern societies in its own, inhuman image. Fears that a value-neutral science would destroy cherished cultural values and social institutions did a great deal to solidify that understanding of science itself.[25]

LEGACIES

For a century, then, influential groups of American commentators have argued that science anchored a faulty cultural understanding of human beings and social relations—and, many added, reinforced the power of a dominant liberal elite in the process. This fact has mattered a great deal. In the mid-twentieth century, especially, anyone who attended a college or university in the United States, or read magazines, or listened to congressional leaders, or engaged in other ways with American public discourses, heard numerous versions of the charge that science represented a moral threat to civilization, due to its corrosive effects on humanity's self-conception. Our contemporary understandings of science, and even of our social and political worlds, reflect the potent impact of the critical tradition described in this book.

Although much more work is needed on specific lines of connection, critiques of science influenced key religious and political developments. During the early Cold War, for example, these challenges fed into anticommunist initiatives, especially but not exclusively on the political right. They also helped shape the eruptions of the 1960s as well as today's post-truth climate. Even the continued strength of Christianity in the United States since 1900—at a time when its power has waned in many other countries— may reflect in part the widespread assumption that the only alternative to religious faith is the dehumanizing worldview of a mechanistic science. Political historians often write as if religious beliefs simply encode social or economic interests, while historians of religion tend to treat faith commitments as primary and uncaused. Yet those commitments emerge in dialogue with other conceptual frameworks. Thus, the vaunted "postwar revival"—a broad surge of religiosity among the general public as well as intellectual, cultural, and political leaders that began during World War II and accelerated through the late 1940s and 1950s—clearly took some of its shape from Cold War–era fears about science's cultural influence. As we will see, many Protestant leaders modified their theological views in response to the perceived secularization of American culture at the hands of scientists and their theologically liberal allies. Public images of science and its cultural implications likely influenced the religious commitments of the rank and file as well. Overt concerns about science's cultural meanings accompanied innumerable expressions of public piety in the late 1940s and 1950s.

During the same period, in the middle decades of the twentieth century, growing religious diversity also made social critics less likely to assume that mainline Protestantism set the tone for American public culture and more likely to trace the roots of contemporary developments to the scientific disciplines and research universities. If a single religious tradition had prevailed in the United States, it would have been more difficult to argue that creeping scientism had produced a society-wide moral vacuum. Rather than viewing the Protestant establishment as the dominant force in American culture, postwar critics favoring a range of alternative cultural programs increasingly saw science in the driver's seat—and often charged that mainline Protestants themselves had replaced Christian teachings with a worldview drawn from modern science.

The tradition of argumentation described in this book may also help explain a broader trend in the religious domain. For many decades, mainstream forms of Protestantism have been losing ground to fundamentalism and other biblically based, theologically conservative approaches that conflict with Darwinism and other scientific theories. Most scholarly interpreters of that phenomenon are sharp critics of theological liberalism who contend that it has declined due to its innate incoherence—its allegedly impossible combination of biblical faith and morality with scientific empiricism. Both religious and secular historians, in fact, have argued that theological liberalism proved unstable, forcing adherents to choose either religion or science. Yet this analysis often rests on an underlying assumption about science: that it is incompatible with theism, or indeed any understanding predicated on moral freedom. In this rendering, science eliminates free will by asserting that human action operates in a strictly lawful, mechanical fashion and is fully determined by antecedent causes. The increasingly vigorous circulation of that critique of science since the 1920s may help explain why so many Americans have abandoned the liberal theologies that view of science challenged.[26]

This critical tradition has shaped American politics as well, by challenging the legitimacy of welfare liberalism. The mid-twentieth-century complex of ideas and institutions that historians call the "New Deal order" suffered from numerous practical and conceptual weaknesses. For example, many white Americans' distrust of racial minorities made them unwilling to devote tax dollars to promoting social equality and produced sharp disparities in employment and housing behind the scenes. But surely it also mattered that vocal critics at every point on the political spectrum—including many mainstream liberals themselves, as well as prominent religious leaders—argued over the years that the American welfare state was dangerously technocratic, bureaucratic, and dehumanizing. They contended that social science, a foundational resource for the New Deal agencies, was ideological rather than neutral and threatened humanity by corrupting its self-understanding. Such critics identified the welfare state as the product of a "disintegrated liberalism," resting on "the illusion that scientific observation and logic alone will suffice in the treatment of human affairs." In so doing, they tied the New Deal to undemocratic, even totalitarian projects of social engineering that turned autonomous individuals into raw material for experts to manipulate.[27]

Across the past century, meanwhile, the style of criticism described in this book has also led critics on the left to repeatedly slide from economic to cultural understandings of power—and often to shift the blame for prevailing social conditions from capitalist elites to scientific elites, from political economy to rationality. In the late nineteenth and early twentieth centuries, social critics assumed that big business held the reins of power, buying the policies it desired while using its cultural influence to sustain a free-market ideology that disabled political opposition. By the postwar years, populist critiques of concentrated power increasingly turned away from big business toward experts. In this view, real power in modern America lay in the hands of secular, liberal professors, not business leaders, preachers, or politicians.[28]

Since then, the emphasis on science as a threat to human values has taken new forms, even as moral commitments have become central to political identities and party affiliations in the United States. In recent decades, public debates have increasingly revolved around a series of competing declension narratives that posit a moral deficit in the nation's public life—and often trace that deficit, in part or in whole, to science's influence. Conservatives have long identified the New Deal as the moment of decline, when the United States lost its moral compass—to some, because a relativistic, naturalistic, and technocratic mentality took hold in American culture and reshaped public institutions and practices accordingly. New Leftists often located that technocratic turn in the years after World War II, while neoconservatives and the Christian Right focused on the 1960s. Proponents of each narrative have discerned a pervasive sense of moral aimlessness that they often linked to the cultural influence of science.

Through all of this disputation, the core image of science as a value-neutral, and thus innately amoral, enterprise has sunk ever deeper into the cultural bedrock. Generations of commentators have taken for granted that science entails a morally detached approach to the world, even as they clashed bitterly over its applications and implications. The consequences, though hard to measure, have been substantial. As the third decade of the twenty-first century opens, a potent new disease is spreading and the planet is lurching toward environmental disaster. Responding effectively to these threats will require us to think much more clearly and precisely about the configurations of scientific expertise that surround us—and often shape our lives in minute detail. Turning our attention from science's champions to its critics can help us do just that.

1

MENTAL MODERNIZATION

YOU STROLL THROUGH MANHATTAN on a late summer afternoon in 1929. As you traverse the right-angled streets, you move from bright patches of sunlight into the lengthening shadows of massive skyscrapers—huge geometrical shapes looming above, the products of innumerable hours of labor and immensely complex processes of planning, design, financing, contracting, scheduling, and construction. Crossing from neighborhood to neighborhood, you encounter banks, groceries, warehouses, sweatshops, tenement houses. Through open windows, over the growl of passing automobiles, you hear snatches of radio, voices in many languages. Newspapers hawked on street corners announce the rise and fall of the stock market, the latest merger. Crowds descend into train and subway stations, waiting to be whisked away to the neat lawns and gadget-stocked kitchens of the boroughs and the mushrooming suburbs. As the sun sets, street lamps flicker to light all over the city, joined by neon signs and movie palace marquees. *Where did all of this come from?* you wonder to yourself. *What holds it together? What kind of future does it foretell?*

The answers that come to mind likely differ greatly from those offered by earlier generations of Americans. Through the nineteenth century, the culturally dominant Anglo-Protestants—and many others—looked at the world around them and saw the results of hard work, self-sacrifice, thrift, and ingenuity. Asked what kind of culture inculcated those virtues, most credited some combination of Protestantism and republicanism. They argued that the autonomous, virtuous individual anchored any healthy polity. From this perspective, Protestantism and republicanism represented the religious and political expressions of individualism, neatly complementing American enterprise and inventiveness. But that story strikes you as out of date. *Isn't something else at work here?* There's a force, a drive, behind the city's pell-mell growth that feels somehow larger than any individual, no matter how thrifty, clever, or hard-working.

Ducking into a bookseller in search of wisdom, you find the shelves bursting with popularizations of modern knowledge: philosophy, psychology, anthropology, history, biology, medicine, and more. Perusing these fat but readable volumes, you find that most of them speak with a single voice. They announce the arrival of an utterly new thought world, as different from nineteenth-century thinking as day is from night. A thoroughly scientific "modern mind," the authors explain, built the bustling metropolis around you—and that mind will singlehandedly shape the future to come.

But as you venture into one of the city's many university buildings and begin eavesdropping outside lecture halls, you hear other voices, other portraits of the present age. A professor of art history laments the philistinism and cultural leveling of a standardized society: the destruction of cherished, organic values by "the machine." Two doors down, an anthropologist shocks the sensibilities of his students by announcing, with evident satisfaction, that human beings remain savages. We are Stone Age relics, he asserts with glee, transported into a Machine Age whose contraptions and institutions we can hardly comprehend, let alone operate effectively. Gliding down the building's marble steps and crossing the street to a neighborhood church, the pugnacious sermon immediately pulls you into a maelstrom of conflict between religious modernists, who welcome science and the machine, and fundamentalists, who protest that newer is not always better—especially when it comes to the revealed word of God.

As you escape into the night air and begin to sort through this welter of perspectives, your mind flashes back to a time, not so distant but before the Great War, when economics, politics, and foreign affairs dominated the public conversation. It strikes you that something essential has changed. Questions about culture, about beliefs and values, have come to the fore in American public life. Yet you also realize that all of the voices you just heard—all the competing accounts of the relationships between science, religion, philosophy, the humanities, and the arts—took for granted that an unprecedented new form of civilization, based on the application of machine techniques, had recently emerged. The questions dividing the commentators started from there. Does the Machine Age hold great possibilities or reflect immense losses? If there is potential to be tapped, what forms of belief and valuation can unlock it? Can humanity gain control of its machines and turn them to the good by thinking in new—or old—ways?

KNOWLEDGE AND MORALITY IN FLUX

Such collective self-descriptions were indeed quite new in the 1920s. Through the nineteenth century, the mainstream Protestants who dominated American public culture defined their country as a product of republican politics and Protestant ideals. Of course, economic progress also figured prominently in the era's portraits of American life. Yet most observers identified growth and prosperity as products of, and aids to, political and religious individualism. They also assumed that science went hand in hand with Protestantism and republican politics, as part of a single complex that American Protestant leaders called "Christian civilization" or simply "civilization." In this understanding, economic freedom reflected a deeper pattern of political freedom, and below that lay the spiritual freedom common to Protestantism and science alike. Science, too, embodied individualism: the liberation of individuals from dogmatic authority and other mental shackles, so that they could truly discern God's truth. From this perspective, Europe still groaned under oppressive, authoritarian institutions: monarchy, aristocracy, the Catholic Church. But Americans, freed from the dead hand of tradition and prideful, self-interested exertions of human authority, followed God's dictates alone.[1]

For the vast majority of nineteenth-century American Protestants, these divine strictures included binding, absolute moral laws that operated beneath the comings and goings of everyday life and alongside the causal patterns revealed by the natural sciences. Of course, individuals could break the moral laws, unlike the physical ones. But they would suffer the inevitable consequences. Far from threatening morality, in this view, science powerfully aided it by proving the existence of God, the creator and lawgiver. Among Protestants, a long tradition of "natural theology" held that scientific findings would square perfectly with biblical teachings, because God had authored both Scripture and the "book of nature." Investigating God's creation could never reveal conflicts between its various elements. Thus, science would inevitably advance in tandem with industry, democracy, and Christian ideals as part of a single, integrated civilization.

Two developments shook this widespread understanding of the intimate connections between science, religion, and morality in the late nineteenth century: Charles Darwin's theory of natural selection and the impact of industrialization on American public culture. At first, the Civil War and Re-construction blunted the force of Darwin's materialistic account of natural history in the United States. Although the theory set professional scientists against one another and was hardly unknown to ordinary citizens, the mainstream Protestant leaders who played such central roles in American culture did not take up the issue in earnest until the 1870s and 1880s. When they did, a few concluded that Darwin's theory utterly negated the prevailing Christian understanding of the world. These figures either reas-serted the Bible's primacy over mere sensory knowledge or threw off their faith entirely. Most leading Protestants, however—and many natural scientists—reinterpreted evolution as a matter of steady, divinely ordained progress, with the human person as its biological endpoint and a fully Christianized civilization as its ultimate goal. This framework of theistic evolutionism preserved the harmony between science and religion that the tradition of natural theology required.

Below the surface, however, aligning Darwin's theory with Christianity subtly altered both, in ways that eventually undermined the natural the-ology tradition. In the churches, as in the new research universities that began to took shape in the 1860s and 1870s, debates over Darwinism accelerated a nascent shift toward theological liberalism. Protestant ministers and theologians—unlike many of their followers in the pews—increasingly followed avant-garde German thinkers in viewing the Bible as a historical product, cobbled together over centuries from preexisting materials rather than handed down by God in its entirety. They interpreted its verses as a set of parables that pointed to universal moral principles, not a literal, blow-by-blow account of either human history or natural history. These liberal tenets also permeated the emerging research universities, whose leaders presented their work as continuous with liberal Protestantism. Natural theology, with its claim that the book of nature and the book of Scripture contained the same divine truths, gradually gave way to a "separate spheres" approach holding that science and religion could not conflict because they did not overlap at all. In this view, scientific and religious thinkers used entirely different methods to address separate sets of questions. Religion provided intuitive and scriptural evidence of the moral law and the need for faith, whereas science used sensory evidence to reveal patterns in nature. Each reigned supreme in its own domain, with no possibility of contradiction.[2]

As it turned out, however, the separate spheres approach generated new kinds of conflicts. Many of its advocates in the sciences reserved potent terms such as "knowledge" and "truth" for the products of their distinctive methods of investigation. In response, a number of otherwise science-friendly commentators, such as Princeton University president James McCosh and Yale University president Noah Porter, insisted that certain foundational principles of Christianity—moral freedom, immortality, God's existence—were matters of cognitive truth, not mere faith, intuition, or poetic insight. Christians, they argued, could not simply cede the mantle of genuine knowledge to scientists investigating the natural world.

Another major axis of conflict took shape when some scientists and philosophers, seeking to understand the implications of Darwin's naturalistic approach, redefined truth itself in ways that challenged even liberal forms of Christian faith. The conventional view of scientific inquiry, associated with the writings of Francis Bacon, held that scientists discovered authoritative, unchanging facts and built their theories on those solid foundations. Empirical facts offered the basic units from which scientists painstakingly reconstructed the universal laws of nature. In the antebellum years, leading American Protestants drew on the tradition of natural theology to accommodate themselves to this Baconian understanding by giving divine warrant to its assertion that sensory evidence offered knowledge of the spatio-temporal dimensions of God's creation.

Yet Darwin's theory seemed to violate Baconian tenets. It ranged far beyond sensory data, trafficking in broad, speculative generalizations about the distant, unobservable past. After the Civil War, as theorists of science worked to accommodate this feature of Darwinism, some adopted versions of positivism. In line with the separate spheres model, positivism held that science revealed only the observable surface of reality; it could not address a set of "ultimate" questions that included the nature of causation and the constitution of reality itself, as well as cosmological questions such as the universe's origin and purpose. Still, positivists assumed that science could find reliable, stable patterns of causation in the world.

Late in the nineteenth century, the pioneering theorists of pragmatism added a further stipulation, based on a Darwinian view of the human mind as an evolutionary product. Following Charles S. Peirce, they concluded that scientific knowledge could never be viewed as permanent and unchanging, even in the restricted domain assigned to it by positivists. Indeed, the pragmatist philosopher John Dewey ruled out the assertion of absolute, timeless truths in any domain, including religion and politics. In a world of constant flux, he argued, all judgments based on human experience remained inescapably fallible and revocable, subject to change as new evidence emerged—or as the contours of reality itself shifted.

Such post-Darwinian theories of knowledge codified the principle of methodological naturalism that defines science today: Investigators must not propose supernatural causes for natural phenomena. At the same time, these understandings of science implicitly questioned the existence of a divinely ordained body of moral laws. Clearly such laws did not reside in the spatiotemporal universe, alongside the other constraints on human behavior. What was the status of Christianity's moral absolutes—if indeed they existed at all?

The answers given by Dewey and his science-minded contemporaries often took their shape from the second key development of the late nineteenth century: rapid industrialization and the dramatic social and cultural changes it fostered. Social scientists, philosophers, and reformers clashed over the political meanings of industrialization. Some insisted that the familiar principles of economic individualism and laissez-faire governance still applied to the new industrial economy. Others, however, contended that the growing size and power of industrial corporations had fundamentally changed the rules of the economic game, requiring new policy responses. The emerging labor movement put a point on the question, especially when its radical wing advocated Christian or Marxian forms of socialism.

Although the fear of violence and class struggle remained strong among middle-class professionals, the 1880s saw a new generation of Christian leaders, philosophers, and social scientists reject free-market orthodoxy, styling themselves "Progressives" in politics and taking up the associated Social Gospel in Protestantism. Dewey and other Progressive theorists argued that industrialization represented the opening of an entirely new phase in human history, wherein the old laws of political economy no longer applied. The massive size of industrial corporations gave them immense cultural and political power, allowing them to run roughshod over individual citizens and smaller competitors alike. As a result, unregulated competition no longer guaranteed widespread prosperity. The Progressives reasoned that industrialization and corporate consolidation demanded new forms of economic intervention, to protect the interests and freedoms of workers and consumers. And in a democratic context, where the people theoretically reigned supreme, such changes in the realm of political economy could only arise from the spread of a new social morality—a new understanding of what the inhabitants of an industrial society owed one another. Progressive theorists worked assiduously to define and promote this framework of social morality, which rested on new understandings of economic causation as well as personal and collective obligation.

The Progressives thus abandoned a major component of the divinely ordained moral law as Protestant leaders had long understood it—namely, the principles of classical political economy. They argued that such allegedly absolute rules, governing the mutual obligations of citizens and the proper scope of state authority, were actually time-bound and no longer applied under industrial conditions. In practice, Progressives usually defined the new social morality as an application of universal Christian principles to unprecedented conditions. But their arguments, when combined with positivist and pragmatist understandings of knowledge, raised the possibility that other elements of the moral law could, and should, change as well. By the 1920s, such challenges to moral absolutes in the name of modern science grew increasingly loud in the universities and the culture at large. In response, critics of modern science argued that its understanding of the world ruled out the possibility of morality itself: either moral principles were absolute and unchanging or they were meaningless. As a result, the entrenched assumption that science and moral progress reinforced one another began to unravel in the wake of World War I.

DEFINING THE MODERN MIND

What was the evidence that an amoral science had spawned an amoral public culture in the 1920s? Understandings of science as a cultural category take their shape from perceptions, not unvarnished facts or pure logical deduction. Those perceptions, in turn, can reflect many factors, including not only scientists' own writings but also claims about science by many other kinds of commentators, as well as economic, social, and political changes that come to be considered scientific in origin or character. Above all, two developments in the 1920s convinced a small but vocal group of critics that science had poisoned the well of American public culture. First, the emergence of a full-blown consumer culture, however limited its actual reach, led many social critics to conclude that Americans had become shallow, materialistic, acquisitive individuals who retreated to suburban enclaves and cared for nothing except their social status. Sinclair Lewis's 1922 novel *Babbitt* provided a popular shorthand for the type: enamored of watches, radios, and other consumer gadgets; obsessed with keeping up with the neighbors and the latest trends; and devoid of moral ideals or even aesthetic preferences beyond those installed by advertisers. Second, popularizations of modern psychology flourished in the 1920s, bringing to the masses what critics considered a profoundly materialistic, reductive interpretation of human behavior. Drawing a causal connection between these two shifts, some observers reasoned that psychology and other expressions of modern science had turned Americans into self-seeking hedonists.[3]

Meanwhile, the writings of a long line of science enthusiasts seemed to reinforce this causal connection by contending that Americans *should* overhaul their thinking in accordance with modern science and industry. Decrying the "cultural lag" between industrialized production and society's self-understanding, Dewey and many other 1920s commentators argued that industrialization, as a near-total rupture in human history, demanded comprehensive changes in morality and religion as well as new political and economic ideals. Although these advocates of mental modernization believed that their work had just begun and the fruits of their labor lay many years in the future, some critics took their writings to indicate that the alarming cultural changes of the era reflected the widespread adoption of scientific thoughtways.

Philosophers, social scientists, and even some biologists argued in the 1920s that preindustrial thought was unsuited to the industrial world in every regard, not just in its economic dimensions. A handful of these mental modernizers targeted the content of Christian morality, rejecting virtues such as charity, beneficence, and justice. The vast majority, however, embraced such broad principles and simply questioned their conventional applications to specific issues such as sexual mores or church-state relations. Still, these figures, like their more iconoclastic counterparts, offered new conceptions of what philosophers call "metaethics": namely, the legitimation of ethical principles, addressing the "why" rather than the "what" of morality.[4]

On what grounds do we follow moral teachings? Not because God decrees them, the mental modernizers argued. A thoroughly scientific answer would require new conceptions of metaphysics, epistemology (the theory of knowledge), and social causation. As Dewey and other mental modernizers worked to place their understandings of humanity on empirical foundations, they challenged core Christian teachings—on the soul, divine purpose, and free will—that even most theological liberals and modernists endorsed. Protestant leaders who had long worked to accommodate Darwinism and other scientific theories sometimes responded with alarm to the 1920s campaign for mental modernization. "Evolutionism does not necessarily exclude a religious interpretation of the world and of human life," noted the Boston University theologian Albert C. Knudson, "but a thoroughgoing psychologism and historicism apparently does."[5]

The mental modernizers did not speak for the scientific community as a whole. Indeed, their ranks included philosophers and even a few literary scholars as well as psychologists and biologists. Meanwhile, numerous researchers, especially in the physical sciences, continued to embrace more or less conventional forms of faith. Yet the statements of the mental modernizers represented new interpretive possibilities that provoked alarm among critics of many kinds. Their writings raised four different kinds of nightmare visions for those observers who increasingly feared science's cultural effects. Three of the leading approaches to mental modernization explicitly disavowed the possibility of meaningful moral action, and many critics believed the fourth version—that of Dewey—did so as well.

The first variety of mental modernization was quite rare but gained enormous notoriety in the 1920s. Anticipating existentialism, its advocates held that human beings were utterly alone in a vast, mechanical universe that was impervious to their hopes and purposes. The British philosopher Bertrand Russell had long argued that modern science required this view. His 1903 essay "A Free Man's Worship" remained an important point of reference in the 1920s, especially after Russell elaborated on "Why I Am Not a Christian" in 1927. Two years later, in 1929, the literary critic Joseph Wood Krutch echoed Russell's vision of an "alien and inhuman world" in *The Modern Temper*. Krutch's bleak book ended on a note of metaphysical resignation and existential hope: "Ours is a lost cause and there is no place for us in the natural universe, but we are not, for all that, sorry to be human. We should rather die as men than live as animals." In truth, both Russell and Krutch feared the social effects of understanding the world in purely mechanistic terms. Yet they thought it intellectually dishonest to locate any anchors for human morality in the world portrayed by science. Russell and Krutch did insist that human beings could act freely and perhaps even morally, in some limited sense of that word. But they cautioned that the universe offered no support and no reward for such action.[6]

A second version of mental modernization, rooted in behavioristic psychology, presented a rather different spectacle in the 1920s. The psychologist John B. Watson and his followers dismissed not only God and the soul but also the human mind. Indeed, the behaviorists appeared to completely eliminate free will—and thus the possibility of moral choice—by identifying environmental conditioning as the sole cause of human behavior. Here, the threat to conventional understandings took the form of a thoroughgoing determinism, not a portrait of existential freedom within a structure of cosmic meaninglessness. Watson and his allies eliminated the concept of mind or consciousness—any sort of entity that processed information and made decisions—because it could not be seen in the laboratory. They instead defined human action as a series of essentially automatic reactions to forces in the surrounding environment. What, then, could freedom mean, if chains of association between external stimuli and programmed responses accounted for human behavior in its entirety—if human beings were no different from Pavlov's dogs, salivating when a bell rang, or the white rats that Watson preferred to study? Where was agency or the self, let alone progress? Even a psychologist who believed that glandular secretions produced individual personalities could complain that "Watsonianity" had become a comprehensive substitute religion in the 1920s.[7]

A third style of mental modernization built on Sigmund Freud's psychoanalytic framework. Despite the major theoretical differences, many critics saw Freudianism as of a piece with Watson's behavioristic psychology—as just another expression of the deterministic bent of modern science. To be sure, Freud elaborately theorized and categorized mental expressions, whereas Watson eliminated the mind altogether. Moreover, Freud directed attention to the special capacity of childhood experiences to shape unconscious behavior and stressed the importance of sexual drives, trauma, and symbolization. His therapeutic interventions also rested on the possibility of examining and redirecting unconscious impulses. But for many critics, Watson and Freud were essentially interchangeable. Each offered a variant of what Knudson called "psychologism": the view that human behavior could be reduced, in a deterministic fashion, to a reflex of essentially material causes in the surrounding social environment.[8]

By contrast, Dewey and his growing cadre of admirers offered a fourth vision of mental modernization that specifically rejected the deterministic tendencies of behavioristic and Freudian psychology, aiming to reconcile moral freedom with scientific inquiry within a broadly empiricist frame. Indeed, the naturalistic but nondeterministic understanding of human behavior that Dewey championed spread far more widely among American philosophers, social scientists, and biologists after World War I than did the other species of mental modernization. Rather than redefining human behavior to fit the mechanistic tenets of the physical sciences, Dewey and his allies defined science and empiricism more expansively to encompass moral freedom and other distinctive features of the human world.[9]

These figures flatly rejected the mechanistic assumptions of the behaviorists. Some of them argued that Watson's brand of materialism applied in the physical sciences but did not capture other aspects of reality—especially human behavior, where moral freedom and value-driven choices predominated. (One argued that the "tight little world" of materialism was almost as "mythological" as medieval philosophy, because it abstracted bare facts from the human values that really mattered.) Other nondeterministic naturalists argued that even physics had now risen above materialism, a reductive philosophy that bore no relation to any of the sciences in practice. In fact, a surprising number of interwar social scientists and philosophers, including Dewey, denied that there were strict, deterministic laws in any science. These critics of determinism interpreted the term "science" quite broadly. They included nonmechanistic interpretations, nondeterministic processes, and nonquantitative assessments under that rubric.

For these theorists, science encompassed virtually any form of replicable, empirical inquiry—any instance of reasoning collectively from publicly verifiable evidence. When they called for extending scientific methods to human affairs, they did not mean viewing persons and social relations through the same interpretive lens that classical physics applied to spatio-temporal phenomena. Rather, this group of mental modernizers, which included many of the leading social scientists and philosophers of the 1920s, argued that human freedom was a thoroughly natural phenomenon. Science, for these naturalistic thinkers, was neither amoral in itself nor destructive of moral behavior. Quite the opposite; it placed moral freedom at the heart of a modern, empirically grounded worldview.[10]

Although Dewey and his allies aimed to reconcile science with familiar understandings of free will, their naturalization of morality threatened to eliminate any distinctive role for religion in human affairs, insofar as religious believers continued to insist on absolute moral truths. Like the closely associated philosophy of pragmatism, the naturalization of morality challenged the idea of absolute truths in any domain. Most of the biologists who argued that nature included values, choices, and purposes worried little about deep epistemological questions; they simply assumed that empirical investigation revealed the contours of the surrounding world. But philosophers and social scientists argued vociferously about the meanings of knowledge and truth. Darwin, many contended, taught that human minds, like human bodies, had emerged through the evolutionary process, to help individuals navigate their worlds. So far, so good; it was easy enough to say that the human mind facilitated survival by providing reliable knowledge of the world. But Dewey and others saw a problem: If all human judgments rest solely on the experiences of particular, subjective, embodied individuals, we can never really know when our supposed knowledge mirrors external reality, or even maps onto it in a direct, one-to-one fashion. No matter how many subjective experiences we gather, share, classify, and analyze, we will never attain an external, God's-eye view that allows us to compare the resulting judgments with the actual world from a vantage point outside human subjectivity.

Dewey and his allies drew two conclusions from this assumption, both of which provoked conflict with religious believers. The first was that adopting science's empirical orientation—either by undertaking an investigation oneself or by heeding the findings of professional scientists—was the best way to ensure the success of any project in the world. Naturalists of this ilk assumed that science's empirical approach, reliable in the past, would likely remain reliable in the future. Certainly it had achieved more success than other intellectual practices in analyzing patterns of correlation in the world of experience, and thereby enabling people to predict the future outcomes of their actions. Therefore, science, while always ultimately fallible, should be the highest source of authority wherever it had spoken.

The second conclusion was that moral judgments were as subjective and fallible as any other products of human experience and reasoning. All talk of transcendent, absolute principles, moral or otherwise, was false—and highly dangerous. Human beings could no more discern whether their moral pronouncements matched the external world than whether their scientific theories did. As products of subjective experience, moral tenets, like scientific concepts, remained essentially conjectural tools for navigating the experiential world. These principles possessed no basis of validity beyond their ability to successfully steer behavior toward the desired real-world outcomes. In short, Darwin's theory of evolution required Dewey's pragmatism: the view that all human judgments, including moral judgments, were ineradicably fallible and could be improved only through the empirical, intersubjective practices characteristic of modern science. Any alleged "law" of human behavior worked only a particular social context and took its justification solely from the practical successes it enabled within that context. This meant that religious believers were deluded and potentially dangerous when they claimed that they possessed timeless moral absolutes, or even that they had based moral judgments of any kind on nonempirical sources—for there were no such sources, given the impossibility of transcending subjective experience.

Dewey's approach resonated widely among politically progressive philosophers and social scientists, especially the generation that launched their careers in the 1910s and early 1920s. But the belief that industrialization required some form of mental modernization, as part of a comprehensive process of "social reconstruction," also took strength from a converse proposition: that conventional, free-market views of economics and politics rested on traditional understandings of religion, morality, and truth. Laissez-faire theorists often claimed that the principles of Adam Smith and David Ricardo functioned as iron laws, valid for all societies. Many scientific thinkers argued that classical economics and religious orthodoxy went hand in hand, as part of a coherent movement of reaction. Dewey identified the common element between the two as the postulation of philosophical absolutes: absolute economic rights in the former case and absolute moral principles in the latter. In economics as in religion, he argued, self-interested leaders had convinced the masses that certain values were true for all time—indeed, dictated from on high—and thus prevented citizens from intervening and changing the conditions of their shared existence for the better. According to Dewey and like-minded thinkers, one was either on the side of the moderns, embracing purely functional scientific and moral judgments, or hopelessly stuck in the past.

From Dewey's point of view, only theological modernism, religious humanism, and naturalism were viable options for those building the infrastructure of a new world. His perspective tied the resurgent forces of Protestant fundamentalism and Catholicism to economic laissez-faire: all were bygone expressions of an age before modern thinkers had applied science to economic production and then discovered that it ruled out all supernatural appeals and absolute laws, whether economic or moral. This rendering of modern history, which neatly linked philosophical and economic changes, made education a particularly important site of controversy. Yet Dewey and his allies saw evidence all around—the slowing of Progressive reform after 1920, the postwar Red Scare, the resurgent Ku Klux Klan, a swelling antievolution campaign—that adults and children alike had failed to adopt a modern mentality.

The naturalization of moral freedom did not align science with ethical action in American public culture, as proponents had hoped. Rather, an array of critics argued that Dewey's approach, like the other varieties of mental modernization, destroyed morality altogether. Watson and Freud seemed to eliminate free will, and Russell and Krutch rendered the universe impervious to moral meanings. But to many observers, Dewey's naturalistic philosophy also obliterated morality. In their view, reformulating morality as a set of conditional, empirically grounded guidelines for ensuring functional success in the experiential world meant eliminating what made it morality in the first place: its transcendent, objective, extrahuman authority. All manner of critics—from fundamentalists to religious liberals, along with humanities scholars, political commentators, and others—objected that genuine moral freedom required the prior existence of absolute, non-subjective moral principles. To say that we craft moral judgments on the basis of everyday experience, they reasoned, is to say that we make up our own rules, reflecting our own viewpoints and interests. If this is the case, then the most powerful individuals and groups will simply impose the rules they construct. And what is this, Dewey's critics concluded, but the nihilistic doctrine that might makes right? Moral freedom, they argued, does not truly exist if it means the freedom to define morality oneself. A conditional, contextual morality is no morality at all.

Critics charged that Dewey and his counterparts destroyed religion as well as morality. Mental modernizers in Dewey's vein argued that religious doctrines, like scientific theories and moral principles, were human creations—hypotheses that helped individuals get around the world. One could say that such doctrines had worked in practice so far, but not that they corresponded to the actual structure of the world—in other words, that they were "true," in that word's usual meaning. Not surprisingly, this view alarmed many of religion's champions, including some liberals and modernists as well as more traditional believers. By the 1920s, it seemed to many observers that philosophers and social scientists had declared it impossible to know anything about the world, or even to believe anything on faith. Surely such a relativistic approach would undermine all that Americans held dear. The gap between academic and popular understandings of truth and knowledge widened substantially after World War I, fostering mutual distrust.

CULTURAL SIGNS

Many critics concluded that figures such as Russell, Krutch, Watson, Freud, and Dewey reflected the deepest cultural implications of a science that had come unmoored from moral strictures and familiar understandings of human behavior. In its modern form, they argued, science categorically denied the moral freedom of the individual and the possibility of purposive action. Still, even these critics were slow to see science as the dominant cultural force in their society. Through the 1920s, most worried about the inroads of a "business culture," characterized by greed, selfishness, and a materialistic emphasis on the things of this world. This understanding of cultural power in the United States was slow to change, even among religious leaders. But a few began to argue in the 1920s, based on evidence from popular culture and the surrounding society, that modern science had expelled morality from American public life. Even as some of the mental modernizers assumed that traditional philosophical and religious understandings anchored the business culture of the day, a number of critics concluded that the business culture took its shape from scientific teachings. Among these figures, changing social institutions and cultural practices fueled a growing perception that science endangered religion, morality, and common sense alike.

Even before the consumption-obsessed 1920s, World War I had brought a host of alarming developments. In 1914, the industrialized world—led by Germany, its most civilized and scientifically advanced representative— had fallen upon itself in a war of unprecedented severity. Much of the bloodshed, which increasingly claimed civilians as well as combatants, stemmed from technologically sophisticated weapons: machine guns, tanks, airplanes, and, most frightening to many, the deadly chemical gases used by the Allies as well as the Germans. Meanwhile, one explanation of the war itself traced its origins to German leaders' belief that Darwinism authorized a brutal struggle for survival between the nations. This claim, stoked by writings such as the prominent biologist Vernon L. Kellogg's book *Headquarters Nights* (1917), soon became a central exhibit in the case against Darwinism pressed by William Jennings Bryan and his theologically conservative allies.[11]

Developments in the 1920s offered further evidence to those who thought that science fostered widespread immorality—and convinced some of them that science had come to dominate American society as well as German culture. One key factor was the middle-class consumer culture that emerged after World War I. This phenomenon did not take shape until decades after the rapid industrialization process began in the United States. Of course, new mass-production techniques accelerated the flow of goods after 1920. But many observers also believed a cultural shift was afoot, fostering a massive new demand for consumer items.[12]

A noticeable turn toward vocational goals in the colleges strengthened the charge that Americans had begun to lose their moral bearings. College professors often lean on their perceptions of their own students when they assess the state of the surrounding culture. In the 1920s, a thoroughgoing shift was afoot. The public universities swelled with students seeking an economic leg up, as second- and third-generation immigrants—especially but not exclusively Jewish youth—sought to improve their vocational fortunes. Even private municipal institutions such as Columbia University and the University of Chicago witnessed a strong push toward vocational aims among their students, not least in the sprawling extension schools and summer programs that such universities built up in the 1920s.[13]

Meanwhile, college students, like other young Americans, also seemed to have succumbed to temptations far worse than vocationalism. In gin joints and fraternity houses, the sexual experimentation of the new generation shocked the sensibilities of most Progressive reformers as well as their more conservative counterparts. More innocuous pursuits such as football and the Charleston likewise reinforced the image of a generation shorn of higher ideals, chasing economic gain or perhaps just immediate, hedonistic pleasures.[14]

What accounted for these dramatic cultural shifts in the 1920s? It was easy enough to argue that consumerism flowed naturally from the massive expansion of productive capacity fostered by technological innovation. Did the changing values of the youth signal something more at work, however? Perhaps counterintuitively, advocates of mental modernization often traced the spread of materialism to the persistence of traditional religiosity. Figures such as Dewey argued that the incoherence and irrelevance of an antiquated, unscientific faith had allowed big business to commandeer religion itself. They cited evidence such as President Hoover's bland piety and the ad executive Bruce Barton's famous 1925 account of Jesus as history's most successful salesman. But other critics argued that scientific thinking had fatally weakened the nation's moral fiber, producing materialistic individuals alongside mechanized production. They traced recent cultural changes to the marriage of technology and commercial enterprise—personified by Hoover, a former mining engineer—with the "modern temper" of a scientific culture. The high-tech character of many leading consumer goods, such as watches, radios, and automobiles, further strengthened the perceived association of science with mindless consumption and conformity. In this view, the endemic moral failures of 1920s Americans represented the fundamentally amoral—even antimoral—quality of modern scientific thought.[15]

A range of cultural and institutional changes fed into this new argument. For one thing, the public was quite literally buying the new arguments about science and its implications. In the middlebrow book market, readers snapped up "surveys" and "outlines" that compressed vast swaths of material into digestible, book-size chunks. Inspired by the success of H. G. Wells's *Outline of History* (1920), authors and publishers churned out surveys of history, philosophy, medical discoveries, and much else. The philosopher Will Durant's tour through the history of his field was even serialized for newspaper syndication. The authors of these surveys typically advocated mental modernization, celebrating the intertwined progress of science, technology, and social liberalism—and often secularism or religious liberalism as well. Their popular treatments brought controversial views to the masses, at a time when many university professors also sought to modernize their students' minds through introductory survey classes and other courses in the natural and social sciences.[16]

Indeed, the historical narratives offered by the popularizers helped to constitute the image of modernity as a scientific age, rooted in the Enlightenment. None of these terms—modernity, scientific age, or Enlightenment—would have made sense to most Americans in 1920. Over the next two decades, however, our familiar categories for describing the last several centuries of American and European history coalesced in the United States, as they had somewhat earlier in Europe. These categories included the West, the Scientific Revolution, the Enlightenment, and the overarching concept of a modern age, or modernity. As we will see, critiques of science's cultural impact helped stabilize the meanings of this cluster of new categories. But those critiques gained strength from the fact that science's advocates also constructed the West, the Scientific Revolution, the Enlightenment, and modernity as linguistic categories, albeit in a much more celebratory manner. The historian James Harvey Robinson and the philosopher John Herman Randall Jr. were among the many scholars and popular writers who foregrounded science in their narratives of modern history, while decrying the lingering influence of traditional religions and other "forces of reaction." A host of other authors, journalists, and professors also declared nonscientific thinking hopelessly outdated in the Machine Age and sought to educate the public in the ways of modern science.[17]

What would a scientific age look like? The popular treatments that proliferated in the 1920s contained potent descriptions of human behavior as well as historical narratives. As usual, critics fastened on the most shocking proposals, especially from behaviorists such as Watson. Behaviorism gained visibility through numerous avenues in the 1920s, including Watson's writings for popular journals, popularizations by other psychologists, and a "Battle of Behaviorism" against the Harvard psychologist William McDougall, held at the Psychological Club in Washington in 1924 and later published as an affordable volume. Watson himself toiled to disseminate his idiosyncratic views on parenting in the 1920s. He insisted that parents should treat children with a flat emotional affect, seeking to avoid strong emotional attachments. They were to prevent children from sitting on their laps and to kiss them only in the rarest instances: once each night at bedtime, and occasionally also when they cried. In an ideal world, Watson wrote, children would be given to a new set of parents every four weeks until they turned twenty. For many commentators, Watson's approach revealed the ultimate tendency of modern science, as applied to human affairs. His writings reinforced the impression that scientific thinking ran counter to everyday morality and simple common sense.[18]

So, too, did the proliferation of advertising and public relations techniques in the 1920s and the systematic use of insights from modern psychology to improve these techniques. In an emblematic development, Watson went to work in the advertising industry after an affair with a graduate assistant cost him his faculty position at Johns Hopkins. Behaviorism's emphasis on creating unconscious associations through repetitive conditioning could inspire not only progressive programs of social reform but also efforts to drum up consumer demand for products, new and old. These initiatives left the impression that ad agencies, like the psychologists on which they drew, saw Americans in exactly the same terms as Pavlov's dogs, who salivated on cue when a bell rang—and that a growing number of Americans actually fit the mold, having abandoned age-old conceptions of rationality and morality.[19]

New theories in the physical sciences also ran afoul of common sense in the 1920s, as historians have noted. Nature itself had proven increasingly confusing and counterintuitive since the late nineteenth century, when mysterious forces such as radioactivity showed their faces. Of course, atomic theory and germ theory had already challenged conventional understandings in the nineteenth century, insofar as the phenomena they posited could not be seen with the naked eye. But those theories made some sense on a metaphorical level, because they posited entities and behaviors that roughly mirrored human-size counterparts. In the 1920s, however, new approaches stretched the popular imagination to the breaking point. Albert Einstein's theory of relativity, postulating a single matrix of space-time that curved in four dimensions, shattered all comparisons of physical nature to the world experienced by human beings. The term "relativity" itself reinforced the association of modern science with a loss of stable moorings. So, too, did Einstein's specific arguments: that no absolute measures of time or space existed, or that time could move at different rates for different observers. Yet Einstein also became a popular celebrity in the 1920s, providing the visual model that would henceforth come to mind when Americans heard the word "scientist." For those attuned to even more cutting-edge ideas, meanwhile, the emergence of quantum theory and Werner Heisenberg's uncertainty principle further eroded the sense that the world was meaningful and predictable, even at a subatomic level. The physical sciences, which many observers identified as the source of industrial production, also became linked in the 1920s to avant-garde rejections of the most fundamental principles of human experience.[20]

Biology, for its part, was also widely associated with a counterintuitive framework—in this case, a new understanding of human behavior. Through the 1910s, biologists and many social scientists embraced biological determinism, the view that human behavior stemmed from underlying biological drives or instincts. Biological determinists argued that behavior widely seen as the product of free will—of some combination of logical reasoning, moral commitment, and emotional sentiment—was entirely beyond human control, stemming from innate physiological structures. In the academic disciplines themselves, biological determinism largely faded in the 1920s, due to pushback from Watson, Dewey, and a growing number of others who rejected the "instinct" concept and stressed the environmental factors shaping human behavior. However, it continued to structure popular understandings of science's meaning through the 1920s and well beyond. Indeed, the most visible application of biological determinism, the eugenics movement, spread like wildfire after World War I. A host of popular eugenicists advocated the conscious, systematic breeding of human beings to remove from the population a set of undesirable traits that they traced to purely biological causes. The array of maladies ranged from physical handicaps to alcoholism, criminality, and even "pauperism": the endemic tendency to remain poor. Despite its growing distance from mainstream biological thinking, especially by the end of the 1920s, the eugenics movement represented one of the most striking outgrowths of the scientific enterprise in the public square.[21]

Critics attuned to theoretical shifts in the social sciences encountered additional challenges to conventional understandings of knowledge and morality in the 1920s. In that decade, many social scientists adopted a more detached, neutral stance and sharply distinguished scientific inquiry from social reform. This empiricist style affected the content as well as the form of scholarship. Realists in political science, for example, chronicled the messy and often alarming details of political systems as they actually worked in practice, rather than defining democracy as an ideal type and seeking to move society toward that ideal. At the University of Chicago, which became ground zero for such approaches in the 1920s, establishment moderates such as Charles E. Merriam joined iconoclasts such as Harold Lasswell in employing this solely descriptive approach. Moreover, these scholars often sought to replace ordinary language with quantitative measures and built their descriptive models on the assumption—drawn from Freud, among others—that human behavior was largely irrational. Outside the universities, the widely read political journalist Walter Lippmann adopted this realist style in a series of well-known books arguing that popular sovereignty could never work because citizens could neither understand nor constructively address the problems of the modern world.[22]

Other groups of social scientists likewise rejected prevailing cultural norms in the name of science. In the law, for example, Oliver Wendell Holmes Jr. and other advocates of legal realism and sociological jurisprudence denied that human laws should, or could, mirror higher ideals. They also argued that judges and juries alike were fundamentally irrational and sought to replace commonsense principles with social-scientific theories as the guideposts for legal decisions. Meanwhile, cultural anthropologists such as Franz Boas, Margaret Mead, and Ruth Benedict developed the principle of cultural relativism. They emphasized the central role of cultural conditioning in human behavior, minimizing the roles of autonomous thought and valuation. Boas and his allies also insisted that no culture could be compared invidiously to another. Each had to be taken on its own terms—and many primitive cultures outdid those of the United States and Europe in important respects. These anthropologists, like many other social scientists in the 1920s, rejected conventional understandings of race, class, and sex, as well as the nature and sources of human behavior more generally.[23]

Although natural and social scientists embraced a wide range of religious views in the 1920s, the writings of a few high-profile commentators proved to many observers that science required a form of militant atheism that rejected moral freedom as well as God. One was the philosopher Russell, who had done important work in logic and mathematics before setting out to undermine traditional religious understandings of the world. After Krutch echoed Russell's view in *The Modern Temper,* the journalist Frederick Lewis Allen's 1931 blockbuster *Only Yesterday* identified Krutch's bleak view of the world as definitive proof that embracing modern science meant abandoning God, morality, and human purpose. Another notorious public figure in the 1920s was the celebrity lawyer Clarence Darrow. Closely attuned to biology and the social sciences, Darrow emphasized the environmental determinants of behavior—including criminal activity—and ridiculed traditional understandings of free will and morality alongside Christianity.[24]

Significant changes in the scale, scope, and justification of state power reinforced the perception that science had become culturally dominant in the 1920s. We are accustomed to thinking of the American state as small and ineffectual prior to the New Deal of the 1930s, but the 1920s actually saw important new governmental interventions, many undertaken in the name of science or with the aid of scientific knowledge. These developments strengthened the association of science with social changes that alarmed many Americans, especially in theologically and politically conservative circles.

Education, as it so often has, became a key battleground for competing cultural visions in the 1920s. By the end of that decade, stepped-up enforcement of truancy laws and the growing economic stability of many immigrant families had made high school attendance typical among Americans. As urban school systems expanded rapidly, their administrators drew heavily on new techniques for sorting and classifying students. The standardized intelligence tests created by psychologists figured especially prominently. Developed before the war, such tests came to the public's attention with American intervention in World War I, when their widespread application to nearly two million Army draftees pegged the recruits'—and thus the wider population's—average mental age at under fourteen years. Changes to the high school curriculum during the 1910s and 1920s also reflected controversial interpretations of modern science. Courses in sex education raised firestorms in Chicago and elsewhere, while social studies and civic biology courses that aimed to both Americanize and liberalize urban youth spread even more widely. The increasingly bureaucratic and specialized character of American schooling itself seemed to reflect the de-personalizing tendency of scientific thought, as urban districts grew and used new forms of testing to place students into vocationally specific tracks. The continuing secularization of public school curricula rang alarm bells among many religious believers in the 1920s as well.[25]

State agencies also drew on medical expertise and biological theories to undertake unprecedented new impositions on the very bodies of citizens during the 1920s. Mandatory vaccination raised the ire of some political conservatives, who saw such programs as a dramatic amplification of state power that portended a draconian, authoritarian future. Meanwhile, bio-logical determinism increasingly influenced public policy as well as popular culture. Even as the intellectual tide turned against that framework among social scientists and then most biologists, it continued to underwrite eugenics, one of the most visible and controversial forms of "scientific" policymaking. Justice Holmes, already notorious for his dismissal of a morally meaningful legal system, firmly aligned the Supreme Court with eugenics in 1924. The 8–1 *Buck v. Bell* decision allowed—indeed, strongly encouraged—state-level medical institutions to sterilize all inhabitants that doctors deemed unfit to reproduce.[26]

In other areas, by contrast, science seemed to recommend inaction on matters that had traditionally been public concerns. In the 1920s, develop-ments in the social sciences were closely connected to the emergence of civil libertarianism, a view that identifies the freedoms guaranteed in the First Amendment as the core of democracy and defines their scope as broadly as possible. Civil libertarians focused especially on freedom of religion, freedom of speech, and freedom of press, often shorthanding these as "freedom of thought" or "freedom of expression." Although the new American Civil Liberties Union (ACLU) emerged to defend pacifists and political radicals during the war, the group was widely thought to have lined up with science against religion in the "Scopes monkey trial" of 1925. There, the ACLU eagerly took up the defense of Dayton, Tennessee, teacher John T. Scopes after he ran afoul of his state's antievolution statute in 1925. The fact that Darrow held a prominent place in the ACLU and led its cadre of defense lawyers down from Chicago strengthened the as-sociation of civil libertarianism with modern science's willful refusal to uphold foundational moral truths in the world. Like cultural relativism, its academic cousin, civil libertarianism seemed to indicate that distinctions of good and bad or right and wrong had no meaning whatsoever—or at least, that science said they did not. Darrow actively fostered that perception by helping to turn the Scopes trial into a spectacular showdown between traditional Christianity and modern, scientific agnosticism. A stream of acerbic newspaper articles by the science-minded cultural critic H. L. Mencken further cemented the idea that science and religious skepticism stood together against traditional religiosity in Dayton.[27]

Even before that trial, moreover, Darrow was infamous for promoting a new understanding of criminal behavior that denied perpetrators' moral responsibility for their actions. He had worked with well-known defendants—and often made controversial claims on their behalf—for three decades, including the socialist labor leader Eugene V. Debs during the 1894 Pullman Strike. But Darrow was best known for his defense of the young murderers Nathan Leopold and Richard Loeb in the 1924 "trial of the century." There, he notoriously argued that these spoiled, wealthy teenagers, who had kidnapped and killed their fourteen-year-old neighbor for pure sport, should not be held morally responsible for their actions, which instead reflected a combination of innate drives and environmental conditioning. Darrow's defense of Leopold and Loeb cut to the heart of the criminal system itself, with its assumptions of moral freedom and consciously chosen action. If science meant that criminals were no longer responsible for their offenses, then how could civil order prevail?[28]

DEWEY AND MANY OTHER ADVOCATES of science protested that it meant nothing of the sort. Yet they struggled to convince Americans at large that a naturalistic view of the world made room for morality, human purpose, or even conscious thought. In the eyes of the public, science entailed the mechanistic, reductive approach of the physicists—and Darrow—rather than Dewey's open-ended, morally inflected naturalism. The burgeoning charge that modern society was awash in materialism made it all the more difficult for Americans to imagine Dewey's nonmechanistic science. Meanwhile, a growing number of literary scholars and religious leaders ignored or flatly denied the assertion that science comported with moral freedom. They powerfully reinforced the widespread assumption that scientific explanations were by definition materialistic, deterministic, and quantitative. To the extent that Russell, Krutch, Watson, and Freud shaped public images of science—and they were far better known outside the universities than anyone in Dewey's vein—their writings, too, buttressed the belief that science was a thoroughly reductive, mechanistic enterprise, deaf to human concerns. If science had indeed reshaped modern culture, some commentators began to argue in the 1920s, then a string of horrors lay ahead.[29]

2

RESISTING THE MODERN

DESPITE THE ALARMING DEVELOPMENTS outlined in Chapter 1, relatively few commentators in the 1920s concluded that science had become culturally hegemonic. As we have seen, the assertion that science dominated Western culture and had squeezed out morality would have sounded absurd to most Americans before World War I—and so it remained for many of them in the decade that followed. Through the 1920s, conventional interpretations of American culture and identity, sustained by potent structural factors, still rendered it difficult to view science as a culturally dominant force. Above all, the new mode of critique ran afoul of the widespread assumption that the United States was a Christian nation with the Protestant churches at its cultural heart.

Yet the overlap between the popular vogue of psychology and seemingly congruent intellectual, cultural, social, and political changes inspired some groups of observers to argue that a scientific worldview had spread outward from the universities and captured the popular mind. Some critics of Darwinism began to question their long-standing assumption that American institutions and public culture remained thoroughly and reliably Christian in character. Numerous Catholic leaders and a few of their Protestant counterparts came to view science as a dominant cultural force. And small but vocal cadres of literary and artistic figures charged that the spread of scientific thinking explained the practical, utilitarian sensibility (and resulting aesthetic sterility) of modern society. Each of these groups concluded that the country's deepest problems lay in the realm of philosophical beliefs—in particular, beliefs about human beings and their capacity for moral action—rather than institutions, practices, or personal morals. In this environment, concern about science's cultural effects began to spread, along with claims about its cultural dominance.

LITERARY CRITICS

The celebratory portraits of a dawning "scientific age" that flourished in the 1920s were alarming to many critics, but also rather transparently aspirational. Advocates of mental modernization defined science as the keynote of modern thinking, the wave of the future. But they also granted that science's influence in the wider culture was anything but established. Indeed, it would require conscious, concerted effort to spread modern thoughtways. Even decades later, this assumption often persisted. "The genuinely modern has still to be brought into existence," John Dewey argued in 1948; it would take "the resolute, patient, cooperative activities of men and women of good will, drawn from every useful calling, over an indefinitely long period." Scientific thinkers, in this view, still needed to overcome an immense mass of ignorance and foolishness in the public domain.[1]

Of course, educated Americans also knew of a long tradition of European thought, dating back to the Enlightenment, which claimed that humanity had entered, or was about to enter, an age of science. But few of the mental modernizers followed European theorists such as Auguste Comte in arguing that Christianity would give way entirely to science as a source of guidance. Indeed, most of science's American champions actively prided themselves on avoiding such European aberrations. Militant and anticlerical varieties of atheism did gain some purchase in the United States during the 1880s and 1890s, but these largely gave way to less confrontational forms of naturalistic thought after 1900, especially in the science-centered universities. Even in the 1920s, most champions of the modern mind said it was thoroughly compatible—and perhaps even continuous—with liberal Christianity.[2]

By then, however, a small but growing group of critics had begun to see the scientific threat to morality as a defining feature of their time—and to think in new ways about science's relation to other cultural practices, especially religion and the humanities. Many joined the mental modernizers in viewing industrialization as a massive rupture in philosophical and religious thought as well as economics and politics. They agreed that age-old moral pillars had eroded alongside bygone economic patterns. Unlike the champions of the modern mind, however, these figures sought to re-build conventional structures of moral authority. Fearing that a scientific mind-set had begun to wreak havoc all around them, they argued that established moral tenets were more necessary than ever in the Machine Age, given the corrosive force of industrialization and the ways of thinking it had unleashed. Without reasserting timeworn beliefs and values, humanity could never tame the machine. Among the innumerable cultural structures dissolved by the "acids of modernity" (in Walter Lippmann's influential 1929 metaphor), these commentators feared above all for traditional conceptions of morality. Critics in both the humanities and the churches worked assiduously to shore up popular belief in the possibility of moral freedom and the existence of stable moral guideposts.[3]

This style of criticism took shape, for example, among groups of literary and artistic figures who decried the cultural effects of industrialization and modern science and held considerable influence among educated elites in New York and Boston. Such figures joined sympathetic philosophers in defining the humanities disciplines as, in effect, the antisciences: fields of study and modes of thought that actively promoted a morally and aesthetically robust understanding of the human person in order to counteract science's devastating influence on American culture at large.

These critics in the humanities inherited a lament about bourgeois civilization that dated back to the rise of the commercial middle class and flourished in Europe during the industrialization process, which seemed to portend the conversion of entire societies to bourgeois patterns. But in the United States prior to the Gilded Age, only a handful of poets, novelists, and philosophers worried about the culturally and spiritually deadening potential of commercial life. Even advocates of the classical model of higher education framed it as an aid to success in business and the professions, although they trumpeted the superior taste and virtue of the Greeks and Romans as well. But as the research universities and disciplines took shape after the Civil War, practical and scientific subjects elbowed aside classical learning. In response, some early theorists and practitioners of the humanities, led by Charles Eliot Norton at Harvard, picked up the concerns of Europeans such as Matthew Arnold and Alexis de Tocqueville about modern society's inclination toward cultural leveling, philistinism, standardization, and conformity. In the early decades of the twentieth century, commentators such as the medievalist and architectural critic Ralph Adams Cram likewise charged that mass production and consumption, combined with social equality, had eroded hard-won canons of taste and reduced culture to its lowest forms. Still, these champions of the humanities did not blame science for the cultural maladies they decried.[4]

However, the developments of the 1910s and especially the 1920s changed the contours of this critique, while also putting new wind in its sails. Although mass democracy and bourgeois values remained common targets, some critics in the humanities now began to target other social and cultural changes. A few identified technology as the culprit. In fact, even some of science's champions lamented in the 1920s that human beings had become machinelike under the cultural influence of modern technology. To some humanities scholars, likewise, the shallow, amoral consumerism of that decade signaled that the machine's cultural impact had come into full view.

But many artistic and literary critics identified science as more than the source of the machine. A growing number saw direct forms of scientific influence behind the cultural malaise of the age. These critics came to view science's cultural impact in terms of its effects on moral reasoning and views of the human person rather than the pressures of the machine on work, consumption, political economy, and cultural tastes. They found science's deepest meaning in the deterministic, materialistic visions of Russell, Krutch, Watson, and Freud and saw Dewey's pragmatic, functional morality as simply another form of scientific materialism. Equating science with a mechanistic, quantitative outlook, these critics traced modern culture's faults to the extension of that reductive approach beyond science's proper domain of physicochemical interactions. Critics in the humanities recoiled, especially, at attempts to develop a science of human behavior, which they believed would eliminate human purposes and moral freedom by explaining individual actions solely in terms of external causes.

Looking back on the arts in the 1920s, the champions of literary modernism dominate scholarly and popular memories. "Lost generation" writers such as F. Scott Fitzgerald and Ernest Hemingway deplored the cultural tenor of the era and sought an antidote in formal experimentation and psychological realism. Yet these figures did not wield the greatest influence in their day, despite being the best-remembered interwar critics of cultural philistinism now. Like Poe and Hawthorne before them, the modernist authors of the 1920s often gained their fame in retrospect, through the canonizing efforts of a later generation of humanists disaffected with the cultural tendencies of the post–World War II era. At that time, the best-known critics of the Machine Age were far more conventional in their tastes and values. Indeed, many of these 1920s commentators equated literary modernism itself with middle-class consumption and sexual rebellion, calling it yet another amoral, self-destructive form of instant gratification. These critics deemed the formal experimentation of a Stein or a Pound profoundly selfish—a perverse rejection of the social responsibility of artists and critics to uphold societal standards, both aesthetic and moral.

This was certainly true of the New Humanists, some of the earliest cultural critics in the United States to ascribe a direct, negative cultural influence to science. That group, led by Harvard's Irving Babbitt and the Princeton-based writer Paul Elmer More, combined the anticommercial sentiments of Babbitt's teacher Norton with a sharp rebuke of modern psychology. The New Humanist movement gained steam in the 1920s and crested in 1930 with a host of articles on its merits and shortcomings, edited volumes debating its tenets, and even a Carnegie Hall discussion. The movement's advocates brought questions about science's cultural implications to the fore and defined science's meaning and influence in ways that would also characterize many subsequent challenges.[5]

Babbitt, More, and their New Humanist allies identified psychology as the central site of conflict between modern science and a healthy view of morality. They defined the human person in dualistic terms, arguing that the higher, conscientious self properly reigned supreme over the lower, willful self. More variously called this higher faculty a "law of taste" or a "faculty of the soul," while Babbitt spoke of a "higher will or power of control." However defined, the higher self was the universal aspect of the human person—the truly human element, which all individuals shared and which would bind them into a web of mutual service if recognized, cultivated, and properly valued. The existence and primacy of this higher faculty, wrote Babbitt, constituted a "human law" that could be conclusively proven. To validate that law, however, the investigator needed to follow "tradition" or "intuition," looking beyond mere sensory evidence to "the something in himself that is set above the flux and that he possesses in common with other men." Babbitt's "ethical positivism" would run parallel to scientific empiricism and stem from viewing "the data of experience" through the lens of "happiness" rather than "power and utility." In short, he argued that equally authoritative forms of moral and psychological knowledge operated alongside the natural sciences, and that these pursuits indicated the human meanings of merely empirical conclusions.[6]

On the basis of this psychological and methodological dualism, the New Humanists defined democracy as the equal opportunity to "measure up to high standards": a laborious process wherein humanistic elites patiently retrained the masses to resist their base instincts and prevailing fashions alike, and instead to follow universal principles distilled from long human experience and captured in great works of Western literature. All of the work of civilization would be undone, said Babbitt and More, if the masses failed to heed their higher faculties; if the humanistic elite shirked its task of cultivating such faculties; or if the educational system failed to produce a humanistic elite. But at present, they argued, psychologists and educators in "an advanced stage of naturalistic intoxication" were busily undermining all three pillars of civilization.[7]

Even as the New Humanists excoriated modern psychology and empiricism more broadly, they also mobilized scientific authority themselves. More's younger brother Louis Trenchard More, a physicist and historian of science at the University of Cincinnati, served as the New Humanist movement's resident expert on science. Ignoring the protests of the prominent journalist Lippmann and other critics of the movement, the younger More strictly limited the term "science" to the mechanistic, mathematical approach of classical physics. The "aim of all science," he repeatedly asserted, is "to express its laws in the language of mechanics." Yet he also argued that the scientist's reductive approach represented the application of a particular, contingent interpretive framework. Mechanistic analysis was merely one intellectual tool among many, and hardly capable of grounding a full-blown philosophy of reality. Far from providing a universally valid picture of the world, scientists simply filtered human experience through their distinctive, utilitarian lens.[8]

Just as they drew on the authority of science, Louis Trenchard More and the other New Humanists also wielded the language of empiricism against the "pseudo-science" of the philosophical naturalists. They argued that scientific materialists had not discovered a world with merely spatiotemporal dimensions, but had simply imposed a narrow metaphysical frame onto a far richer reality and ignored its other features. What could be more dogmatic and antiempirical than arbitrarily excising the many important elements of the world that were empirically discoverable but could not be expressed in science's mechanistic vocabulary? According to the New Humanists, scientists failed to live up to their own empiricist standard if they denied the existence of phenomena that the physicist's interpretive lens obscured. Fortunately, they believed, genuine scientists were beginning to throw off the "metaphysical dream" of modern materialism and recognize that its "mechanistic hypothesis" represented a valuable technique rather than a stable worldview—that this framework was "relative and provisional," not "absolutely true."[9]

Most importantly, the New Humanists argued, scholars were coming to see that scientific methods were unsuited to studying "the phenomena of the subjective world" and thus could say nothing about "the problems of life." Indeed, they contended, modern thinkers were gradually recognizing the "primordial" fact of a higher faculty and an associated human law. Louis Trenchard More agreed with Babbitt that there were multiple, equally valid forms of knowledge, each attained by a distinctive method. Sensory evidence, he allowed, offered a valuable, if hardly definitive, window onto spatiotemporal phenomena. But he urged students of human subjectivity to resist the behaviorists' materialistic dogma altogether and embrace the alternative method of "self-examination" that introspective psychology had long employed. Only in this manner could investigators conclusively prove the existence of the higher will, with its unique capacity to discern universal human laws. The younger More's analysis promised that psychologists could rejoin the humanistic elite by turning back from experimentation to introspection. They, too, could sustain the cultural standards and broadening experiences that the democratic masses needed to subordinate their lower, instinctual selves to their higher, civilized selves.[10]

Crucially, Babbitt, the More brothers, and the other New Humanists identified modern psychology's reductive, materialistic view of the human person as a clear and present danger, not a distant threat. All around them, they saw educational and cultural initiatives reflecting a "naturalist conspiracy against civilization" that had produced the conflagration of global war, among other frightful outcomes. The problem was simple: Many working scientists, psychologists, and educational theorists failed to grasp the purely instrumental character of science's mechanistic framework and interpreted it as a comprehensive worldview instead. That was especially true, the New Humanists argued, of the misguided "humanitarians" controlling education, with their faulty psychological approach, their naturalistic "gospel of service," and their "deterministic denials of man's moral freedom."[11]

As would so many other critics in the coming years, the New Humanists placed Dewey, an increasingly influential theorist of philosophical naturalism, social psychology, and education alike, at the head of the "prophets of the flux." Ignoring the distance between Dewey's formulation of naturalism and the reductive approach of figures such as Watson, they read his attacks on psychological dualism as evidence that Dewey reduced human behavior to purely material causes—and falsely assumed that human beings were intrinsically good. In reality, Babbitt argued, human goodness succumbed easily to base motives. A citizen armed with advanced technology but no moral control became "an efficient megalomaniac." For Babbitt and the New Humanists, the only way forward in a technological, democratic age was to resist the campaign for mental modernization at all costs, adopting a dualistic view of the self and patiently training oneself and others to "like and dislike the right things," with the aid of time-tested literary works.[12]

Whereas the New Humanists equated science with the provisional, instrumental use of materialistic presuppositions and called for alternative, nonscientific ways of understanding human behavior, other literary and artistic critics sought to elevate a genuine science—one that encompassed moral and aesthetic judgments—over the ersatz, mechanistic version that they thought dominated American culture. Among the most visible of these critics were the Southern Agrarians, the Nashville-based group led by John Crowe Ransom and Allen Tate that issued the well-known volume *I'll Take My Stand* in 1930. Like many champions of literary and artistic modernism, these figures combined an avant-garde style with cultural traditionalism. They looked back to an imagined past when artistic innovation seemed to have flourished organically in Western societies. Many of the chapters in their 1930 collection argued that the science of their day, like all other forms of cultural expression at present, reflected the faulty social philosophy of industrialism rather than a genuinely human understanding of social relations. Although these writers believed that the results of scientific research itself were unimpeachable, they argued that the corrupt, inhuman social environment created by industrial development had badly skewed the practical applications of science.[13]

The Southern Agrarians, like the New Humanists, drew on the authority of a working scientist—in this case, the Vanderbilt psychologist Lyle H. Lanier. In a chapter attacking Dewey's philosophy, Lanier argued that science's promise would be realized only when its practitioners consciously pursued the "right relations of man-to-nature" underlying true religion and art and the "right relations of man-to-man" underlying the social graces. Indeed, Lanier called for "far-sighted 'social engineering'" to redirect social resources from industry to agriculture, in order to rehabilitate "the agrarian economy" and thereby restore the "old individualism" it had fostered. In the volume's introduction, the Agrarians traced the needed humanistic culture to an "imaginatively balanced life," "lived out in a definite social tradition" such as that of the preindustrial South and sustained by a comprehensive "standard of taste" that encompassed social and economic conditions alongside the arts. Scientific applications guided by this human standard would inevitably redound to the good, the Southern Agrarians argued.[14]

Another critic of modern science, Lewis Mumford, adopted a more forward-looking social vision while sounding many of the same themes. He urged modern thinkers to abandon their fixation with physics and draw out the cultural possibilities latent in recent biological theories. Like fellow "Young Intellectuals" Van Wyck Brooks, Harold Stearns, and Waldo Frank, Mumford looked back on recent history with dismay. He argued that the combination of pell-mell growth with a mechanistic worldview derived from physics had dramatically impoverished American culture since the Civil War. Meanwhile, Mumford joined Lanier and the New Humanists in identifying Dewey's pragmatic naturalism as the epitome of modern thought's sterility. Dewey's deflationary account of moral claims as practical, fallible judgments, he argued, eliminated all genuine considerations of value. Mumford described Dewey and the historian Charles Beard as the leaders of the "New Mechanists," who believed that "physical science will one day serve as a key to every aspect of experience" and that wisdom entailed nothing more than adjusting all human behavior to "external facts."[15]

Whereas the New Humanists called for introspection and the "Twelve Southerners" emphasized an agrarian way of life, Mumford saw the needed alternative to mechanistic, materialistic approaches in another corner of the sciences themselves: modern biology. In particular, he seized on the new concept of emergent evolution as the philosophical basis for the "organic humanism" of the future. Taking hold among many philosophers and biologists in the 1920s, emergent evolutionism held that there were multiple phases or levels of reality, each with distinctive dynamics that could not be understood by reducing them to the terms of lower levels. In Mumford's rendering, the levels of reality, in ascending order of complexity and importance, were the physical world, the organic world, the social world, and the individual personality. With each step up the ladder, he asserted, interpreters necessarily moved away from strict certitude and quantitative measures toward subjective judgments of value and quality. Ever since the breakup of the medieval synthesis, however, scientists had wrongly treated the other levels of reality as extensions of the purely mechanistic, quantitative domain addressed by the physical sciences. By so doing, they had torn asunder values from facts, art from science, and human purposes from the conditions of their achievement. Unlike the New Humanists and Southern Agrarians, who connected the cultural maladies of their day to the recent industrialization process, Mumford joined Catholic critics in arguing—albeit in a secular idiom—that the mechanistic, materialistic worldview of the era reflected age-old philosophical errors and had dominated all of Western civilization for centuries.[16]

RELIGION AND THE MODERN MIND

Of course, Protestant leaders still held pride of place in the popular culture of the 1920s. Although the writings of literary and artistic critics reached influential segments of the population and reflected important conceptual shifts, the portraits of science offered by religious leaders would ultimately set the tone of the public conversation in the United States. In the churches, many carried forward the conventional association of science with moral progress in the 1920s. This perception took strength from the continued power of theistic evolutionism—the view that the evolutionary process was God's preferred means of perfecting humanity, both physically and morally—among mainstream Protestants (and many Catholics) in both the churches and the universities. But a number of these theistic evolutionists began to worry that the spread of scientific materialism threatened the divine plan by fostering atheism and reversing hard-won moral progress. These figures abandoned the long-standing assumption that a pious Christian majority controlled American society, despite the presence of a vocal cadre of atheistic naturalists. As in the humanities and artistic circles, groups of religious critics came to believe that the avatars of science enjoyed substantial influence over the minds of the masses. They blamed modern science's mechanistic approach to the world for a lengthy roster of social problems, from the immoral hedonism of the rising generation to the mindless consumerism of the suburbanites to the public's robust appetite for outlines of modern psychology.

Religious critics of science's influence were less likely than literary and artistic commentators to stress the homogenizing impact of the machine, although a few sounded that note as well. Rather, they tended to focus on philosophical questions and perceive the kinds of direct cultural harms from scientific thinking that the New Humanists, the Southern Agrarians, and Lewis Mumford had detected. Science, they argued, had burst its bounds, pushing beyond its proper domain—the interpretation of natural phenomena—and badly eroding religion and morality in the process. These critics, too, identified modern science as a materialistic, deterministic mode of explanation that ruled out not only supernatural forces but also the entire domain of human subjectivity: purposes, values, emotions, meanings. Like their literary and artistic counterparts, religious critics typically allowed that scientific materialism provided the proper lens through which to view certain phases of reality. But they insisted that it did not apply to human beings, except in their basic physiological contours. Understanding humanity required very different interpretive methods, based on evidence beyond that available to the senses.

In a process that began in the 1920s and stretched into the 1950s, the spread of this analysis of science's cultural impact steadily led groups of mainline Protestants toward a much more skeptical view of the scientific enterprise as a whole. The separate spheres model crafted by nineteenth-century theological liberals gave religion a prominent intellectual and social role as the conservator of moral absolutes, justified by nonempirical means. But now, it seemed, scientific thinkers had crossed the line, invading religion's special province by challenging traditional understandings of morality. Dewey attacked the very idea of absolute truth, while Russell, Krutch, Watson, and Freud made a mockery of moral freedom. During the decades after World War I, multiple generations of religious thinkers who had painstakingly adapted their theology to Darwinism and defended modern science against theological conservatives came to believe that their scientific allies had betrayed them.

By the 1940s, mainline Protestants who identified religion primarily with moral truth increasingly joined theological conservatives in arguing that scientific worldviews threatened American society at large. Looking around them, they charged that scientists controlled precincts of American culture far outside their assigned sphere, with particularly disastrous effects in the moral domain. They variously blamed social ills on the faulty thinking of natural scientists seeking to extend their insights to the social world, on social scientists failing to recognize the inapplicability of scientific methods in their fields, on naturalistic philosophers offering science-based worldviews, and on the journalistic, political, and educational allies of these academic cadres.

These Protestant critics, like their counterparts in the humanities, demonstrate clearly that scientific authority is not a pill one swallows whole. They never questioned the legitimacy of physics, chemistry, or even evolutionary biology as interpretations of the natural world. But they rejected associated forms of philosophy and social science—above all, modern psychology. In fact, these critics redrew the boundary between the separate spheres of science and religion. In their new separate spheres model, science provided reliable, empirical knowledge of the spatiotemporal world, as it had before. However, religion also provided knowledge—genuine, validated *knowledge,* not just intuitions, leaps of faith, or biblical parables—within its own domain of moral and spiritual truths. This redefinition of core religious tenets as truths rather than faith commitments decisively shaped what historians have called the "Protestant theological renaissance" of the interwar and post–World War II years.[17]

In turn, of course, changing views of science's character and cultural roles—especially the perception that modern culture embodied a scientific rejection of morality—fed into the new understanding of religion as a source of genuine truths. This dynamic is evident in the earliest writings of Reinhold Niebuhr, who later taught for decades at Union Theological Seminary in New York. The preacher-turned-theologian Niebuhr played a central role in the theological renaissance by developing a highly influential version of "Christian realism" that directly challenged science as a source of guidance in human affairs. Niebuhr's famed assault on theological liberalism reflected his understanding of science as a culturally dominant and thoroughly materialistic philosophy. By welding this understanding of science to a sharp critique of capitalism, he also rejected the long-standing reliance of socialists, both Christian and secular, on science as an interpretive resource. A native German speaker who kept close tabs on European thought, Niebuhr picked up key arguments about science and modern social institutions from Max Weber and other commentators.[18]

Although Niebuhr's mature themes of sin, irony, and power would emerge by the 1930s, much of his earlier work focused squarely on what he considered naturalistic thinkers' pathological obsession with "the forces of external existence." His 1927 book *Does Civilization Need Religion?* equated science with determinism and then argued that its cultural and institutional influence had destroyed the "personalization of the universe" that characterized a genuinely religious culture. Science only comported with religion, Niebuhr wrote, when its atomizing tendency was embedded in a broader, integrative philosophy that upheld moral freedom and ruled out all deterministic pretensions. This was also the only way to prevent science from destroying society, he added. According to Niebuhr, science's application to the real world had fueled a relentless "depersonalization of civilization" since the nineteenth century. With metaphysics weakened as "a coördinator of the sciences" and human relationships robbed of all personal, moral dimensions by the "mechanization of society," science now had free rein to continue its depredations.[19]

In particular, Niebuhr wrote, "mechanistic psychiatry and psychoanalysis" had usurped religion's claim to unify the self, though they could never actually deliver. Meanwhile, the "absolute determinism" of a "purely naturalistic ethics" left a brutal struggle between individuals, groups, and nations as the only means of "social reconstruction." As "absolute standards of truth, beauty and goodness" sank in a mire of materialistic greed, Niebuhr summarized, "Western civilization is enslaved to its machines and the things which the machines produced." The ruthless battles between "Nietzschian and Marxian cynics" that dominated modern life were "lazily witnessed by vast hordes whose main purpose in life is to gratify their senses and who give their sympathy to one or the other side according as it offers least hindrance to their enjoyments." Niebuhr foresaw total social collapse unless a new cadre of "spiritualized technicians" could grasp modern society in its divine dimensions and take charge of machines without becoming machines themselves.[20]

Throughout the rest of his career, Niebuhr would continue in various ways to indict scientific naturalists for modern ills. By the early 1930s, the outlines of his mature interpretation were clear: ultimate blame lay with naturalistic social scientists and philosophers and their theologically liberal allies in the churches, not the natural scientists whose methods and outlooks they fruitlessly aped. Niebuhr's explosive 1932 book *Moral Man and Immoral Society* argued that a pair of hopelessly unrealistic and mutually reinforcing assumptions had systematically undercut campaigns for social and economic justice in the United States. These two pillars of bourgeois liberalism, each reflecting its "romantic overestimate of human virtue and moral capacity," were scientific materialism and sentimental Protestantism. Liberal Christians, Niebuhr argued, sought to address social conflicts through love, while scientific naturalists such as Dewey relied solely on rational argument. Neither strategy could possibly work, whether alone or in tandem. Niebuhr held that the self-interest of individuals, and still more the self-interest of groups, inevitably corrupted their judgments, both moral and rational. He excoriated Dewey and his allies in the universities and education schools for believing that "our social difficulties are due to the failure of the social sciences to keep pace with the physical sciences." In fact, Niebuhr argued, the social sciences could never attain the objectivity of the physical sciences because reason was subservient to interests in the social domain: "The will-to-power uses reason, as kings use courtiers and chaplains to add grace to their enterprises."[21]

The liberal Protestants that Niebuhr flayed were hardly uncritical celebrants of science, however. Even self-proclaimed "modernists" such as the Baptist writer Shailer Mathews, who explicitly embraced both Darwinism and the empirical findings of the social sciences, often had their own concerns about science's cultural influence. Mathews defined theological modernism itself as "a phase of the scientific struggle for freedom in thought and belief" against dogmatism and authority. Yet he insisted that it had nothing to do with "the naturalistic, mechanistic interpretation of nature given by those who deny the existence of personality in the universe." The modernist, said Mathews, joined hands with the fundamentalist in denying that "evolution is a Godless impersonal process." Meanwhile, he portrayed science as an utterly dominant force in the modern world, lying behind "advertising and meat-packing, oil-finding and automobile-building, radio concerts and a million other things in which the human mind has grown accustomed to think in terms of facts and inferences rather than authority." But no amount of knowledge could generate the virtue that religion alone provided, Mathews warned. Class conflict further ensured that the "deep motives of human life are diseased," he contended, and the "naturalism of some men of science only makes for deeper distrust of mankind," leaving "pagan enjoyment of animal life" as the sole motive for behavior. Like Niebuhr, Mathews presented modern civilization as the joint product of science-based technologies and scientific materialism, which together had ousted all moral considerations from human affairs.[22]

Much of the criticism by mainline Protestants in the 1920s centered on behavioristic psychology in particular. Concern had grown sufficiently widespread by the end of the decade that a host of Protestant leaders issued a collective rebuttal to Watson and his allies in *Behaviorism: A Battle Line*. As the title suggests, these Protestants welcomed the natural sciences but drew the line at applying science's naturalistic outlook to human affairs. One contributor after another asserted religion's primacy in the interpretation of human behavior.[23]

The liberal Congregationalist minister John Wright Buckham became a leading Protestant critic of modern psychology. That field, he wrote in 1924, had transformed itself from "a respectable and harmless department of philosophy" into "a flood, inundating religious life, business, therapeutics, art, literature, education." What Buckham called the "New Naturalism" of modern psychologists denied the existence of the human self as a moral person, undermining the core principles shared by all moral and religious systems. But Buckham reassured his readers that this antihuman, antiethical portrait of persons as "mere mechanisms, automata, products (or by-products) of natural forces" was not in itself a scientific concept. Rather, it was a speculative extrapolation from the natural sciences—and one that natural scientists themselves actively worked to destroy. According to Buckham and like-minded Protestants, modern psychology did not simply turn individuals away from the divine law. It ruled out of court not only God but also the very concept of free will—the capacity to choose whether or not to follow the moral law. This interpretation rendered traditional morality fictitious and impossible, not merely foolish or irrelevant.[24]

Even when they identified psychology as the root cause, most mainline Protestants were vague on how, in institutional terms, scientific naturalism had infected the wider society. By contrast, theologically conservative Protestants often singled out a particular social institution: education. Of course, the fundamentalists, Pentecostals, and other theological conservatives who populated the ascendant antievolution movement in the early 1920s aimed most immediately to establish popular control over high school biology classes, as witnessed in the "Scopes monkey trial" of 1925 and the state-level statute that occasioned it. But antievolution leaders such as William Jennings Bryan decried Darwinism's influence on American higher education as well as primary and secondary schooling. Bryan traced the content of high school curricula to the machinations of a "scientific soviet" that already controlled the universities and was extending its reach into lower levels of education. He made much of the psychologist James Leuba's 1916 study, which showed that only 42 percent of American natural scientists believed in a personal God and only 51 percent believed in immortality—and that these numbers declined substantially among the top tier of scientists. Citing Leuba's results and other evidence, Protestant antievolutionists sought to oust Darwinism from universities and high schools around the nation.[25]

Although antievolution activists worried about the salvation of individual students and insisted that local communities should control education, they also lamented the cultural damage caused by modern science more broadly. The politically progressive Bryan, for example, saw Darwinism's emphasis on ruthless material struggle as the cause of competitive capitalism, the eugenics movement, and German militarism alike. Indeed, he told the evangelist Aimee Semple McPherson in 1925 that the world's troubles stemmed from a "combination of all unbelievers against Bible Christianity." Identifying the mechanistic outlook of modern science as a comprehensive threat to Western civilization, Bryan interpreted Darwinism itself as simply one more in a long line of misguided attempts to extend physical scientists' materialistic framework into a field of analysis—the history of human life on earth—where it did not apply. Theological conservatives such as Bryan argued that the vogue of Darwinism and the historical criticism of the Bible in the late nineteenth century had put society on a philosophical path to ruin. Fortunately, these Protestants believed, the scientific disease afflicting American culture dated back only a few decades and the solution stood at hand in the "old-time religion"—the doctrinal rectitude that most living Americans had learned in their youths and many still upheld.[26]

Catholic thinkers also centered much of their criticism on the schools. However, they often embedded their specific concerns in a far broader historical account encompassing developments across the centuries, extending all the way back to the breakup of medieval Christendom. Most other critics in the 1920s focused on intellectual tendencies in the United States since the nineteenth century, such as evolutionary thinking, theological liberalism, and "industrialism." By contrast, Catholic theorists defined what came to be called "modernity"—a concept still typically captured in adjectival constructions such as "the modern world" or "modern thought"—as a centuries-long phenomenon that characterized Europe as well as the United States.

Catholic leaders, and some laypeople as well, tended to view American events through a European lens, with particular reference to church-state battles in France. Indeed, the French Revolution and its secularizing inheritors came to figure centrally in Catholic theorists' perception of American culture during the 1920s. Of course, Catholics had long argued that Western societies had been unraveling since the Protestant Reformation. In their view, the fateful embrace of the private interpretation of the Bible had destroyed the basis for legitimate social and political authority. But this argument had obvious limitations for an immigrant-based community seeking cultural acceptance in a Protestant-dominated country. Over the course of the 1920s and 1930s, American Catholics such as the historian Carlton J. H. Hayes began to locate the source of modern ills in the eighteenth-century Enlightenment, not the earlier Reformation. The problem, in this view, was the concerted attempt to organize Western culture and institutions around science, and to ignore their true, religious sources.[27]

Hayes and many other Catholics of the 1920s highlighted the militant anticlericalism of French rationalists and secularists, drawing a direct line to the writings of Dewey and his naturalistic allies. They charged that recent changes in American education had brought the schools into alignment with a tradition of secular rationality that the Church had been fighting since the earliest phases of the European Enlightenment. In this regard, many American Catholics criticized the teaching of evolution in the 1920s, although they found the attempts of antievolution activists to install Protestantism in the public schools equally alarming. But they embedded the educational changes of the 1920s in a longer historical frame, seeing them as continuous with other expressions of a materialistic, self-divinizing form of rationalism that had plagued the West for centuries.[28]

Not that Catholics had any problem with human reason, when it was properly applied and bounded. Indeed, the neo-Thomistic philosophy that spread rapidly among Catholic intellectuals in the early twentieth century held that scientific thinkers had abandoned rationality itself when they decided to rely solely on sensory evidence. In the United States, neo-Thomism emerged as a powerful force in the mid-1920s, with the launch of *The Modern Schoolman* at St. Louis University in 1925 and the founding of the American Catholic Philosophical Association in 1926. Neo-Thomists presented themselves as champions of genuine reason against a culture of despair and drift, offering a stable source of social guidance for an intellectually unmoored citizenry. Indeed, they identified their philosophy as the perfect complement to other features of modern, industrial societies. One neo-Thomist, for example, proposed mobilizing "all the mind-reaching media that American genius has so far developed" in the campaign against scientific irrationality. Modern civilization, in this view, could be redeemed only by reversing its empiricist drift.[29]

Largely bracketing the question of priestly authority, Catholic neo-Thomists argued that individual minds could access the natural law—an underlying structure of permanent, divinely ordained moral truths that could guide them in any and all situations. By contrast, they charged, the empiricism of modern science—its refusal to acknowledge anything beyond sensory data—led interpreters to overlook the moral order, which operated in another non-sensory domain of experiential reality. Like many other critics at the time, the American neo-Thomists were particularly exercised by modern psychology and flatly denied that a narrow, empiricist approach applied to questions about human behavior. This was particularly true for the leading proponent of "everyman's Thomism," Fulton J. Sheen. In a series of books beginning with *God and Intelligence in Modern Philosophy* (1925) and stretching into the 1960s, Sheen defined his neo-Thomistic position against pragmatism and other forms of modern philosophy. These systems, he contended, all shared the foundational defect of starting with human experience rather than divine truth. After 1930, Sheen's message reached many millions of people through his popular radio show; he made the transition to television in 1952.[30]

Other Catholic thinkers of Sheen's generation engaged more sympathetically with secular discourses. Rather than rejecting American thought in its entirety, figures such as Michael Williams and George N. Shuster positioned themselves as interlocutors of Mumford, Brooks, Lippmann, H. L. Mencken, and other leading thinkers of the day. At the same time, however, they discerned little promise in American culture, despite the presence of this handful of insightful observers. Shuster, for example, lauded critics of modernity such as Cram, the historian Henry Adams, and the poet Louise Imogen Guiney for their interest in medieval themes. But he argued that "official America" was now firmly in thrall to a "practical realism" or "here-and-now positivism" that faced few substantive challenges. The Civil War, Shuster explained, had released a flood of practical expansion that killed off philosophy entirely and produced "a new rationalism" that picked up the science-worship of the eighteenth century but jettisoned its humanitarianism. The late nineteenth century's "newer positivists," Shuster allowed, had "believed in human progress with a rare firmness, and stroked their constantly increasing machines with affection." But the endpoint was inevitable: "a landscape so denuded of beauty that it bespeaks the barbarian; an almost infinite natural wealth bottled up and possessed by a few; the rise of industrial cities and the frank acceptance of class warfare; and, ultimately, the weakening of the appeal of religion to such an extent that sixty out of every hundred citizens would profess no attachment to any creed." Shuster added that pragmatist philosophers such as William James had ratified the practical, amoral outlook endemic to modern society. "James simply noted the universal prevalence of a mental disease or state of inactivity," Shuster declared, "and because it was normal he called it health." As would critics of Alfred Kinsey's reports on human sexuality after World War II, Shuster held that scientific thinkers replaced substantive moral tenets with merely descriptive guidelines that enjoined individuals to do what everyone else was doing.[31]

From the other side, literary and artistic figures also explored the many points of overlap with religious critics of modern science. The New Humanists engaged religion in several different ways. Paul Elmer More spoke in a Christian idiom, equating the higher self with the soul. Babbitt, by contrast, sought a nonsupernatural, psychologically grounded framework that he sometimes described as a secular religion. He found resources in the teachings of the Greeks and Buddha. Still, many supporters saw more theologically conservative implications in Babbitt's work. The Catholic scholar Louis J. A. Mercier, an ally of the New Humanists, argued that Babbitt actually embodied a broad move toward Catholicism among American intellectuals. For his part, the poet and cultural critic T. S. Eliot, who had studied under Babbitt, denied that an independent humanist tradition could exist and argued in 1926 that his former mentor faced an either-or choice between naturalism and supernaturalism. Faced with Eliot's challenge, Babbitt agreed and aligned himself with Christianity's recognition of a higher nature in the human person.[32]

Yet Babbitt's rhetoric pointed forward in time, not backward to a centuries-long Christian tradition. He portrayed the New Humanism as a path through and beyond the modern mind-set, writing that it foreshadowed "a point of view so modern that, compared with it, that of our smart young radicals will seem antediluvian." Although the New Humanists' emphasis on cultural leadership sounded more Catholic than Protestant, they employed the same language of liberation from arbitrary human authority as American Protestants and scientists alike. They also described their "scriptures," the great works of Western literature, as human achievements rather than revelations from on high. Following such texts, in their view, represented a reasoned deference to the sifted wisdom of the ages, not a slavish devotion to tradition.[33]

A melding of artistic and religious concerns also characterized the work of many traditionalist Episcopalians (often called "Anglo-Catholics") in the 1920s. Eliot, of course, exemplified that pattern. Converting to Anglicanism and taking up British citizenship in 1927, he grew increasingly critical of mass democracy, calling himself a classicist and royalist. In the 1930s, Eliot famously championed a Christian society as well. Back in the United States, Bernard Iddings Bell, who was more of an American-style conservative in his politics than Eliot, wrote extensively against naturalism and connected it to the collectivistic tendencies that he believed were swamping true individualism in the modern world. From his position as head of St. Stephen's (later Bard) College, Bell excoriated the leveling tendencies of modern education and advocated an intellectual aristocracy grounded in foundational moral principles. At the end of the decade, the Anglo-Catholic fiction writer and musician Harvey Wickham, who had moved to Rome, reached broad American audiences with a trilogy of blasts against modern culture in *The Misbehaviorists, The Impuritans,* and *The Unrealists.* In Anglo-Catholic circles, literary and religious concerns tended to fuse into forms of cultural (and often political) conservatism that set the authority of Western tradition against the widespread misapplication of mechanistic models drawn from physics.[34]

OVERALL, A HOST OF PHENOMENA struck these Anglo-Catholics and other 1920s critics as evidence of science's deleterious influence. Heading the list were the new consumer culture, the moral and aesthetic rebellions against Victorian norms, changes in American education, and the popular vogue of psychology and other social sciences. Although some of the critics viewed industrial technology as the main avenue through which science had impacted modern culture, most discerned more direct lines of cultural transmission from the sciences into the culture at large—above all, through the field of modern psychology. All of them firmly rejected the idea of mental modernization, as formulated by science's advocates. Various religious leaders and humanities scholars, as well as cultural critics, politicians, and ordinary citizens, sought paths forward through the Machine Age that were continuous with the best elements of bygone eras. They rejected the assumption that economic changes had consigned older, nonscientific thoughtways to the scrap heap. Industrialization certainly represented a sharp break in history, these critics allowed. But what was new was the intensity of the danger posed by a scientific mind-set in a society that increasingly embodied science's depersonalizing tendency. The solution lay where it always had, in traditional understandings of the human person and moral truth.

Such arguments gave rise to a distinctively modern form of the classic jeremiad genre. By the 1920s, many commentators elaborated on the usual lament that citizens had fallen away from genuine faith by tracing those secularizing tendencies to a particular source: science's cultural influence. Outside religious circles, scholars in the humanities—or anyone else claiming possession of firm moral truths—could also employ this updated mode of jeremiad, because it centered on morality rather than theism per se. The charge against science was less that it denied the existence of God or the truth of the Bible than that it destroyed the basis for moral judgment.

The new jeremiad thus entailed a more pragmatic defense of religion than had been the case in earlier years. It deemed religious teachings (or the humanities, or both) valuable primarily because they demonstrated the existence of moral absolutes and delineated their content. In the decades to come, a remarkably broad array of religious leaders, along with many other kinds of critics, would press the charge against science. In the Cold War years, especially, they would use the new jeremiad to attribute all manner of social problems to the philosophical errors of a scientific culture.

3

SCIENCE AND THE STATE

WHY DID THE APPARENT LOSS of moral guideposts in American public culture matter? Some critics of the 1920s listed a range of specific outcomes, from class conflict to eugenics to uninspiring architecture. For the most part, however, their arguments operated at a higher level of generalization. Indeed, they sometimes took a rather circular form: a loss of moral absolutes is bad because moral absolutes are good. But the political developments of the 1930s changed the context for such arguments and made them far more persuasive to many Americans. By the end of that decade, the rapidly growing body of critics had mostly converged on the charge that the modern era's scientific delusion portended a very concrete outcome: the transformation of the United States into a totalitarian dictatorship. A host of changes in the 1930s, both at home and abroad, convinced new groups of observers that a scientific mind-set was running rampant in the modern world and directly undermining the cultural foundations of American democracy as well as morality more generally. These critics of science's influence increasingly argued that social and political institutions took their shape from an underlying public philosophy and that scientists were busily remaking that philosophy in their own, materialistic image, with dire consequences for democracy's future.

As developments overseas loomed larger in public debates, it became an article of faith for innumerable Americans that the rise of totalitarianism in Europe and Russia reflected the secularization of Western societies at the hands of scientific thinkers. In this view, the Nazi regime rested on a modern form of paganism that was as relentlessly atheistic—indeed, as actively antireligious—as Soviet Marxism. Critics argued that both of these frameworks grew out of the materialistic, deterministic mind-set of the sciences. The total state was the inevitable result of applying science's amoral perspective to politics, because destroying traditional understandings of God, society, and self had left a vacuum that the champions of totalitarian state-worship rushed to fill. By the end of the 1930s, many Protestants of various persuasions had joined Catholics in defining totalitarianism as the quintessentially scientific form of politics.

Political changes at home often struck these 1930s critics as proof that science's pernicious influence had spread throughout American public culture as well. Criticism of secular public education accelerated in the 1930s, for example. But the main event was the emergence of the American version of the welfare state under Franklin D. Roosevelt. The New Deal brought unprecedented economic interventions, mushrooming bureaucracies stocked with social scientists, the transfer of welfare functions from local and religious bodies to the federal government, and much more. These intertwined developments led many observers to conclude that a scientific worldview, radiating outward from the universities, held sway over the minds of American policymakers or the political culture as a whole. Indeed, some feared that the apparent merger of social science and government, under the rubric of planning, would produce a brutal totalitarian regime in the United States.

TECHNOCRACY

As before, many critics of science's political influence worried mainly about its apparent destruction of morality. But a new issue also came to the fore as Roosevelt's administration undertook to manage the American economy and build a social safety net. A number of conservative and centrist opponents of the New Deal charged that economic regulation and even social welfare programs embodied a basic error that permeated the modern social sciences: namely, an overestimation of the predictability of human behavior. The New Deal state, they argued, opened itself to disastrous failures insofar as its agencies used social-scientific knowledge instrumentally to inform ambitious planning projects.

Various economic and political individualists worried in the 1930s that Roosevelt's New Deal would lead down the path to Soviet-style collectivism. This critique appeared among a range of figures, from the libertarian writer

Albert Jay Nock, who penned *Our Enemy, the State* in 1935, to the liberal journalist Walter Lippmann, whose distaste for economic laissez-faire was matched only by his fear that the New Dealers would trample sacred political freedoms in their rush to develop a comprehensive planning regime based on social-scientific knowledge—knowledge, he said, that could hardly support such reckless initiatives. Such figures often blamed science for the vogue of economic planning and new forms of administration by far-off, unaccountable federal bureaucrats.[1]

This group of American critics was fairly small, but they significantly expanded the scope of critiques of science's influence, in ways that portended a large body of post–World War II thought. Through the 1920s, most of the critics who worried about the cultural influence of science had defined it as a faulty cognitive orientation that had spread into key domains: popular understandings of psychology, say, or high school curricula. Even when these critics targeted institutions, such as the schools, they focused on philosophical teachings. But in the 1930s, individualistic thinkers began to see the New Deal state itself—its actual practices and institutions, not just the philosophy inspiring it or theories about its proper scope—as scientific in character. In certain regards, they echoed 1920s figures such as George N. Shuster, who had argued that the machine and its social products—cities, buildings, consumer goods, and many more—embodied the dangerously depersonalized orientation of modern science. In the 1930s, however, Lippmann and a number of other critics defined key features of the federal government itself as essentially—and improperly—scientific. In their view, the American state under Roosevelt was becoming more scientific insofar as it was increasingly bureaucratic, secular, and active in regulating economic affairs and other social dynamics.

Since the early years of the twentieth century, a growing cadre of what one historian has dubbed "rational reformers" had sought to redefine politics as a form of engineering, portraying the state and perhaps even society itself as a vast machine that could be used to achieve any number of ends. The sociologist William F. Ogburn and the political scientist and historian Charles Beard were among the many American thinkers, dating back to the Progressive period, who used such political metaphors. But before 1933, few commentators believed that such a mode of governance had actually emerged. It was only when leading rational reformers vaulted into Washington bureaucracies—including the famed "Brains Trust" advising Roosevelt on broad policy issues—that some critics began to view the American state itself as an outgrowth of science's depersonalizing tendency.[2]

To be sure, not all of the New Dealers fit the mold of rational reform. Many rejected the idea of a value-neutral social science and explicitly sought to yoke empirical investigation to progressive social ideals. In fact, these figures often worried about the cultural effects of science unmoored from values. In the mid-1930s, for example, Roosevelt and Department of Agriculture head Henry A. Wallace traced technological unemployment—the replacement of human workers by machines—to engineers' deficient sense of social morality. In an October 1936 letter to the heads of American schools of engineering and technology, Roosevelt assigned them much of the blame for the Depression. A few critics at the time actually proposed a moratorium on scientific research, so that society could catch up morally with its inventions. But Roosevelt and Wallace instead advocated retraining engineers so that they would attend to the social implications of their work. The "remorseless discipline of higher mathematics, physics and mechanics," Wallace explained, produced a singularly heartless, amoral perspective that seeped into other areas of engineers' lives, including their social and political views. To counteract this tendency and keep science's corrosive potential in check, Wallace and Roosevelt wanted to immerse engineering students in "imaginative, non-mathematical studies, such as philosophy, literature, metaphysics, drama and poetry." They believed that exposure to the arts and humanities would confine the scientist's relentlessly value-neutral approach to its proper domain—namely, the laboratory—and prevent it from affecting how engineers thought about the practical applications of their innovations.[3]

Other features of the New Deal, however, reinforced the nascent view that Roosevelt had created a scientific state. Although the New Deal was a thoroughly ad hoc affair in both its public justifications and its policies, administration officials often portrayed it as a massive exercise in applied science. Of course, Roosevelt hired hundreds of social scientists to run and staff his expanding bureaucratic agencies. Moreover, significant aspects of the New Deal's public face reflected the specific approach of the University of Chicago's Ogburn, who was well known in intellectual circles for framing both science and technology as thoroughly instrumental, amoral pursuits. Ogburn had come to the attention of many observers in 1922, when he coined the term "cultural lag" in his book *Social Change, with Respect to Culture and Original Nature*. That book laid out a theory of technological determinism. The relentless march of industrial innovation, Ogburn argued, inevitably forced corresponding changes in every other area of society and culture. However, there was often a significant gap in time between technological change and the eventual responses in other domains, as the cultural process worked itself out. Ogburn's cultural lag concept spread rapidly among theorists of mental modernization because it captured what they considered the failure of political institutions to catch up with technological changes—to undertake the regulatory initiatives and welfare provisions that Progressive social scientists identified as the needed response to industrialization and corporate consolidation.[4]

Ogburn gained additional notoriety after 1929, when he turned his presidential address to the American Sociological Society into a manifesto for scientific inquiry as a rigorously value-neutral pursuit. To become truly scientific, Ogburn argued, sociologists would need to ruthlessly "crush out emotion" and "taboo our ethics and values" in order to attain the "pure gold" of reliable, value-free knowledge. Ogburn certainly did not intend to limit the participation of scientists in reform efforts, but he drew a sharp line between their activities as scientists and their very different pursuits as reformers. Methodologically, Ogburn held up quantitative analysis as the ideal. "My worship of statistics had a somewhat religious nature," he later observed. "If I wanted to worship, to be loyal, to be devoted, then statistics was the answer for me, my God." As critics routinely noted, the social science building at the University of Chicago, home to Ogburn and many other champions of social-scientific realism and value-neutrality, featured an inscribed motto attributed to Lord Kelvin: "When you cannot measure, your knowledge is meager and unsatisfactory."[5]

Ogburn's technological determinism, cultural lag framework, and value-neutral approach circulated widely in Washington during the 1930s. He himself oversaw two major federal reports that brought social-scientific research to bear on the causes of the Depression. The first, *Recent Social Trends in the United States* (1933), had been commissioned by the Hoover administration; it followed Ogburn's belief that sectors such as "agriculture, labor, industry, government, education, religion, and science" should all advance in a coordinated fashion, with national planning employed to prevent asymmetries between them. In 1937, *Technological Trends and National Policy* went further by attempting to predict the future interactions of technology and social institutions. Like all of Ogburn's work, these reports contended that social change started with technological innovations, which then forced changes in the economy, which in turn transformed social and political institutions, finally altering the "social beliefs and philosophies" of the masses. In Ogburn's view, the technological change associated with industrialization would, sooner or later, produce a planning-oriented state and a thoroughly scientific attitude toward political institutions.[6]

To Lippmann, however, such talk of planning reflected the dangerous, authoritarian potential of applying science to politics. In his 1937 book *The Good Society*, the famed political commentator turned against Roosevelt, whom he had previously supported. He now painted a nightmare vision of the New Deal's eventual outcome. Following the lead of planning enthusiasts, Lippmann contended, would inevitably produce what Hilaire Belloc, a Catholic traditionalist in Britain, had called the "Servile State."[7]

The decades-long dialogue of Lippmann, a secular Jew, with Catholic political theorists reveals the capacity for shared critiques of science to create strange bedfellows. That convergence had begun with Lippman's publication of *A Preface to Morals* (1929), as he puzzled over a question that was at once deeply theoretical and eminently practical: What could lead individuals to discipline their desires, and thereby minimize social conflict, in a world where the "acids of modernity" had dissolved the existing foundations of social authority? Lippmann's book immediately ruled out the answer given by the Catholic thinkers he engaged over the next few decades. In his view, there could be no return to theological orthodoxy, given the inroads of democracy, church-state separation, and other modernizing forces. Instead, Lippmann looked to a formally secular tradition of humanism that stretched through figures such as Aristotle, Buddha, Confucius, and Spinoza, teaching an inner discipline based on the adjustment of one's desires to the extant conditions of the world.[8]

Through the early 1930s, Lippmann identified his humanistic liberalism as an antidote to the principle of unfettered economic freedom, which he had despised since his early days as a socialist. He stressed the importance of treating other individuals interchangeably with oneself, and thereby providing the conditions under which all could flourish. On this basis, he welcomed the New Deal and roundly condemned the philosophy of laissez-faire in works such as *The Method of Freedom* (1934). At the same time, however, Lippmann distrusted Roosevelt. When the president tried to "pack" the Supreme Court with sympathetic justices in February 1937, Lippmann finally broke with him. In *The Good Society,* published later that year, he contrasted his version of liberalism, not with laissez-faire, but rather with the collectivism that he saw on the march all across the modern world—including in Roosevelt's America. Lippmann now identified the New Deal as a form of "gradual collectivism," featuring arbitrary rule by political pressure groups rather than genuine self-governance. Lippmann still found free markets unappealing, but *The Good Society* drew on the critiques of economic planning developed by the conservative Austrian economists Friedrich Hayek and Ludwig von Mises.[9]

The Good Society also linked science firmly to collectivism. Liberal democracy, Lippmann argued, stood everywhere under threat from a resurgence of authoritarian political ideals, including rationalistic versions. In his view, the survival of an industrial civilization depended on its commitment to liberalism, defined in terms of an ongoing social process: the "gradual encroachment of true law upon willfulness and caprice." This liberalism, he averred, was the "necessary philosophy" of any society featuring an extensive division of labor. And its main competitor, on display all over the world, had grown out of modern science. In the wake of traditional religion's collapse, Lippmann argued, a new faith had arisen in its place and attained dominance through the chaos of World War I. This technocratic, collectivist framework centered on science's application to politics. "Out of the union of science with government," he wrote, "there is to issue a providential state, possessed of all knowledge and of the power to enforce it." In a scientific state, Lippmann continued, political officials would possess "all the authority of the most absolute state of the past." However, they would not act of their own accord. Rather, they would blindly follow the "consecrated technicians," the "engineers, biologists, and economists who will arrange the scheme of things." This scientific authoritarianism would obliterate all of humanity's hard-won freedoms, swapping in the arbitrary authority of scientists for that of priests and despots.[10]

Lippmann discerned a second threat to democratic polities from the prevailing understandings of science as well. This stemmed from the determinism and materialism of modern thought. In his final chapter, "On This Rock," Lippmann restated liberalism as a "higher law" that stood behind existing, positive laws and gave them their legitimacy. This law encompassed a categorical "denial that men may be arbitrary in human transactions," Lippmann wrote. As he formulated the principle elsewhere, "one human being shall not be the instrument of another, shall not be a thing like the carpenter's axe or his saw." Even more simply, liberalism meant the absolute primacy of what Lippmann called the "humanist ideal": namely, "the inviolability of the human person," as expressed in the concept of human rights. But this core liberal insight, he wrote, had been casually tossed aside by a host of materialists—"the Hegelians, the Marxians, the pseudo-Darwinians, and the Spenglerians"—who had thrown out the baby with the bathwater and "brought down the humanist ideal in the crash of the supernatural order."[11]

Lippmann traced the problem to the materialistic orientation of modern science itself. Scientific thinkers, he explained, "could not find the human soul when they dissected their cadavers; they could not measure the inalienable essence." And so they had simply tossed out the concept. Since then, Lippmann declared, Western intellectuals had cultivated "a radical disrespect for men," treating "justice, liberty, equality, and fraternity" as simply "old superstitions along with God, the soul, and the moral law." The results were as terrifying as they were predictable. "With man degraded to a bundle of conditioned reflexes," Lippmann wrote, "there remained only an aimless and turbulent moral relativity"—the perfect breeding ground for the "revivals of tyranny" in Russia, Italy, and Germany. Lippmann ended his book by portraying the modern world as the scene of a bitter struggle between dictators and the "men of deep religious faith" who alone defended the human person against technocracy and totalitarianism. Although Lippmann himself grounded moral freedom in a form of humanism rather than supernatural religion, he agreed with Catholic theorists and other religious thinkers that totalitarianism reflected the characteristically modern denial of moral freedom in the name of scientific materialism. Indeed, he saw religious traditionalists as his main allies in the defense of human ideals against a thoroughly science-obsessed world, even though he did not share their theological views.[12]

CULTURAL FOUNDATIONS

Most critics of science's cultural implications in the 1930s focused almost exclusively on the second threat Lippmann identified: the influence of materialistic, mechanistic theories of human behavior. Rather than the contours of the state or the planning ideal, they emphasized broad principles of political philosophy. Indeed, most of these critics welcomed the New Deal in its general outlines, if not all of its specifics. They worried instead about science's deleterious impact on the American public's capacity to firmly commit itself to democracy, over and against totalitarianism. The existing charge of moral relativism took on new forms, as critics argued that science-minded citizens had no reasonable basis on which to choose between freedom and dictatorship.

On both sides, in fact, debates over the proper response to totalitarianism tended to recapitulate 1920s struggles over mental modernization. The disputants repeatedly portrayed fascism and communism as outgrowths of their opponents' faulty philosophical tenets. Conversely, each party now identified its own cultural program as the basis of democracy itself, as well as the needed complement to industrial organization. Had totalitarianism emerged because Western nations had foolishly tried to develop a modern, secular alternative to traditional religion, or because they had failed to do so? Should the West move toward a culture based on science, or away from it?

John Dewey, like many other mental modernizers, redoubled his insistence on the need for scientific guidance as totalitarianism loomed. "A culture which permits science to destroy traditional values but which distrusts its power to create new ones," he famously declared in 1939, "is a

culture which is destroying itself." Meanwhile, Dewey and many of his allies argued that upholding absolute truth claims, in morality or anywhere
else, represented the hallmark of totalitarian thinking and the source of the
total state. Sooner or later, they reasoned, the adherents to such absolute
principles would succumb to the temptation to impose them on doubters
by any available means, fair or foul. Democracy, by contrast, rested on
fallibilism: a recognition of the experiential grounding and the tentative,
experimental character of all judgments, including moral principles. Only
such a view of human thought, Dewey and his compatriots argued, would
ensure that the individuals and groups in power refrained from forcibly
imposing their conclusions on others. For these mental modernizers, democracy was a matter of civil liberties, especially freedom of thought
and expression, while totalitarianism's key feature was its effacement of
those liberties.[13]

On the other side of the emerging battle over democracy's cultural prerequisites, many religious thinkers and a number of important literary and
artistic figures rooted democracy in robust, universal moral truths. They
viewed totalitarianism as the inevitable product of the moral relativism that
scientists had injected into Western society. In an important sense, they defined the problem as too much freedom of thought: Scientific thinkers had
taught citizens to apply the critical, empiricist approach of the modern sciences to morality, where it emphatically did not pertain. Now, that expansion of science's domain threatened the fabric of civilization itself. As totalitarianism loomed in Europe and drove a flood of illustrious émigrés to
American shores, the debate over democracy's cultural foundations began
to exert an enormous gravitational pull on American intellectual life.[14]

Not surprisingly, given the perceived stakes, a newly aggressive tone
could be heard on both sides of the debate in the 1930s. Among scientific
thinkers, a number of particularly pugnacious modernizers declared that
the forces of reaction should no longer enjoy free rein to sow discontent.
George S. Counts, a radical educational theorist trained by Dewey at Columbia's Teachers College, argued in the early 1930s that American teachers
should "indoctrinate" students in the basic principles of a new social order.
Later in the decade, Harry Elmer Barnes, another Columbia figure, brought
a particularly strict form of determinism into the modern mind literature.
In *Society in Transition* (1939), Barnes identified human behavior as a
product of environmental conditioning. On that basis, he drew what many
critics saw as the logical conclusion of all naturalistic thinking: that moral
freedom was entirely illusory. "No one can be held personally responsible
for his actions," Barnes declared, for these stemmed from "hereditary and
social conditions": "A man who commits a multiple murder is no more re-

sponsible for his behavior than an amiable and generous philanthropist."
Barnes's textbook also featured many other controversial recommendations: centering school curricula on the social sciences, delegating the administration of national policy to experts, and replacing traditional religions
with "a secularized social religion" featuring "a new and much needed
moral code, founded upon natural and social sciences." Even relatively
mild-mannered progressives such as Eduard C. Lindeman, a theorist of social work and adult education, argued in the 1930s that a "technological
age" could not "afford to have its values set by persons unfamiliar with
the foundations of science and technology." Scientists, Lindeman recommended, should aid not only in designing welfare programs but also in
"making reasonable tests of all proposed values."[15]

These mental modernizers also sounded a central Deweyan theme: that
moral absolutism fostered political absolutism. Dewey's own rhetoric grew
more militant in the 1930s, when he sensed a final battle brewing against
the defenders of absolute truth. Rather than the bourgeoisie and the working
class, Dewey believed, the modern struggle featured two factions of the
middle class: scientific thinkers, representing the forces of progress, and re-
actionary, prescientific thinkers who would destroy democracy by clinging
to the patterns of an outmoded era. All supernatural religions produced
"habits of mind at odds with the attitudes required for maintenance of democracy," he declared forthrightly. Moreover, for Dewey, supernaturalism
included any postulation of a nonempirical source of knowledge whatsoever. In the late 1930s, signs such as Reinhold Niebuhr's growing popularity and the apparent embrace of supernaturalism by Lippmann—a cofounder of the *New Republic,* a longtime associate of many social scientists,
and an erstwhile champion of pragmatism and moral fallibilism—led Dewey
and many other theorists of the modern mind to fear that the American
intelligentsia was embracing authority and seeking to turn back the cultural clock.[16]

On the other side of the struggle over democracy and culture, critics of
mental modernization argued that a scientific outlook was congenitally in-
attentive to values and did not even allow citizens to justify their prefer-
ence for democracy over its political competitors. A remarkable number
and variety of American commentators traced Nazism, as well as communism, to the spread of a secular, scientific culture. "A nation without God
and morality must end in tyranny," ran the typical argument. Catholic commentators were particularly inclined to draw causal links from scientific
thinking to moral relativism to totalitarianism, as they continued to shift
their attention from the Reformation to the Enlightenment as the main
source of modern ills. Figures such as Carlton J. H. Hayes, Monsignor

John A. Ryan, and prominent *Wall Street Journal* contributor Thomas F. Woodlock argued in the 1930s that the fruits of science's cultural domination included Nazi Germany as well as Soviet communism. The moral vacuum created by a scientific culture, Hayes contended, cleared the way for a totalizing religion of nationalism: the worship of the state as a substitute god, the sole source of meaning and value. Catholic leaders continued to worry about the secularization of the schools in particular, charging that Dewey and his allies in education circles had dynamited the foundations of democracy.[17]

Among Protestants, meanwhile, the accelerating theological renaissance reflected, in part, growing fears that the improper extension of scientific thinking had leached the moral content out of American public culture. Niebuhr, a central figure in the shift of the late 1930s and 1940s, penned a series of influential books, beginning with *Moral Man and Immoral Society* (1932), that decisively shaped the turn toward Christian realism—and away from the social sciences—among many mainline Protestant thinkers. Niebuhr argued that theological liberals actively, if unknowingly, abetted the destruction of society's foundations by taking their cues from science. After all, scientific understandings of the social world did not and could not acknowledge the patent fact of moral freedom, let alone humanity's endemic, irreducible tendency to succumb to pride and self-interest in using that freedom. Much like T. S. Eliot with the New Humanists, Niebuhr insisted that religious leaders needed to choose between the fundamentally incompatible principles of naturalism and theism. Meanwhile, he portrayed the modern world as utterly in thrall to a reductive, secular outlook. Even the churches were shot through with scientific understandings, Niebuhr contended. He increasingly argued that religion provided genuine knowledge of human affairs—a symbolic or mythical form of knowledge that no empirical endeavor could ever reveal. In the wartime lectures that became *The Nature and Destiny of Man,* Niebuhr traced the philosophical errors of the modern world back across the centuries, ascribing to them an immense historical sweep.[18]

As the 1930s progressed, a growing number of Niebuhr's counterparts in the mainline churches joined him in targeting both theological liberalism—especially insofar as its advocates allied themselves with social science—and proposals for mental modernization. By the end of the decade, these Protestants increasingly identified "secularism" as the leading threat to religion, both at home and around the world. Pioneered by Catholic critics in the 1920s, the category of secularism indicated something rather different from the heathenism, paganism, or atheism targeted by past religious leaders. Bridging theory and practice, it encompassed both personal commitments and social conditions—the influence of naturalistic outlooks as well as the evacuation of moral content from daily life. Indeed, the new charge of secularism drew a direct causal line between the two.[19]

Protestant critics of secularism found much common ground with Catholic counterparts such as Hayes. Among the former were figures engaged with the missionary enterprise, such as the Harvard philosopher William Ernest Hocking and the Quaker Rufus Jones. Attentive to developments at home as well as abroad, they argued that the various denominations and confessions, rather than competing with one another, should instead join hands against a secular mind-set that threatened all religions equally. Niebuhr famously set himself against secularism as well, at a major 1937 conference in Oxford. Other contributors to the theological renaissance likewise asserted the dire need for religion in a thoroughly secular, scientific culture. Even the interfaith movement, which historians typically associate with an inclusive, cosmopolitan sensibility, featured a concerted movement of resistance to secularism in the late 1930s. Within the National Conference of Christians and Jews, for example, Hayes sought to build a cross-confessional alliance of Protestants, Catholics, and Jews against the outgrowths of secularism—including liberal theologies, which he found too friendly to science. Protestant (and Jewish) listeners also tuned in as the neo-Thomist Fulton J. Sheen denounced naturalism and secularism on his popular radio program.[20]

The influx of émigré scholars from Hitler's Europe after 1933 further sharpened the terms of the cultural struggle over science. Those émigrés who advocated a scientific culture, such as Rudolf Carnap and other logical empiricists in philosophy, typically defined science in starker, more rigidly value-neutral terms than did many of their American counterparts. On the other hand, émigré critics of science's cultural impact identified value-neutral "positivism" as a virulent social disease, corrupting individual personalities and social institutions alike. This camp included not only humanities scholars and religious thinkers but also a group of political theorists who treated politics as an autonomous realm with its own, nonempirical norms and values. Even many of the émigrés who identified as social scientists decried the extension of physical-science methods into those disciplines. Thus, Frankfurt School leaders such as Theodor Adorno blamed Nazism and modernity's other ills on an Enlightenment-inspired "instrumental rationality" anchored in positivistic social science. On a more individual level, the Gestalt psychologist Wolfgang Köhler called positivism "an evil that destroys young energies as surely as does tuberculosis."[21]

THE WAR YEARS

During the early 1940s, arguments over the cultural prerequisites for democracy continued unabated. Indeed, they remained largely unchanged in their basic contours. As before, most exponents of the modern mind held that moral absolutes undermined democratic tolerance, while many critics of science insisted that moral relativism sapped democratic commitment. Specific events, such as the 1940 struggle over the appointment of Bertrand Russell to a visiting position at the City College of New York, served as flashpoints for these tensions and further heightened suspicions on each side about the other group's motives. So, too, did a series of clashes at the annual Conference on Science, Philosophy, and Religion in their Relation to the Democratic Way of Life (CSPR), launched by Louis Finkelstein of the Jewish Theological Seminary in 1940. The University of Chicago philosopher Mortimer Adler, a fellow traveler with neo-Thomism who emphasized metaphysical first principles and deplored the empiricist bent of modern social science, famously blasted Dewey and his ilk at that group's first meeting. Americans, said Adler, should fear "the positivism of the professors" more than "the nihilism of Hitler." Prior to the conference, Adler had proposed requiring all participants to "repudiate the scientism or positivism which dominates every aspect of modern culture."[22]

Beneath these fireworks, however, subtle shifts were occurring. Most importantly, additional numbers of mainline Protestants soured on their long-standing alliance with social scientists during the war. As they did so, some embraced the "neoorthodoxy" of Niebuhr and his allies. Others, such as F. Ernest Johnson at Columbia's Teachers College, developed new formulations of liberal Protestantism that were more confrontational toward science and secularism. In this regard, the CSPR's second meeting in 1941 was especially revealing. Although the event offered few of the explosions provided by Adler and his opponents the previous year, a joint paper written by a group of Protestant scholars from Princeton University and Princeton Theological Seminary highlighted the growing fear of a scientific culture among theological liberals. Any naturalistic view of the world, argued the Princeton group, was "essentially materialistic" because it portrayed the individual as "simply a highly developed animal." Naturalists could not sustain the belief in human dignity that anchored democracy because their approach definitively ruled out such a quality. Hocking spoke more sharply in 1942, deploring the "death's-head world-view" of naturalistic thinkers.[23]

Two years later, the Episcopalian political scientist John H. Hallowell followed many émigrés in giving the imprimatur of social science itself to the widespread charge that secularization explained Hitler's rise. In an influential study, he traced Nazism back to nineteenth-century changes in German liberalism. How had the famously liberal German middle class suddenly become a mass of antiliberal zealots in the early 1930s, with hardly a "murmur of dissent"? Hallowell mobilized a raft of empirical data to show that German liberalism had been hollowed out morally by the elimination of its religious foundation in the previous century. Much of the blame fell on scientific thinkers, he explained: the "normless factualism" of a positivistic approach had undermined the necessary belief in a transcendent moral order. Indeed, Hallowell wrote that "the decline of liberalism parallels the degree to which liberal thinkers have accepted positivism." Of course, he noted, capitalism, with its worship of progress and purely monetary measures of value, had also contributed to eliminating liberalism's moral core. The scientist and the decadent middle classes converged on the impersonal standards of "technical efficiency and mechanical certainty," he declared. But Hallowell ultimately gave pride of place to positivism, with its morally deadening effects on legal theory and other forms of thought. Shorn of all "objective values and truths," he explained, liberalism retained only its "subjective and anarchical elements." These remnants could not check arbitrary exercises of will, whether by the Nazis or by any other powerful group. Democracy, Hallowell concluded, required a firm belief in "the essential moral worth of human beings" that flowed only from "the Christian concept of individual souls" and could never be justified in scientific terms.[24]

The political commitments of many scientific thinkers were prominently on display in the early 1940s as well. During those years, the question of America's broad war aims, and later its specific goals in reconstructing Europe and Asia, brought an array of arguments about science's cultural implications to the fore. American commentators had no desire to relive the betrayal that so many of them had felt when Congress rejected the League of Nations in 1919. Indeed, they began planning for peace long before the United States even entered the war. By September 1942, so many plans for postwar reconstruction were already circulating that a book reprinted nearly thirty proposals. These ranged from E. U. Condon's "A Physicist's Peace" to the *Christian Century*'s "Winning the Peace" to statements by Jews, Quakers, and even theosophists. Astute readers might have noticed that comparatively few of the scientific thinkers—especially the natural scientists—wanted to remake the postwar world along purely scientific lines. Rather than a culture organized around the modern mind, they proposed various alternatives. Natural scientists, for example, often argued that a universal system of ethical values could provide the basis for world organization. But a number of prominent commentators did take a Dew-eyan line, identifying global democracy with fallibilism and resistance to moral absolutes. The postwar peace, wrote the philosopher Max Carl Otto, should not only provide for "the most livable life for all" but also make each individual "the final judge of what 'most livable' means for him." Meanwhile, even centrists such as the political scientist Charles E. Merriam contended that the modern world itself took its shape from the rise of science.[25]

Among those advocating versions of mental modernization during the war aims debate, psychologists and psychiatrists offered particularly breath-less visions of "world mental health." They called for widespread cultural retraining through "education for scientific thinking in human relations." Ninety-nine percent of the psychologists asked to sign a 1944 statement on international affairs agreed that exposing children to "symbols of unity and an international way of thinking" could eliminate both war and prej-udice. Faculty groups discussing postwar reconstruction predicted that the Germans and Japanese would require particularly extensive interventions—perhaps, one psychologist proposed, "alternating sequences of 'active' and 'passive,' 'direct' and 'indirect' techniques of preventive mental hygiene and psychotherapy." These views intersected with the development of a "culture and personality" approach whose chief patron, the foundation executive Lawrence K. Frank, defined modern societies as neurotic mental patients requiring therapeutic interventions and whose best-known advocate, the widely read anthropologist Margaret Mead, had called for a thorough overhaul of American culture, perhaps along the lines of the "spectacular experiment in Russia," during the 1930s.[26]

Other ambitious proposals for scientifically informed cultural change flourished during the war. Some critics feared that the common tongue it-self was being refashioned in the mechanistic mold of science. General se-mantics, a kind of cognitive-neurological therapy with messianic overtones that sought to eliminate social conflicts by retraining individuals to ap-proach language from the "engineering point of view"—to recognize the concrete, referential content of their words—became a national sensation by late 1938. College courses on general semantics appeared, and two pop-ularizations featured as Book-of-the-Month Club selections in 1941. The founder of general semantics, Alfred Korzybski, told audiences that the lay-person should refrain from challenging experts and "abolish his old fash-ioned 'private opinions'" within the territory covered by science.[27]

Expressive language seemed to be under assault from other quarters as well. Roosevelt's assistant attorney general, Thurman Arnold, contended that all talk of ideals was meaningless except as a device for securing loyalty to one's own projects. The British psychologist C. K. Ogden and the Harvard literary scholar I. A. Richards proposed reducing the English language to 850 words: just the number needed to describe physical movements in space. And in philosophy, logical empiricists such as Carnap were widely, if inaccurately, believed to have categorized as "nonsense" all sentences that could not be rewritten in the mechanical terms of physics. Meanwhile, the philosopher W. V. O. Quine was among those lauding the possibility of translating everyday language into symbolic logic, a notation system that intimidates readers even today. Applying this "unambiguous language" to the "political, economic, and social fields," he predicted, would "conveniently reflect the structure of these fields and make discussion and analysis easy." To Quine, symbolic logic would finally allow a nonquantitative science, providing "explicit techniques for manipulating the simplest ingredients of discourse." Indeed, another enthusiast contended that the adoption of symbolic logic foretold "a stage where all the language of thought will be calculable like mathematics." More generally, wartime commentators often agreed with Carnap that logicians could play a prophylactic role in society, healing the social body by rooting out illogical propositions in public discourse.[28]

Although these proposals fueled outrage, the broader political situation continued to inspire most of the wartime challenges to science's sway. The association between science and totalitarianism grew ever deeper, especially among religious leaders. On Christmas Day, 1942, the American Institute of Judaism issued the familiar charge against scientific materialism. "Misreading the findings of the sciences, both physical and social," the group declared, "men have given their allegiance to false philosophies, spiritual and moral values have been divorced from human life and materialism has been made supreme in the affairs of men." A month earlier, in a talk at St. John's, the Catholic philosopher Sheen tied world events directly to the fortunes of Dewey's philosophy. It had been Japanese exchange students trained in the pragmatic, utilitarian American universities who had incited that nation's aggression, he charged. In Europe, too, Sheen declared, the Allies faced the task of "stamping out the barbaristic barnacles of misdirected progress" that pragmatism had generated, namely "scientism and relativism." On the other hand, Sheen also saw a major positive outcome from the war. "Pearl Harbor," he wrote, "saved us from being another France, because that one swoop had the effect of killing every pragmatic philosopher in the United States."[29]

In his radio addresses, as in his numerous writings, the prolific and highly visible Sheen offered some of the wartime era's most striking attacks on scientific thinking—as he did before and after that conflict as well. "Science today is the enemy of man," he declared in one publication. Sheen went on to identify scientific materialism, embodied by Dewey, as one of the leading superstitions of a secular culture and to argue that it had "ruined higher education in the United States by prostrating itself before the god of counting." Applying a mechanistic framework to the world, he argued, led individuals not only to "ignore purposes in our laboratory" but also to "eliminate purposes from the universe." Sheen thus called for a comprehensive rejection of "the way the modern world thinks": "The modern man wants back his soul!" He warned Dewey and like-minded thinkers that the American soldiers would sooner or later "come to hate not only the enemy they meet in battle, but the intelligentsia who told them they were only animals. . . . And when they come marching home there will be a judgment on those who told them they had no soul; they will live like new men and they will give a rebirth to America under God."[30]

AS SHEEN PREDICTED, the immediate postwar years would witness a tremendous efflorescence of religious commitment in the United States. And a significant portion of that energy would reflect the belief that scientific materialists had commandeered American public culture for their own purposes. But it was not simply intellectual trends in the universities that fueled such a perception. The political changes of the 1930s also played a key role. To many observers, Hitler and Stalin emblematized the political implications of modern scientific thought. Back at home, meanwhile—and paralleling aspects of state growth abroad—the expansion of federal regulatory power and the creation of a secular welfare state in the 1930s provided new opportunities for critics to conclude that science had left its mark on society.

The struggle over science's political meanings in the late 1930s and 1940s tore through the increasingly tenuous alliance between religious and secular progressives, giving rise to new alignments defined by structures of cultural authority rather than political commitments. Since the 1880s, American intellectuals had routinely reached across the science-religion divide, joining hands to advance shared political projects. By the 1940s, however, they increasingly united across the political divide to champion the cognitive claims of science or religion. The intensity of these battles reflected a growing belief that culture, and above all philosophy, shaped politics. On the scientific side, Dewey called political institutions largely "an effect, not a cause," of the underlying cultural substrate. Niebuhr concurred on this matter, if on little else: "In the end, the logic of a system of ideas becomes the pattern of human action."[31]

Behind the scenes, meanwhile, legions of natural and social scientists had joined forces with the federal government to create entirely new instantiations of science. Heading off to Washington, or Cambridge, or Los Alamos, these figures played key instrumental roles in the campaign against Hitler. As they did so, they built up institutions and personal networks that would give rise to the postwar military-industrial complex, with its phalanxes of contractors, academic research centers, and funding channels. After the war, as fascism receded into memory and the Soviet enemy took shape, these new research efforts and the theories of human behavior they generated would combine with other forces to create an unprecedented degree of fear about the cultural effects of scientific thinking. In the postwar years, the idea that Americans lived—for better or for worse—in a scientific age would become a simple matter of common sense.

4

SOCIAL ENGINEERING

WHAT IS SCIENCE DOING TO US? How is it changing our world? These questions have loomed large since the eighteenth century. But they have rarely pressed more insistently than in the post–World War II United States. The period from the late 1940s to the early 1960s brought profound uncertainties and fears. In the minds of many Americans, bombers patrolled the skies, communists lurked under the bed, and the nuclear family could hardly restrain the seething anxieties of suburban life. Waves of cultural panic arose, targeting perceived dangers that ranged from communism to homosexuality to comic books. But underneath it all, many observers discerned an even deeper threat: an unprecedented, epochal shift in the locus of cultural authority. Science, in this view, had become the new public philosophy—or perhaps an established religion in its own right. It had replaced not only religious beliefs but also political ideals, artistic sensibilities, and even commonsense understandings. Frightful metaphors proliferated: science was a genie escaped from a bottle, a car without a driver, a servant or slave turning on its master. Relentlessly expanding its reach, science represented a profound moral danger, not a reliable force for good. It had escaped its proper domain—the value-neutral analysis of physical nature—and invaded the realms of human meaning and morality.

As concerns about science's cultural influence spread in the postwar years, the terms of the debate continued to shift. For example, disputes over the validity of Darwinism receded even further from center stage than they had in the 1930s. On the side of Darwin's champions, American philosophers largely withdrew from public disputation after World War II, occupying themselves with technical questions and employing a highly specialized form of discourse, even in the burgeoning field of value theory. By the time John Dewey died in 1952, American public life no longer featured an organized and highly visible cadre of naturalistic philosophers who defended a worldview grounded in modern biology. Meanwhile, the theologically conservative Protestants who inveighed against evolution increasingly occupied their own, insulated discursive world by the 1950s. Those religious leaders who accommodated Darwinism tended to dominate mainstream public debates.

Of course, the activities of a prominent group of natural scientists—nuclear physicists—profoundly shaped cultural understandings of science and its social impact after World War II. The physicists' contributions to weapons research helped to foster a belief that the typical scientist was "cold, calculating, and without social interest or moral standards." As the arms race accelerated, science advocates argued that public perceptions of scientists had grown so negative as to produce a widening gap in "scientific manpower" between the United States and the Soviet Union. Citing a substantial decline in the number of high school and college students studying science since the early twentieth century, the chemist Charles Allen Thomas worried that the scientific community could no longer replicate itself across the generations. The minds of both students and parents had been "been poisoned by the insidious cloud of anti-intellectualism which hangs over this country like a great shroud," Thomas lamented. "Somehow, science has become identified in the minds of a great many people as a sort of super 'Svengali,' responsible for all our dilemmas."[1]

In this climate, even fear of nuclear destruction often reflected a deeper sense that moderns lacked the moral resources to turn their newfound power to the good. As a result, much of the postwar disputation about science's broader meanings focused on the social sciences. Critics of science's cultural influence dissected what they took to be the social scientists' substantive claims, implicit moral assumptions, cultural and political goals, and steadily growing areas of jurisdiction. In this regard, Cold War–era conflicts over science were more extensive, if rather different in structure, than the earlier rounds. Whereas conservative Protestants had been largely alone in challenging the actual findings of the natural sciences, critics of myriad persuasions now opposed scientific interpretations of human persons and societies. New terms of opprobrium arose as these critics sought to analyze the precise target of their concerns. By the end of the 1950s, the pejorative term "scientism" had gained popularity as a label for the illegitimate extension of scientific methods into domains where critics said they did not apply. Behind the rather abstract, bloodless specter of scientism, however, lay a much more concrete and frightening phenomenon: "social engineering." Whereas critics viewed scientism as a philosophical error that had turned into a cultural sensibility, they saw social engineering as a devastating social force in the everyday world: Scientism's advocates now held the levers of power and pursued a total transformation of human existence. Social engineering was not just misguided, said the critics, but also fundamentally undemocratic—even un-American—because it involved manipulating human beings without their consent.[2]

All manner of postwar social changes led critics to conclude that science's detached, depersonalized, and implicitly manipulative perspective had invaded the precincts of modern culture and transformed itself into social engineering. Ironically, attempts by liberal reformers to *weaken* constraints on individual behavior—namely, the traditional moral strictures that shaped parenting, schooling, criminal punishment, and sexual norms—triggered many of the accusations of social engineering. To be sure, critics also saw scientism and social engineering behind other phenomena, from advertising to bureaucratization to weapons research. However, they often targeted a "permissive" orientation that they associated with the growing authority of modern psychology. Taken together, the entire array of postwar developments struck many critics as the early phases of a comprehensive project of social engineering and convinced them that science itself had relativistic and authoritarian tendencies—indeed, that modern science potentially served as the amoral ideology of power-hungry experts.

POLITICAL CHANGES

Postwar fears of science and social engineering reflected, in part, a perception that American society was undergoing rapid secularization. Far-reaching changes in church-state relations buttressed this view. In its *Everson* and *McCollum* decisions of 1947 and 1948, the Supreme Court abandoned the prevailing "accommodationist" theory of church-state relations, which held that governmental agencies could aid all religions equally so long as they did not favor any particular church. The justices, invoking Thomas Jefferson's metaphor of a high "wall of separation" between church and state, now declared that governmental agencies could not privilege religion of any kind over unbelief. In practical terms, the outcomes of *Everson* and *McCollum* were mixed. Moreover, in *Zorach* (1952), the Court famously declared, "We are a religious people whose institutions presuppose a Supreme Being." But *Everson* and *McCollum* enshrined a strict separationist reading of the First Amendment that fueled intense anxieties about secularization among many religious leaders, politicians, and citizens. New groups of Protestants now joined Catholics in charging that the secularization of public institutions actually entailed the forced imposition of secularism as an official, antireligious faith. Reinhold Niebuhr and the Jesuit theologian John Courtney Murray offered a variation on this argument, seeing a watery "religion of democracy" behind the new, secular establishment.[3]

Everson and *McCollum*, like other apparent signs of secularization, caused such outrage because huge numbers of Americans believed that substantive religious principles underpinned democracy itself—and that secularism led inexorably to totalitarianism. In this regard, the category of secularism served to connect postwar anticommunism with charges of scientism and social engineering. Not all anticommunists focused on the atheistic character of Marxism; some deplored the Soviets' abridgement of civil liberties, others their destruction of free enterprise. But the Soviet regime's official atheism loomed large in American understandings of communism. And what was atheism, many critics reasoned, if not an outgrowth of scientism—the improper extension of a materialistic orientation from the natural sciences into philosophy and politics? The right-wing anticommunist Whittaker Chambers thus wrote in his 1952 best seller *Witness* that Stalin's worldview did not actually stem from Marx's writings, but rather from the physicist's injunction, "All of the progress of mankind to date results from the making of careful measurements." Chambers identified democracy as "a political reading of the Bible" and totalitarianism as the logical endpoint of any naturalistic philosophy. Like many other Americans, he feared that science also threatened democracy and human freedom at home in the United States, insofar as it relentlessly spun off atheistic philosophies.[4]

Numerous developments abroad reinforced the widespread belief that secularization led to totalitarianism. For example, the poor treatment of church leaders in Eastern Europe and elsewhere became a cause célèbre among anticommunist Christians. Critics also decried the emergence of strong communist parties in Europe and linked their influence to the official secularism of the United Nations. Suspicion of the United Nations' philosophical orientation deepened in 1946, when the British biologist Julian Huxley was chosen to lead its Educational, Scientific and Cultural Organization (UNESCO). Huxley, the grandson of Darwin's associate T. H. Huxley, was an outspoken naturalist (and unrepentant eugenicist) who had laid out his controversial views in readable books such as *Religion without Revelation* (1927), *A Scientist among the Soviets* (1932), *If I Were Dictator*

(1934), and *Evolutionary Ethics* (1943). As the United Nations crafted its International Declaration of Human Rights in 1948, an array of Christian leaders lobbied heavily to include an affirmation of God in the document; they were gravely disappointed when the campaign failed.

Huxley was hardly the only scientist to gain access to the corridors of power in the wake of World War II. In the United States, too, physicists, engineers, and other natural scientists continued to exert significant influence after 1945. As the defeat of fascism gave way to the arms race against the Soviets and then a hot war in Korea, science became ever more deeply intertwined with the military—and both became central to American governance in a way that they had never been before. The dense network of university departments, research institutes, military contractors, and funding agencies that President Eisenhower famously dubbed the "military-industrial complex" grew steadily in both size and influence, driven by an unprecedented flood of federal money and US military leaders' embrace of high-tech warfare.

In turn, the military-industrial complex transformed the scientific enterprise itself. Universities focused heavily on research that would draw federal grants and scientists organized themselves into large teams for huge, collaborative projects reaching across numerous institutions, both public and private. Although physics and engineering led the way, social scientists, among others, quickly adapted to the new model, launching ambitious surveys and other projects that required teams of researchers and addressed national security concerns. As the natural and social sciences became embedded in the burgeoning military-industrial complex, what Alvin Weinberg dubbed the "big science" model made scientific research itself a highly organized and increasingly bureaucratic practice. When the Soviets' launch of Sputnik in 1957 prompted national soul-searching about the state of American science, Congress' passage of the 1958 National Defense Education Act seemed to reorient the entire educational system toward the systematic production of scientists and engineers.[5]

What kind of culture did this unprecedented national mobilization of scientists portend? We know now that many elite scientists—including leading members of the Manhattan Project—responded with anguish to Harry Truman's decision to drop atomic bombs on Hiroshima and Nagasaki in August 1945. They launched a series of public initiatives, aiming to subordinate atomic energy to moral principles by bringing its immense power under the control of an international body dedicated to ensuring peaceful applications and preventing an arms race. But the impact of these figures on the public conversation was limited. Ordinary citizens did not read the *Bulletin of the Atomic Scientists,* wherein prominent scientists repeatedly elevated moral guidelines over technical considerations and urged their peers to take responsibility for the practical uses of their research. What the public saw, mainly, was that American scientists wanted to hand over their knowledge—and therefore atomic power—to the Soviets. In the early 1950s, American citizens also witnessed a bitter struggle between Edward Teller, demanding more and larger bombs, and the aloof, patrician, cryptic, and allegedly communistic J. Robert Oppenheimer, who called for limiting the nation's nuclear arsenal. The wave of political activism among American natural scientists after 1945 might have mitigated the suspicion that science exerted a harmful moral influence, had its full contours been known to the public. In many ways, however, scientists' postwar activities actually heightened that belief.[6]

So, too, did a series of popular works by natural scientists that challenged conventional understandings of politics, morality, and religion. According to the biologist George Gaylord Simpson in his widely read 1949 book *The Meaning of Evolution,* science had proven empirically that there were no "absolute ethical criteria of right and wrong." Simpson couched his arguments in expansive, humanistic terms. Indeed, he managed to deduce a sturdy framework of human rights from the moral indeterminacy of the universe by declaring all attempts to restrict individuals' freedom of interpretation scientifically illegitimate. At the same time, however, Simpson sharply bounded that interpretive freedom by arguing that citizens should treat empirical evidence, and empirical evidence alone, as final and authoritative. They should systematically defer to science wherever it had spoken and endorse only those religions and philosophies compatible with the prevailing scientific frameworks—including science's disproof of moral absolutes. Meanwhile, Simpson decried excessive freedom in the economic realm, pushing beyond New Deal liberalism to a form of democratic socialism that he again portrayed as the only logical application of science to politics.[7]

Whereas Simpson anchored his prescriptions in an existentialist view of the human predicament, other natural scientists looked to the human body as a source of wisdom and insight. Yet these works, too, flouted traditional understandings. For example, postwar champions of individual freedom could hardly have found solace in the anthropologist Ashley Montagu's *On Being Human* (1950), which held up infant dependency as a model for adult belonging. At a time when the language of individual rights had come to prevail among even the staunchest champions of a planned economy, Montagu invoked an older, Progressive Era dichotomy between natural, cooperative relations and unnatural competition. Well known for his antiracist writings, Montagu now told his readers that "there are certain values for life which are not matters of opinion but which are biologically determined." Chief among these was universal brotherhood. "Man does not want to be independent, free, in the sense of functioning independently of the interests of his fellows," Montagu declared. "This kind of negative independence leads to lonesomeness, isolation, and fear. What man wants is that positive freedom which follows the pattern of life as an infant within the family—dependent security, the feeling that one is a part of a group, accepted, wanted, loved and loving." Fortunately, Montagu asserted, the "tissues of every organism remember their dependency and interdependency" from early childhood, such that adults sought constantly to maintain or recapture that state. Every fiber of our being, said Montagu, pointed toward loving cooperation, whereas aggressive competition was a learned trait, alien to our biological heritage. In the charged postwar climate, such claims would have sounded positively communistic to many readers.[8]

The Progressive dichotomy of cooperation versus competition and the claim that "humanity has moral goodness built into it as insistently and biologically as hunger" found less frightening but equally strenuous expressions in the writings of other postwar scientists. The Yale biologist Edmund W. Sinnott contended that both Christian ethics and Western political values grew out of human protoplasm; the core democratic tenets of "freedom, progress and the worth of individuals" flowed directly from "the fundamental character of protoplasmic structure and activity" itself. Unlike most of his readers—and a growing number of his colleagues—Sinnott denied that scientific explanation entailed viewing the natural world in deterministic terms. Quite the opposite, he said; science proved that all living entities, from human beings down to sea sponges, pine trees, and even slime molds, guided their own behavior by projecting purposes and choosing ends. On that basis, Sinnott concluded that the structure of biological life underpinned Christian ethics. "Love is the climax of all goal-seeking, protoplasm's final consummation," he wrote. "To love your neighbor as yourself is the only basis for human relationships." Elsewhere, however, Sinnott argued much more specifically that biology authorized (and Western civilization required) "the faith of religious liberalism." He identified Catholicism and traditional Protestantism—the faiths of a large majority of Americans—as fundamentally unscientific, violating the teachings of nature and political experience alike. In Sinnott's writings, as in those of Simpson and Montagu, science buttressed certain conventional understandings but ran wildly afoul of others. There was little in these works to reassure those readers who believed that the American state had hitched its wagon to scientific expertise—and perhaps to a secular philosophy rooted in evolutionary biology.[9]

THE SOCIAL SCIENCES

There was something for everyone to fear in postwar American science. But many critics responded especially strongly to the actions and beliefs of social scientists. Would the scholars charged with applying the lessons of science to human affairs engage in deterministic fantasies? Would they redefine freedom in biological terms that ruled out the religious beliefs of most Americans? Or would they preserve robust, traditional understandings of moral freedom and religious faith? In short, would the articulation and application of the social sciences uphold or undermine cherished cultural values?

Of course, the answers depended on where particular observers looked and how they read the evidence. With the help of archives and databases, we can see in retrospect that postwar social scientists were deeply preoccupied with the question of values, even when their public statements seemed to disavow such a concern. In the 1920s and 1930s, strict behaviorists such as John B. Watson had sought to eradicate all talk of human values, choices, or purposes—indeed, even such familiar categories as "mind" and "thought." Although these figures never constituted anything close to a majority, they were an important presence in the interwar conversation about science's cultural meanings. By the 1950s, however, a similarly titled but significantly different approach called behavioralism prevailed across many of the social sciences. As opposed to psychological behaviorists, behavioralists in fields such as sociology and political science took the existence of values, choices, and purposes as a given—an empirical fact about human behavior. Human beings did, in fact, have minds and values; they were not mere creatures of social conditioning. Indeed, a large part of the task that postwar behavioral scientists set themselves was to explore the interactions of value-laden choices with other, structural forces.[10]

At the same time, however, these scholars touted their own personal and collective detachment from the values under study, at least during the research process. In high-profile contexts—including the congressional debate in the late 1940s over whether the proposed National Science Foundation would fund research in the social sciences—these scholars portrayed their work as strictly value-neutral, just like that of physicists. Leading social scientists such as the sociologist Talcott Parsons argued that social values never influenced research findings themselves, even though they mattered enormously in the real world and could be studied empirically like any other phenomena. In other words, these figures aimed to apply the detached, empiricist outlook and methods of the physical sciences to the distinctive subject matter of the social world. The value-neutral study of value-driven behavior proved a highly attractive ideal for social scientists and foundation executives alike in the postwar years.[11]

Whereas many interwar advocates of mental modernization had argued that the empirical study of human experience, past and present, would reveal a comprehensive, secular framework of values, postwar social scientists, like their colleagues in philosophy, largely rejected that view. Rather, behavioral scientists typically assumed that value systems derived from nonempirical sources and portrayed social-scientific knowledge itself as purely instrumental in character—as a set of theoretical generalizations and associated techniques of intervention that could be used to advance any given value system. Unlike publicly engaged biologists, these social scientists argued that their work represented a necessary complement to religious, artistic, or political frameworks, but was emphatically not a substitute for them. Science advanced values; it did not determine them.[12]

Many postwar social scientists argued that the proper framework of subjective values and norms was already on display in the contemporary world, in the United States and perhaps parts of Europe as well. Some high-profile theorists—including Parsons, whose influence extended across the disciplines, and the cultural anthropologist Margaret Mead, one of the most visible public intellectuals of her day—saw that reservoir of morality in Christianity. More commonly, postwar social scientists identified the needed moral framework as the American "liberal consensus": a set of deeply rooted political ideals that dated back to colonial times and reflected the unique characteristics of American social experience. A third, smaller group of social scientists joined humanities scholars in invoking a sturdy heritage of Western civilization that stretched back to ancient Greece. Each approach, however, held that Americans already had the values they needed close at hand. For moral guidance, citizens and policymakers should look to the American or Western past, not the present-day social sciences. The latter merely enabled Americans to implement more consistently and effectively a framework of moral principles that was already lodged at the heart of their culture.[13]

For all of these figures, as for their colleagues in the physical sciences and many biologists, "science" meant the production of instrumentally rational knowledge: technical means for implementing values derived from other, nonscientific sources. The scientific method required strict objectivity, even when the subject matter at hand was the morally free, value-driven behavior of individuals. At the same time, however, these scholars also linked science to the promotion of broad social, cultural, and political ideals. Few postwar social scientists followed Sinnott and Montagu in arguing that science simply dictated particular forms of belief and action, let alone a comprehensive way of life. Many, however, joined Simpson in reaching essentially the same conclusion by declaring empirically invalid each of the actual alternatives to prevailing liberal ideals. These scientific liberals argued that only the combination of representative democracy, civil libertarianism, a secular public sphere, and a robust, expert-managed welfare state comported with the observable facts of human behavior—and with the prevailing moral framework, whether it was Christianity, a liberal consensus, or Western civilization. From this point of view, postwar American liberalism was the only viable political response to the human condition, at least in an industrial age.[14]

The tendency to "naturalize" postwar liberalism—to describe liberal values as reflections of empirically discoverable structures of reality rather than contingent human choices—also took other forms after World War II. This was especially true in psychology and sociology, the fields most frequently associated with dangerous programs of social engineering. In the collaborative 1950 volume *The Authoritarian Personality,* as in other prominent statements by psychologically minded commentators, the potent concepts of psychological maturity and mental health anchored political visions. That book's authors detailed how various social practices and institutions stymied the emergence of normal personalities—and thereby threatened the social order itself. Meanwhile, numerous sociologists argued that the postwar United States featured a mode of social existence—industrial modernity—that ensured the maximal satisfaction of both material and psychological needs and would eventually spread to every society around the world. Again, they presented the associated guidelines for human behavior as reflections of pre-given features of the world, not contingent political values and choices. Characteristically, the anthropologist W. Lloyd Warner and two of his University of Chicago colleagues urged young readers to "adapt themselves to social reality and fit their dreams and aspirations to what is possible." Just as biologists grounded liberalism in human physiology and numerous other figures based it on a shared American or Western culture, these social scientists used empirically grounded assertions about human beings and social structures to declare certain political visions realistic and others pathological.[15]

Thus, the self-declared avatars of neutrality in the postwar social sciences typically took clear political stances by locating both radicals and conservatives entirely outside the bounds of American political culture, or the Western tradition, or even rational thought. They argued that Marxism and free-market thinking were unrealistic and even delusional, contradicting patent facts and shared values alike. The many social scientists who declared that World War II had brought an "end of ideology" added the force of history to the usual dismissal of radicalism and conservatism, by placing political deviations safely in the disappearing past. By the end of the 1950s, theorists of "modernization" had begun to argue that every country in the world would eventually converge on something like postwar American institutions, if US policymakers helped steer them away from the pathologies of communism. All societies would one day arrive at the same terminus, becoming democratic, secular, pluralistic, and welfarist.[16]

The secularity of these liberal visions reverberated loudly in public debates, even though social scientists rarely stressed that dimension of their work. Like Sinnott, many of them ruled out a wide range of religious positions as well as political stances. In particular, postwar social science strongly buttressed theological liberalism, which comported well with political secularity. This could be seen especially clearly in the pedagogical domain. Across the disciplines, professors spoke casually about "education in liberalism" and the need to change students' value structures in their classes, even though they increasingly disavowed any intention to do so with their research. A survey instrument widely used to measure college students' "mental and emotional maturity" docked respondents for favoring censorship, criticizing Europeans or modern literature, distrusting liberals or American political leaders, or seeking to limit science's reach. Meanwhile, educators contrasted liberalism not only to radicalism and conservatism but also to a belief in absolute truths, especially moral and religious truths. To them, liberalism was as much a spiritual orientation as a political one. Even scholars who invoked an age-old framework of Western principles, celebrated American Christianity, or used the new language of human rights typically disavowed the absolute, unconditional prescriptions and proscriptions that most Americans called "morality." Rather, they usually had in mind a rather thin, largely political, and conspicuously secular set of commitments, especially to civil liberties—including tolerance of all forms of belief and unbelief—and economic security.[17]

Meanwhile, the same science that seemed to be undermining society's moral foundations also became associated with distant, unaccountable forms of power in the minds of many Americans after World War II. On all sides, it seemed, there were influential but shadowy institutions that determined the daily conditions of life but which citizens had little ability to control—and were sometimes hidden from them entirely, as the Manhattan Project had been. The military, of course, created a string of national research laboratories featuring tightly controlled access, not only to the facilities themselves but also to the resulting knowledge. On campus, a mushrooming network of federally funded research centers housed researchers with few ties to the academic departments or the teaching activities of the universities; many of them undertook classified work as well. These centers and nonacademic counterparts such as the RAND Corporation also spun off alarming new conceptions of humanity, with approaches such as game theory, systems theory, operations research, and cybernetics reducing individual and collective behavior to abstract variables, aggregate statistics, or predictable, mechanical dynamics. On the civilian side, the federal government worked to manage the national economy as a whole, through the Council of Economic Advisers and other means, alongside its other regulatory initiatives in the 1940s.[18]

The ongoing proliferation of federal bureaucracies, staffed by experts, represented yet another point of reference for those who feared that science had broken free from the chains of public morality and was steadily reshaping society in its own image. By the 1950s, political scientists had reinforced this image of a scientific and largely unaccountable state—and, critics said, silently imported their own political values once again—by equating democracy with the effective advancement of citizens' material welfare by elites and experts, with as little input as possible from the public. A few went so far as to discount the importance of voting itself. Democracy, according to these interpreters in the social sciences, was a matter of cars and refrigerators, not town meetings or ballots—let alone equality, dignity, and other cherished values.[19]

Overall, then, the postwar social sciences presented two faces to the public: a stance of rigorous objectivity that portrayed science as a thoroughly nonmoral enterprise and a staunch commitment to a secular, expert-driven, antipopulist version of welfarist liberalism that denigrated the foundational beliefs—political and especially theological—of innumerable Americans, even as it sought to exclude them from the political process. But that inconsistency did not inspire the most extravagant fears surrounding the postwar social sciences. Nor did postwar concerns simply reflect scholars' portrayal of scientific investigation as a thoroughly amoral practice. Most consequentially, an array of real-world changes appeared to indicate that the social sciences were bringing to life the lurid nightmare visions portrayed in fiction.

Above all, two futuristic novels gave many postwar Americans their sense of how a thoroughly scientific society would look. One was Aldous Huxley's *Brave New World,* which portrayed a stultifying society characterized by top-down manipulation and a total loss of individual autonomy for the masses. Originally published in 1932, the dystopian novel enjoyed considerable postwar success in the United States and appeared in several new editions starting in 1946. Huxley actually shared a scientific bent with his brother Julian and their famed grandfather. Indeed, only his poor eyesight kept him from a scientific career himself. Yet Aldous Huxley had long insisted that science could not suffice as a cultural guide because it was essentially materialistic and denied the existence of values—indeed, that it had done much to squeeze moral commitments out of modern societies. Science explained sensory experience, said Huxley, but there was a larger reality beyond the senses that he sought to capture in mystical terms. Far from being machines, he argued, human beings were "freedom-loving animals, far-ranging minds and God-like spirits." Science and technology were simply means to the truly human end: "the unitive knowledge of the immanent Tao or Logos, the transcendent Godhead or Brahman." Huxley's book portrayed the use of such resources for very different purposes, however: a global dictatorship based on a rigid caste system that suppressed individual effort, thought, and personality through a combination of eugenic breeding, strict conditioning, and the casual, hedonistic use of sex and drugs. *Brave New World* served as a constant point of reference in postwar American debates about the future of a scientific society, giving many readers a compelling portrait of—and a catchy shorthand term for— science's destructive potential. Its stark vision reinforced the suspicion that social scientists hoped to take control and radically alter American culture and institutions, destroying individual freedom in the name of social peace.[20]

The 1948 novel *Walden Two,* by the arch-behaviorist psychologist B. F. Skinner, appeared to confirm that suspicion. Thanks to a lengthy feature in *Life,* Skinner was already known to the reading public for keeping his infant daughter for long stretches in a glass-sided, climate-controlled box. For many readers, his postwar book revealed the deepest meaning of modern science, opening a window onto the secret hopes and dreams of the multiplying legions of social scientists. Even worse than the actual contours of Skinner's ideal society was his apparent inability to grasp how terrifying it was. It was bad enough if social scientists undertook social engineering for selfish purposes, critics noted. But *Walden Two* seemed to show that scientific thinkers—and the culture they increasingly shaped—could no longer even tell the difference between good and bad, right and wrong, free and unfree. In Skinner's future community, there was only the distinction between social and antisocial behavior. Moreover, the latter was firmly beyond the imaginative capacity of individuals who had been programmed since birth to prioritize society's needs. The children of *Walden Two* were taken from their parents at birth and raised as a group. Constant conditioning to the community's egalitarian rules obviated the need for collective decisions of any kind, whether democratic or otherwise. Here, it seemed, was a world truly *Beyond Freedom and Dignity,* as a subsequent book by Skinner would put it. Openly challenging core political ideals and conventional self-understandings—he flatly denied the existence of free will as well as the soul—Skinner drew new opprobrium to the idea of cultural change unmoored from religious or humanistic guidance. His writings convinced many readers that science entailed not only a mechanistic, deterministic view of the human person but also a manipulative, antidemocratic social vision. Just as Dewey had symbolized the danger of science for an earlier generation, Skinner came to symbolize its danger after 1948.[21]

The reviews of *Walden Two* revealed its resonance with broader fears about science's cultural meanings. Some reviewers found Skinner's vision harmless enough, if implausible and distasteful. But many saw dark portents of ongoing social changes. The philosopher Daniel C. Williams called Skinner the leader of the "scientific samurai" seeking to take "the reins of history." *Walden Two,* said Williams, actually addressed philosophy—which Skinner "tackles with the gaily bellicose irresponsibility of a kitten on the keys"—rather than psychology. Indeed, he wrote, Skinner's "rampant scientism" and "behavioristic hedonism" comprised "a philosophy as convinced as the Inquisition's, though considerably less coherent." Whereas "the ordinary dictator simply makes his subjects behave," declared Williams, "the psychological dictator makes them behave *and like it.*" Another reviewer averred that mid-twentieth-century utopias had become "horror stories," populated by "contented non-political robots." A sociologist charged that Skinner peddled "patent scientism" like snake oil. And the harshest reviewer, who saw in *Walden Two* only "the freedom of those Pavlovian dogs," concluded on a note of fear: "If social scientists share Professor Skinner's values—and many of them do—they can change the nature of Western Civilization more drastically than nuclear physicists and biochemists combined." For many readers, *Walden Two* whitewashed a project of scientific social engineering that would inevitably produce a society like that of *Brave New World.* Skinner's novel powerfully reinforced the growing association of science with "the systematic, planned control of human beings by other human beings."[22]

American social scientists and their counterparts in law and education did publish occasional, if rather benign, calls for social engineering (or "human engineering") in the late 1940s, before McCarthyism made that language largely untenable. Indeed, the onset of the atomic age led some to redouble their emphasis on the need for social-scientific guidance. But the critics were not necessarily reading the *Journal of Social Philosophy, Philosophy of Science,* or even *Science.* Rather, they were mainly responding to a series of practical changes in American public life—in education, the law, and many other domains—that often turned diffuse, if intense, concerns about scientism into the belief that a program of social engineering by manipulative, amoral experts had commenced. To the critics, these social changes indicated that a scientific worldview, with its relativistic or deterministic understanding of human behavior, was systematically replacing traditional moral strictures and a commonsense understanding of moral freedom in one area of social practice after another. Such developments seemed to reveal the endpoint of modern society's scientific drift—and the major fronts on which to fight it. All manner of postwar observers concluded that American social scientists had honed the naïve relativism of the natural scientists to a point and were now using it to manipulate their fellow citizens.[23]

Above all else, the growing influence of psychology and psychiatry in the postwar United States fueled the belief that science was creating a regimented, unfree world. Talk therapy became a mainstream phenomenon in the late 1940s and 1950s, following its wartime adoption by the armed services and the federal government's identification of mental health as a policy goal in the late 1940s. Ordinary citizens, far beyond the walls of mental institutions, increasingly sought out therapeutic interventions. By 1955, the sway of the "psy-sciences" was sufficiently dramatic that EC Comics issued *Psychoanalysis,* whose hero fought neuroses rather than traditional, flesh-and-blood villains. As in the 1920s, many postwar critics identified both behaviorist and Freudian approaches as examples of the deterministic tenor of modern thought. Despite the manifest differences between Skinner and Sigmund Freud, both seemed to eliminate moral freedom by tracing individual behavior to nonrational stimuli, whether it was social conditioning or sexual drives and childhood traumas. The mass marketing of psychotropic drugs, beginning with Thorazine in 1954 and Miltown the following year, reflected yet a third approach to understanding and treating psychiatric conditions. To critics, however, the appearance of these mood-altering drugs again seemed to portend Huxley's brave new world. So, too, did invasive psychiatric techniques such as lobotomy and electroshock, although these practices were relatively slow to impinge on the public consciousness.[24]

New polling and advertising techniques, harmless though they might seem in retrospect, also convinced many postwar observers that Skinner and his scientific allies were gaining power and launching concerted programs of social engineering. To these critics, the manipulation of public perceptions entailed a comprehensive loss of individual freedom—especially when those pulling the levers used cutting-edge knowledge of the human psyche. Although modern polling techniques had emerged in the late 1930s, they suddenly seemed to be everywhere in the 1950s. Observers worried that polling reduced complex opinions to simple numbers and subsumed the individual into a statistical mass. Gallup and Roper described their survey results as antidotes to the anonymity and massification of modern life, but many commentators thought polling procedures actually reinforced such tendencies. Advertising, meanwhile, seemed to allow the creation of new consumer desires from scratch, often in direct opposition to moral principles and even the individual's own good. The technological sophistication of many new products themselves—from refrigerators and automobiles to "hi-fi" audio systems—once again strengthened the association of advertising with science. So, too, did the ubiquity of white-coated figures in the ads themselves. Scientists appeared everywhere in postwar advertisements, hawking toothpaste, cigarettes, and innumerable other products while psychologists worked behind the scenes to tailor the pitches for maximum efficiency.[25]

It was not just the advertisers that seemed to be putting science to work, however. Critics also saw a scientific mind-set behind changes in many spheres of governance. Of course, the sprawling bureaucratic agencies of the federal government still employed many natural and social scientists, as they had since Franklin Roosevelt's day. To some observers, moreover, the bureaucratic mode of organization itself reflected science's highly specialized, rationalized approach to the world. But changes in public policy raised the sharpest concerns. Especially controversial was the push to extend greater leniency to both schoolchildren and criminals, an approach that critics deemed thoroughly relativistic—and hopelessly counterproductive. Many commentators connected the rise of permissive parenting to parallel shifts in the treatment of lawbreakers and argued that these applications of science had fueled massive increases in adult and juvenile crime. In these areas, critics contended, a scientific approach erred twice over: first, by eliminating genuinely moral judgments, and second, by deeming individuals innately good and blaming any deviations on their social environments.[26]

The rise of permissive parenting was no mere figment of the critics' imaginations. Benjamin Spock's 1946 childrearing manual, which told parents to respond to their children's misdeeds—even stealing—with affection rather than shame or punishment, sold millions upon millions of copies. Spock's prescriptions did stem from a belief in the innate goodness of the child, who needed to be loved and appreciated, not scolded and disciplined. (In a similar vein, the hero of the *Psychoanalysis* comic informed his clients, "There are no delinquent children . . . only delinquent parents!") His childrearing book both reflected and advanced a broader cultural shift. Over the course of the 1940s, the number of families that let their babies' hunger determine their mealtimes rose from 4 percent to 65 percent. Other forms of strict discipline and moral judgment likewise declined precipitously in the postwar years. To some critics, this new, scientific approach to raising children threatened to create a world lacking the richness, depth, and complexity of the adult experience by reducing everything to the child's level of instant gratification. A sharper critique identified parenting as a matter of taming "child-beasts," defined by their natural depravity. Meanwhile, Spock's emphasis on helping children fit in with peer groups seemed to replace morality with mere popularity.[27]

Alfred Kinsey's controversial reports on male and female sexual behavior, published in 1948 and 1953, reinforced the fear that scientific thinking advanced a "statistical" or "descriptive" morality, insofar as they not only revealed but also condoned the frequency with which Americans violated sexual norms. A scientific, empirical approach to this topic, said critics, necessarily conflated the normal with the normative. Ignoring genuinely moral considerations, such an outlook simply defined the most widespread practices as the best ones. As one Christian critic put it, Kinsey ignored the distinction "between the idea of norm as a simple report of what people do and the idea of norm as what people *ought* to do." Indeed Kinsey appeared to advocate what many readers considered a degraded, feral sexuality, akin to that of *Brave New World*'s narcotized inhabitants. The visibility and popularity of Kinsey's books—the 1948 study sold nearly a quarter of a million copies, and 20 percent of poll respondents had already heard about it a month before its publication—further reinforced the belief that a scientific mind-set was fueling widespread societal decline. To critics, Kinsey's patent desire to loosen restrictive standards of sexual behavior foretold the destruction of morality altogether. One correspondent decried Kinsey's implication that "a wrong should be indulged because a pathetically large group do it." The *Newark Advocate* agreed: "Moral right and wrong come not from numbers, nor from the deeds even of the majority, but from God himself." For such critics, the failure of Kinsey's subjects to heed established ideals did not in any way invalidate those ideals.[28]

Also alarming, to those already wary of science's cultural effects, was the increasingly widespread preoccupation with social belonging in the postwar United States. Dr. Spock's recommendation that parents help their children blend into peer groups seemed to validate the relentless search for acceptance that permeated the mushrooming American suburbs, at the cost of both moral rectitude and individual autonomy. But worst of all, for many critics, was the apparent elevation of social conformity into the central principle of a reorganized public school curriculum under the rubric of "life adjustment." This was not the only educational change that concerned postwar critics of scientism and social engineering, by any stretch. Many, for example, saw the psychologists' "mechanistic scientism" behind the vogue of standardized testing and other quantitative, statistical methods that threatened to reduce students to mere numbers. However, the life adjustment approach drew most of the critics' ire. An outgrowth of progressive education that lacked the critical edge of early theorists such as Dewey, the life adjustment model jettisoned traditional academic subjects and sought to give students the practical skills they needed to seamlessly integrate into society, both as workers and as citizens. Some exponents even denied that every student should learn to read. Not surprisingly, life adjustment provoked intense criticism from many quarters. Letters and articles from the postwar period are full of blasts at the "professional educationists" and their conformist goals. "To hell with Democracy" if it meant educational leveling, declared one critic. In this arena, he continued, "there is a danger in glorifying Demos." A reader of the New York Herald Tribune wrote, "The true democracy on which this country was founded—that the best in education should be as nearly as possible available to everyone—has been twisted into the perverse theory that the lowest denominator of learning capacity should be the yardstick for all."[29]

Ironically, some of the sharpest and most influential criticism of these cultural tendencies came from publicly engaged sociologists. Their widely read books gave the imprimatur of science itself to the claim that postwar Americans were obsessed with conformity—and that this morally relativistic sensibility reflected the spread of scientific thoughtways. The earliest of these works, David Riesman's The Lonely Crowd (1950), remains the most commercially successful work of American social science ever published, having sold more than 1.4 million copies by 1995. Riesman described a cultural landscape in which the stable internal gyroscope of the inner-directed individual increasingly gave way to the sensitive weather vane of the other-directed type, whose moral judgments extended outward to encompass the needs of others. Readers often reduced Riesman's nuanced analysis to a simple story of moral decline in which Americans had thrown off all ethical constraints and eagerly sought acceptance by adopting the prevailing standards of the group. Although The Lonely Crowd gained credibility from Riesman's social-scientific credentials, it also resonated because it reinforced the widespread belief in a trend—the loss of firm moral guideposts—that many critics blamed on science's cultural influence. Indeed, commentators often argued that the world would be safe if social scientists adopted Riesman's approach rather than Skinner's—which presupposed that they did not.[30]

Well-known sociological works by William H. Whyte and Vance Packard could likewise reinforce fears about the cultural effects of social science.

Whyte's *The Organization Man* (1956) asserted the society-wide replacement of the Protestant ethic by the "Social Ethic," the "secular faith" of the well-adjusted cog in a bureaucratic machine. Only a minority of social scientists, he cautioned his readers, fancied themselves social engineers and avatars of a managerial order. Indeed, their colleagues found these outliers embarrassing. Nevertheless, he continued, this minority group had helped make the promise of "an exact science of man" into a central pillar of the new administrative outlook. "At the present writing," Whyte declared, "there is not one section of American life that has not drunk deeply of the promise of scientism." The following year, Packard's *The Hidden Persuaders* detailed the enthusiastic contributions of social scientists to the latest advertising and public relations techniques. He identified the more extreme examples as moves towards "the chilling world of George Orwell and his Big Brother," based on the assumption that "man exists to be manipulated." Indeed, the book's jacket described it as a "guide to the Age of Manipulation." Like Whyte, Packard ultimately concluded that "we cannot be too seriously manipulated if we know what is going on." (Whyte famously ended *The Organization Man* by teaching employees how to cheat on the personality tests used to measure their group-mindedness.) Yet readers could easily draw much darker conclusions from these writings. Like *The Lonely Crowd*, the works of Whyte and Packard echoed widespread concerns about societal trends associated with scientism and social engineering.[31]

To what degree did these views of science's cultural impact actually take hold among the general public? One can pile up examples, and scientists themselves certainly worried about their public image. As they do today, polls found a mixture of distaste for particular techniques and products—everything from skin whiteners to the prospect of "test-tube babies"—with support for science and technology overall. Respondents endorsed numerous other advances in science and technology and even expressed a willingness to pay higher taxes to fund research. In 1947, almost half of a polling group could think of no invention that they wished had not been created. Only 29 percent of them even mentioned the atomic bomb. Still, there is some evidence that admiration for such technical achievements coexisted with deep unease about science's moral implications. For example, college instructors and educated readers created a steady demand for new editions of *Frankenstein* and *Faust,* the classic Romantic critiques of technology and reason run amok. Similar themes echoed in modern works of science fiction as that genre flourished after World War II. (Packard's account of Madison Avenue bore some resemblance to the world of Frederik Pohl and Cyril M. Kornbluth's 1952 novel *The Space Merchants,* subtitled

A Novel of the Future When the Advertising Agencies Take Over in its single-volume version.) More direct evidence comes from the anthropologists Mead and Rhoda Metraux, who surveyed views of scientists among high school students in the mid-1950s. They presented their results by listing composite descriptions of positive and negative traits, arranged in pairs. However, one negative image of the scientist had no positive counterpart: "He may not believe in God or may lose his religion. His belief that man is descended from animals is disgusting." For this reason, among others, Mead and Metraux's respondents exhibited little interest in becoming (or marrying) a scientist.[32]

THE QUESTION OF RELIGIOUS BELIEF gets at the heart of many ordinary Americans' concerns about the cultural implications of science. Yet the potent concerns about science that circulated in the early Cold War years were hardly confined to religious critics of "godless communism"—let alone creationist critics of Darwin. Indeed, the fear of social engineering—of scientific techniques unmoored from human values—was sufficiently diffuse and flexible to find expression among some of the leading architects of the postwar social sciences, including the sociologist Robert K. Merton and even Skinner himself. Virtually anyone looking at American society in 1945, or 1950, or 1955, could identify numerous institutions and practices that seemed fundamentally out of step with human values and conclude that these phenomena reflected a dangerous decoupling of scientific expertise from shared norms and goals. Throughout the postwar era, numerous critics of varied persuasions would decry what they saw as the scientific transformation of their society—above all, the conversion of a scientific worldview into potentially deadly programs of social engineering.

5

MODERNITY AND SCIENTISM

IT IS OFTEN ASSUMED that the atomic physicists' unprecedented contribution to ending World War II opened the golden age of American science. With Hiroshima and Nagasaki, the story goes, science became a key symbol of power, a source of national pride, and a widely heeded authority. In fact, however, the bomb dramatically reduced the scientist's authority for some observers—and reinforced science's doubtful status for others. "The scientist is not so great a figure as he formerly was in the eyes of the common man," editorialized the Protestant *Christian Century* shortly after the bombings. "His stature is dwarfed by the magnitude of the thing he has done. He is no longer a messianic figure." Catholic commentators, as usual, contextualized the events of August 1945 especially broadly. Here was the endpoint of Western secularization, declared the *Catholic World*'s editor: "ethical anarchy" had finally completed its centuries-long erosion of "the universal and everlasting moral law." Milton Mayer, a Reform Jew turned Quaker pacifist, averred that the bomb's creators "might have been useful men, instead of murderous mechanics, had they been Thomists instead of atomists." These scientists lacked any knowledge of morality, Mayer declared, because they were products of an educational system "sold out to the legend of scientism."[1]

Such critics argued that they, not the scientists, were the true realists, because they recognized the patent fact of moral freedom. The familiar idea that "human beings are affected by the judgments of value to which they give allegiance," explained the progressive Congregationalist Buell G. Gallagher, was "just as much a 'fact of experience' as any other 'fact.'" To "rule out of court all matters of value," he continued, "is to miss significant sections of the causal nexus and to do violence to the fundamentals of empiricism itself by limiting the purview of experience." But that was precisely the method of the pragmatists who had long dominated education, said Gallagher. The result was an "ethical relativism which breeds widespread moral cynicism."[2]

Not surprisingly, fundamentalists and other theologically conservative Protestants issued vocal warnings about the dangers of a scientific outlook after 1945. During the next twenty years, it became de rigueur for these Protestants to embrace young earth creationism, challenging modern science even more thoroughly than in the 1920s by rejecting a day-age theory in which the six days of Genesis symbolized ages or epochs. As before, the response of theological conservatives reflected not only their commitment to a literal reading of the Bible but also their concerns about the impact of scientific materialism on views of human behavior and morality.[3]

As the quotes above suggest, however, challenges to science's authority in the study of humanity appeared across virtually the entire spectrum of confessional commitments and theological persuasions in the postwar period. Even theological modernists, whose religious stance was defined precisely by an embrace of scientific authority in its proper domain, sometimes restricted that domain to the natural sciences and reconsidered their long-standing reliance on the increasingly secular social sciences. Many theological moderates and liberals decried the growing cultural sway of science as well, and often claimed that scientists had no business exploring human affairs. In all of these religious circles, the concept of "modernity" as a troubled age of scientific dominance—perhaps even the scene of a comprehensive "modern crisis"—became widespread in the 1950s. Meanwhile, the term "scientism" took hold as a way of pushing back against the application of scientific methods to the study of persons and societies. Moreover, growing numbers of Protestants joined Catholics in viewing modernity and scientism as centuries-old phenomena. Looking back on Western history, they increasingly discerned a cultural infatuation with the scientific method across the whole period from "Descartes to Dewey," as the evangelical leader Carl F. H. Henry put it. Many added the now-familiar assertion that totalitarianism had arisen in the moral vacuum created by scientism. With Hitler and Stalin, they argued, the West's atheistic chickens had come home to roost.[4]

Importantly, most of these religious critics were political liberals. Seeking a middle course between laissez-faire and Marxism, they generally favored the economic regulations and welfare provisions associated with the New Deal. At the same time, though, they blasted the philosophy on which they believed Roosevelt and his allies had built. Indeed, these critics identified "modern" or "secular" liberalism as the root of innumerable evils in the postwar world. Genuine liberalism had been hollowed out by secularization, they argued, leaving a desiccated caricature to dominate the American political scene in the mid-twentieth century. According to numerous critics, this secular liberalism embodied scientism, with its view of the human person as a mere animal that was either subject to deterministic laws or intrinsically good and completely self-sufficient in the world. In this view, the American polity took its essential shape from scientism's refusal to acknowledge the manifest realities of God, sin, and moral freedom, which would inevitably lead Americans down the same path as the Russians—or perhaps the Germans. Of course, they noted, totalitarianism would be fine for the degraded beings of the social scientists' fantasies, since it could meet their bodily needs efficiently. But those citizens would lack spiritual freedom: their birthright, their essence as human beings, the distinctive contribution of religion to humanity's self-understanding. Here lay the missing element in the secular, materialistic liberalism that dominated modern societies.

THE POSTWAR REVIVAL

Such criticism fed into the massive surge of religious affiliation and expression that followed World War II and extended through the 1950s. During those years, American churches and synagogues swelled, Billy Graham's prayer meetings filled football stadiums, and Congress declared "In God We Trust" the national motto. Almost half of the nonfiction best sellers were religious books and more than 80 percent of Gallup's respondents said they "would refuse to vote for an atheist for president under *any* circumstances." Even the National Education Association, sometimes accused of peddling atheism itself, issued a 1951 report stating that democratic government possessed "a moral and spiritual basis" and found its highest purpose in serving "God's man." Religious themes were equally ubiquitous in the media, in many areas of intellectual life, and in the public pronouncements of all manner of national leaders. "The President, the Cabinet, and the Congress all recognize the priority of spiritual forces," announced Secretary of State John Foster Dulles, a former official of the Federal Council of Churches, in 1953. "Our system demands the Supreme Being," Dwight D. Eisenhower declared repeatedly, because democracy was merely "the translation into the political world of a deeply felt religious faith." The corollary, for Eisenhower, was that truth could not be defined "solely in the narrow terms of mere fact or statistic or mathematical equation."[5]

The official paeans to faith and associated challenges to secularism hardly began with Eisenhower, however. His predecessor, the Democrat Harry Truman, called for "a re-dedication of this nation—individually and collectively—to the unchanging truths of the Christian religion" and identified the country's leading task as "implanting in the child's mind the moral code under which we live": namely, the Ten Commandments. A devout Baptist who often said "my political philosophy is based on the Sermon on the Mount," Truman consistently framed the struggle against communism as a spiritual conflict. In his diary, he called communist countries "Russian Godless Pervert Systems." The connection between religion and politics, Truman explained over and over, lay in their convergence on "the worth and dignity of the individual man and woman," whereas in dictatorships "the individual amounts to nothing" and "the State is the only thing that counts." The totalitarian state, in his view, not only grew from atheism but also sought to protect itself by eliminating criticism on religious grounds; "wherever it can it stamps out the worship of God." At home, too, Truman questioned the democratic credentials of atheists. In a letter to Pope Pius XII, with whom he sought to build a united spiritual front, Truman stated, "I believe that those who do not recognize their responsibility to Almighty God cannot meet their full duty toward their fellow men." Many members of Truman's administration likewise claimed that totalitarian countries went "whole hog on the path of Godlessness," as George Kennan put it. For Kennan, communism was simply the age-old specter of "evil."[6]

These sentiments persisted in the highest levels of government during the Eisenhower years. Dulles, like Kennan, had a one-word explanation for the Cold War: "irreligion." The "great trouble with the world today," he specified, "is that there are too few Christians." Materialism, denying the truly human dimensions of persons, was spreading around the world, just as "the tide of Islam swept over much of Christendom . . . in the tenth century." A few years later, Dulles spoke in a radio address of a moral law "imbedded in the conscience of all men . . . which is just as real as the laws of physics." As for Eisenhower, internal White House memos occasionally called the president the "Defender of the Faith" or "'Pastor' of the Nation." From the campaign trail to the White House, the president consistently defined the Cold War as "a fight between anti-God and a belief in the Almighty," a struggle of "good and evil." National strength was always a religious matter, he insisted; it could never mean "bombs and machines and gadgets."[7]

Such sentiments enjoyed widespread popular approval, as a large majority of Americans defined their political institutions as deductions from

Christian (and perhaps also Jewish) religious tenets. National leaders and ordinary citizens alike also drew the opposite conclusion: that naturalistic worldviews—and perhaps the associated forms of religious liberalism—directly threatened democracy's survival. In this view, the unbelief fostered by modern science eroded democracy from within, leading either to its final collapse or its easy defeat by foreign enemies. The powerful reconsolidation of American civil religion after World War II, often under the sign of the "Judeo-Christian tradition," was routinely linked to a narrow, materialistic understanding of science—and a terrifying vision of its influence and implications.[8]

In religious circles, as elsewhere, postwar challenges to scientism ran along several tracks. Virtually every critic targeted B. F. Skinner, and education was a nearly universal concern as well. But commentators also made other selections from the postwar smorgasbord of scientism. Some highlighted the postwar nexus of social science with sexual behavior or criminal justice. Such critics blamed scholars, politicians, and administrators for peddling a permissive mentality, a deterministic view of human action, or a purely descriptive morality that equated the best behavior with the most frequent. Other commentators carried forward the polemic of the 1930s, decrying the inability of a naturalistic philosophy to ground democracy and its concomitant tendency to produce totalitarian regimes, including "godless communism." These critics often singled out secular education for particular blame, but they often spoke more broadly of "modern thought" as a whole. In this regard, they sometimes extended their critique into the deep historical past—back to nineteenth-century materialism, or the eighteenth-century Enlightenment, or even the earlier writings of Newton, Descartes, or Bacon.

As expressions of public religiosity spread, they grew so prevalent and vehement that they implicitly undercut the narrative of rampant scientism and religious decline itself. Nonetheless, the critics continued to identify themselves as a pious minority under the yoke of a scientific establishment. Academic commentators and theologians often maintained the image of a culture blinded by science by arguing that the religious revival itself was largely secular in character. One version of this argument portrayed most of the postwar religiosity as expressions of a spurious theological liberalism that cloaked the wolf of scientism in the sheep's clothing of biblical language. On this view, only "traditional" or "historic" religion—a category that excluded liberal theologies—offered the normative resources that moderns needed and a scientific culture lacked. Critics also described secularization as a practical matter: even the staunchest believers ignored their faith commitments in the daily rounds of life, unconsciously acting out of secular motives despite their sincere professions of piety. (Thus, the Jewish writer Will Herberg equated secularism with "the practice of the absence of God in the affairs of life.") But either way, it seemed, secularizing forces had invaded the churches and synagogues themselves. In the pews and pulpits, as in the universities and bureaucratic agencies, the avatars of scientism had systematically replaced genuine faith with delusions of scientifically guided social and moral progress.[9]

SCIENCE AND SECULARISM

As critics of scientism and religious liberalism developed their views in the late 1940s and 1950s, they made a number of important conceptual shifts. Conservative evangelicals, for example, often dramatically expanded the scope of their attacks. Beyond Darwin's heresy, many now saw centuries of scientific materialism across the West. Even Henry, a "neo-evangelical" who hoped to reconcile theological conservatives to the twentieth century by squaring religion with science, nevertheless identified scientism as the root of Western civilization's flaws. "In a thousand and one ways," he wrote, "the western world has been taught that nature is the ultimate reality and that man is only an animal." Henry traced this philosophical error all the way back to God's rejection by Renaissance humanists. Ever since then, he declared, science had been "the bondslave of a naturalistic philosophy," its very name "blasphemed by false prophets" as Western thinkers reduced human beings to merely physiological beings in the physicist's "purely mathematical universe." For Henry, as for many others at the time, the two world wars reflected the culmination of the modern crisis. The West's "controlling ideas" had been "contradicted and repudiated by history and experience," he wrote: The "decay of Berlin and even more the atom-bombing of Hiroshima symbolize the disintegration of contemporary culture." And the decline would continue unabated, Henry declared, until Westerners recaptured "the great convictions that nature is not alone real, that man is not merely an animal, [and] that morals are not merely artificial conventions." After World War II, conservative evangelicals such as Henry increasingly joined Catholics, their sworn theological enemies, in defining the modern West—now in its "death coma"—as a product of scientism's devastating philosophical influence.[10]

Catholics themselves often emphasized that the modern crisis had spawned the Cold War, which they viewed as a global battle between the champions of Christianity and the minions of naturalism. Communism threatened abroad, they argued, but a subtler danger at home was in many ways more immediate: the "wave of intellectual and moral nihilism" emanating from the universities. The postwar "infatuation with 'science,'" wrote the Catholic sociologist Edward A. Marciniak, led inexorably toward a form of "scientific fascism." Similarly, the political theorist Thomas I. Cook contended that behavioral science "confuses man the subject and actor with man the physical object," embodying "that scientism which is the most dangerous element in the dominant contemporary outlook." Catholic critics reserved particular scorn for postwar social scientists' vocal claims of value-neutrality. "Men who lack the courage of commitment deserve little better than to be thought of as women," declared the traditionalist philosopher Frederick Wilhelmsen. But these critics also lambasted the other varieties of social science, since each one—"evolutionism, secular humanism . . . historicism, sociologism and psychologism," the Catholic journal *America* enumerated—denied the religious tenets that sustained democracy and civil order. A Seton Hall student thus dubbed secular universities "The Kremlin's Hidden Ally."[11]

Historians have noted that the American Catholic community's ferocious, consistent, and long-standing anticommunism facilitated their integration into the cultural mainstream after World War II. Anticommunism had been a central thread in Catholic thought for decades—perhaps even *the* central thread, according to John Cogley, a rare Catholic critic who deplored communism but thought its threat had been exaggerated. Throughout the postwar period, as earlier, Catholics of all theological and political stripes joined hands in seeing Marxist atheism as a threat to humanity's survival. As a result, Protestants' age-old suspicions of Catholic political loyalties, stoked further in the 1930s by the Church's friendliness toward Mussolini and Franco, began to dissolve in the cauldron of postwar anticommunism. The religious framing of the Cold War, combined with what Herberg called the "over-average religiousness" of Catholics, made them seem quintessentially American—even more so than mainline Protestant leaders, who were sometimes hauled before Sen. Joseph McCarthy's committee in the 1950s. "In the era of security clearances," Daniel Patrick Moynihan later noted, "to be an Irish Catholic was *prima facie* evidence of loyalty. Harvard men were to be checked; Fordham men would do the checking."[12]

But it was not just anticommunism that aligned Catholics with the American mainstream. It was also the particular shape of their anticommunism. Catholic anticommunists did not lament the abolition of private property in the Soviet Union, as economic conservatives did. Nor did they decry the Soviet regime's pattern of suppressing dissent by eliminating civil liberties, as mainstream liberals did. Instead, they focused on a philosophical error that they ascribed to Russell, Watson, Freud, and Dewey as well as Marx,

Lenin, and Stalin: their adoption of an atheistic, materialistic, deterministic orientation that Catholic critics discerned in virtually all precincts of modern thought. When Pius XII codified the Church's tentative embrace of theistic evolution in the 1950 encyclical *Humani generis,* the Catholic position was clear: natural scientists were right about the evolution of human bodies, but evolutionary theory could not be developed into a comprehensive, naturalistic philosophy of existence. That path led straight to totalitarian rule.[13]

This view of science's cultural meanings, like the broader matrix of anticommunism, extended across the full range of Catholic political thought. On the right, Catholic participants in the resurgent conservative movement of the 1950s had much to say about the influence of science in the modern world. Francis Graham Wilson argued that Dewey's twin commitments to progress and Darwinism had provided the basic conceptual framework for the profoundly dangerous behavioral sciences. Other Catholic conservatives likewise decried "the scientific illusion of progressive objectivity," which destroyed the possibility of a sympathetic connection between the observer and the object of study, and a social ideology that identified science and technology as the sources of social improvement—and had produced the Nazi and Soviet regimes, critics declared. Yet the same arguments appeared at the other end of the political spectrum. Even a progressive such as John LaFarge, known for his antiracist work, viewed communism as the inevitable outcome of the modern scholars' "de-spiritualized and grossly naturalistic world." Other Catholic writers saw the final outcome of societal secularization in Truman's use of atomic weaponry. "Nothing remains but nihilism," declared the *Catholic World* after the bombing of Japan.[14]

These arguments reached wide audiences through the writings of prominent liberal Catholics. The Jesuit theologian John Courtney Murray, a fierce critic of secular liberalism, explained with his usual precision that scientists could not rule out "the spiritualist and theist hypothesis" without "illegitimate recourse to some non-scientific *a priori* absolute." Of course, that hypothesis could not be proven through the usual method of gathering sensory evidence, "for the data here are in another, though not less real, order of experience." But it would find adherents once again if scientists would throw off their "arbitrary dogmatism about the nature of man and the limitations of his intelligence" and embrace, in a truly empiricist spirit, "whatever costly truths are encountered in the search." Throughout his postwar writings, Murray also insisted that the spread of secular liberalism had produced both the German and Soviet varieties of totalitarianism.[15]

Jacques Maritain, a lay philosopher and a leading voice of liberal Catholicism in the World War II era, also traced modern ills to the obliteration of "the true image of man" by foolish rationalists and materialists. As a result of this error, Maritain wrote, "Human Reason lost its grasp on Being, and became available only for the mathematical reading of sensory phenomena, and for the building up of corresponding material techniques—a field in which any absolute reality, any absolute truth, and any absolute value is of course forbidden." The political effects, among others, had been devastating: "Modern man believed in liberty—without the mastery of self or moral responsibility, for free will was incompatible with scientific determinism; and he believed in equality—without justice, for justice too was a metaphysical idea that lost any rational foundation and lacked any criterion in our modern biological and sociological outlook." All that remained was "hope in machinism, in technique," and the results were predictable: "Communism is the final state of anthropocentric rationalism."[16]

Similar claims could be heard from Catholic media figures such as Fulton J. Sheen, the neo-Thomist philosopher and a popular radio and television commentator on personal issues. "Once man is identified with nature, so that psychology is nothing more than behaviorism, [and] theology nothing but comparative religion," Sheen wrote in *Communism and the Conscience of the West* (1948), "it is not long until man begins to be treated the way nature is—as a means, an instrument, a tool, and then the Moloch of collectivism swallows up the man of democracy." Sheen saw the materialistic foundation for totalitarianism behind not only Marxism but also its leading competitors: economic laissez-faire and a secular form of welfare liberalism that "denies all standards extrinsic to man himself, measures freedom as a physical power rather than a moral power and identifies progress by the height of the pile of discarded moral and religious traditions." Although Sheen took his wildly popular show *Life Is Worth Living* from radio to television in 1952, the high-tech medium did nothing to temper his blasts at materialism and scientism.[17]

The prominent writer and politician Clare Boothe Luce, who had converted to Catholicism under Sheen's guidance, was one of the highest-profile American Catholic commentators. The wife of *Time, Life,* and *Fortune* publisher Henry Luce, she served in Congress during the 1940s and held several other political offices in later years. Like Sheen, Luce described human behavior as a matter of deduction from first principles. "The basic concept of any man is his concept of his own nature and its relation to the universe," she wrote in *The Twilight of God* (1949). "His laws, morals and ethics, and all his cultural, political and economic institutions, including his international actions, will reflect that concept faithfully." And "the basic concept of scientific materialism," she explained, led to Stalin's brutal regime. Humanity faced a stark choice between two political guideposts: "absolute truth and justice" or naked force. Luce warned in the wake of Hiroshima and Nagasaki that American culture rang with the prayer of science's devotees: "I believe in the Atom, Power Almighty, Substance of Heaven and Earth, once and forever divisible, first split at Oak Ridge, as prophesied by St. Einstein, then dropped over Hiroshima in the form of a bomb, killing millions, still to be split over the whole world, whence it may bring the atomic Kingdom of Heaven on Earth; or come to judge mankind obsolete. I believe in Uranium, Plutonium and the Cyclotron; the Communion of Scientists; the corruption of the body; the relativity of all mind and matter, world with an end, Amen."[18]

In the postwar years, mainline and liberal Protestants, as well as some Jewish commentators, increasingly employed versions of this secularization narrative as well. Such critics typically described scientism's effects in less apocalyptic terms than did Catholics and conservative Protestants. Still, they saw its influence behind the modern crisis and often identified totalitarianism as the inevitable outgrowth of a science-obsessed civilization. Few of these Protestant and Jewish critics worried about modern biology, as did Henry. Most, like Catholics, found the main locus of modernity's error and the primary vector of its spiritual disease in the social sciences, the intellectual fruits of the post-Enlightenment attempt to express political and social dynamics in secular terms.

On the Protestant side, Henry Luce himself was a particularly important critic. Although he never followed his wife into the Catholic fold, his commitment to Christianity—and his distaste for scientism—shone through in his wartime and postwar work. Luce consistently interpreted totalitarianism as an outgrowth of scientific thinking. Penning "A Speculation About 1980" for *Fortune* in 1955, he started with the proposition that science "discloses no 'meaning' or 'purpose.'" What would happen, he asked, if moderns continued to take science as their guide to human relations? "Twenty-five years from now, will men believe that there is in life a 'dignity' infinitely precious—that liberty wagered against death will win?" As the trend of modern thought stood, Luce predicted, they would not. Deploring that prospect, he urged his contemporaries to reunite science and religion, and thereby give each citizen of the modern world "the kind of picture of himself-within-the-universe that can unite his mind, his hope and his conscience in the service of some intelligible, confident and lawful purpose, place him at home in a harmonious universe of truth, and assure him that all his terrestrial tasks have their final consecration from the fact that they further a destiny that is not terrestrial."[19]

Protestant critics decried the views of celebrity biologists such as Britain's Julian Huxley as well as B. F. Skinner and other social scientists.

Huxley, who saw his term at UNESCO reduced from six years to two after American delegates protested his selection, continued to issue controversial endorsements of eugenics and other "scientific" policies thereafter. After Huxley published one such piece in *Time* in 1960, several readers recoiled at his proposals. "According to Julian Huxley," one Episcopal minister wrote, "man has now become his own savior and has complete control of his own future." One needed only to "look at the world situation today," he continued to "see the works of the scientific savior, who only studies man, who cannot love man because to love would destroy the validity of the 'scientific method,' and who, finally, can only point to, but not forgive, the sins of the world." Huxley's scientific god, this reader concluded, "sends not his son but rather a ballistic missile." Another minister, from a Methodist church in Texas, issued a direct challenge to "Huxley and his biological associates": "Let them eradicate this thing we theological boys call sin." If they could do it, he promised, he would give up the ministry and join them in the scientific endeavor.[20]

These claims about science and its cultural impact accompanied the burgeoning theological renaissance among Protestant leaders. Since the late 1930s, a growing number of mainline Protestants had shifted away from their historical alliance with secular and scientific thinkers. Increasingly, they identified naturalism, materialism, or "secularism"—an amalgam, as one historian has noted, of "naturalistic assumptions about the nature of the human being, positivistic-scientistic assumptions about knowing, and popular, utilitarian, and materialistic assumptions about the purposes of learning"—as the leading threat to the faith, and thus to humanity. The pioneering Christian realist Reinhold Niebuhr led the way through the 1940s and early 1950s, continuing to excoriate social scientists for their cognitively, morally, and politically inadequate view of humanity. Although Niebuhr was a former socialist who remained decidedly left of center, his 1952 blockbuster *The Irony of American History* deemed even the most conservative businessmen and politicians preferable to social scientists. There, Niebuhr equated Dewey with Skinner. In their quest for a science of society, he explained, both thinkers inevitably fell into the trap of social determinism by describing human material as akin to the stuff of the physical world and therefore subject to the same kinds of rigid causal laws. Niebuhr called the extension of scientific methods to the study of human behavior "the culminating error in modern man's misunderstanding of himself" and declared that the "spiritual confusions arising from this misunderstanding constitute the cultural crisis of our age, beyond and above the political crisis." Other Protestant participants in the theological renaissance likewise sought a Christian faith that could stand up to scientism, successfully breaking its stranglehold on modern culture. Lambasting theological liberals for capitulating to scientific authority, they identified "traditional" or "historic" religion as the only form capable of challenging scientism. According to Niebuhr's influential version of this critique, social scientists and theological liberals shared the naïve belief that salvation would come from within the world of mere experience—the framework of human history—rather than through God's grace.[21]

As part of the theological renaissance, many Protestants sought to reestablish a Christian framework for American higher education. Taking inspiration from Niebuhr, his brother H. Richard Niebuhr, and the brothers' close ally Paul Tillich, they set out to liberate the universities from what Wooster College president Howard Lowry called, in an influential 1950 book, "the cult of objectivity." These Protestants had a solid institutional core in place by 1950s: student movements, campus ministries, and other religious organizations on campus; the Hazen Foundation, the Danforth Foundation, and the National Council on Religion in Higher Education; and the Commission on Higher Education of the National Council of Churches (whose Department of Campus Life published *The Christian Scholar*), along with the closely associated Faculty Christian Fellowship. Such a movement of restoration was necessary, these critics contended, because the universities inevitably molded the surrounding culture in their image. As went academia, so would go society as a whole.[22]

Many of these Protestant educators argued that the value-neutrality championed by postwar social scientists (and implicitly assumed by most natural scientists) was illusory. All human thought rested on foundational presuppositions, they argued—and only Christian presuppositions adequately explained the facts of human experience. No matter how objective scientific thinkers claimed to be, they always smuggled in an underlying religion from some nonscientific source. "There are no atheists—in fox holes or otherwise," wrote the Episcopalian leader James A. Pike. Elsewhere, Pike declared that "nobody is objective": Each observer had "a perspective, a world-view," that was simply "taken on faith" rather than "proven." And according to Pike, there were only two choices: "the Biblical world-view" or "the secularist world-view." As chaplain at Columbia University from 1949 to 1952, Pike joined Reinhold Niebuhr and his wife Ursula in wresting control of Columbia's religion offerings from Dewey's naturalistic followers in the philosophy department. The Christian ethicist Robert E. Fitch similarly argued that only God "possesses truth as an absolute": "The supreme pride of intellect is to pretend to be without a point of view." Fitch, like Pike, saw only two possibilities: "the perspective of autonomous reason" and "the Christian perspective," which could be proven rationally superior.

Fitch added that Darwinism, despite its creator's intentions, had served "to destroy the religious foundations of morality, and to give scientific respectability to the ancient and evil ethics of Moloch." These Protestant arguments echoed the long-standing Catholic claim that there could be no neutrality in religion, philosophy, or even science, because theological assumptions underpinned all forms of thought, expression, and action. Since value-neutrality was impossible by definition, all individuals and institutions—scientists, schools and universities, the state—needed to actively choose sides.[23]

A smaller group of Catholic and Protestant commentators offered mysticism as the needed alternative to scientism. Aldous Huxley's excursions into Eastern religions echoed a fascination with mysticism among many postwar Americans. By the early 1960s, the Romanian émigré Mircea Eliade had also pushed the study of religion in American universities toward mysticism, which Eliade viewed as a universal category and a universal experience. The writer Joseph Campbell, author of *The Hero with a Thousand Faces,* likewise identified universal mythological categories as pathways to mystical vision. But the best-known mystic of that era was a high-profile convert to Catholicism. Thomas Merton's enormously successful spiritual autobiography of 1948, *The Seven-Storey Mountain,* detailed his journey from the study of English at Columbia—where he imbibed the usual "economic and pseudo-scientific jargon" before coming under the influence of the literature professor Mark Van Doren and the Catholic chaplain Father George Ford—to the monastic life in the Abbey of Gethsemani in Kentucky. Many of Merton's other writings portrayed those who chose the contemplative life as operating both within and against a thoroughly scientistic, technocratic world.[24]

Even theological liberals, routinely accused of sleeping with the scientific enemy, sometimes issued their own challenges to scientism. Indeed, the Unitarian leader James Luther Adams turned the tables by arguing that the deficiencies of religious orthodoxy had fueled the influence of scientific worldviews. He argued that secularism, grounded in scientism, dated back to the Renaissance and took much of its appeal from the obtuse refusal of church leaders to liberalize their views in response to modern social tendencies. For Adams, scientism and theological orthodoxy stood in a destructive, dialectical relationship. A purely scientific worldview, Adams wrote, "does not have the power to resist the temptations of racism, bourgeoisism, and nationalism," and its inevitable failure "often induces a nostalgia for the securities of authoritarian religion." Adams presented a liberalized but firmly antiscientistic form of faith as the only viable source of guidance in the modern age.[25]

Jewish commentators occasionally took aim at scientism as well, though much less frequently than their Christian counterparts. The presence of numerous nonbelievers in their ranks led many American Jews to resist the vigorous postwar campaign against secular worldviews. However, the prominence of secular Judaism redoubled the desire of a few Jewish critics to turn their coreligionists away from science and back toward the historic faith. Figures such as Louis Minsky of the Religion News Service, the writer Waldo Frank, and the theologian-sociologist Herberg joined leading Christians in declaring that faulty, naturalistic philosophies derived from modern science threatened democracy's theistic foundations.[26]

This was especially true of Herberg, a former Marxist who eventually found himself at the conservative *National Review* by the end of the 1950s. Famed for his sociological study *Protestant-Catholic-Jew* (1955), Herberg was also a sharp critic of science's cultural influence. He rooted democracy firmly in the Bible, advocated federal aid to parochial schools and the teaching of religion in public schools, and insisted that secular liberalism could not sustain a democratic culture because it denied God's sovereignty over human affairs and the centrality of religion in public life. Herberg's major theological statement, *Judaism and Modern Man* (1951), limned an apocalyptic crisis of Western civilization: economic weakness, military conflict, moral degradation, social breakdown, and state expansion unchecked by recognition of God's ultimate sovereignty. Behind all these threats, contended Herberg, lay the errors of a "popular philosophy of scientism" that viewed human affairs through the blinkers of "natural determinism" and the "uniform laws of reason and causality" that defined "abstract-objective thought."[27]

Yet Herberg also saw amid the chaos an emerging "post-modern generation, shocked out of its illusions by three decades of unbroken horror." This new generation, according to Herberg, had begun to recognize that only traditional religion could anchor a meaningful worldview and that "ends and values lie in a realm beyond positive science, whose usefulness is limited to devising means for ends already established." The postmoderns, he continued, sensed that the modern path could only bring "further depersonalization, further atomization, further spread of mass standardization, [and] further stultification of man's aspirations toward a worthy and significant existence." Science, Herberg concluded, "may prove an invaluable servant, but when it turns master and savior, it inevitably becomes a brainless monster, imperiling life." Although Herberg was a disciple of Reinhold Niebuhr, his sweeping analysis of scientism's cultural impact across the centuries sounded more like that of a Carl Henry or a Fulton Sheen.[28]

Closer to the main currents of Jewish thought, the Conservative leader Robert Gordis offered a similar account of Western history in *A Faith for Moderns* (1960), despite clashing with Herberg on numerous theological issues in the 1950s. Gordis, too, offered a schematic, idea-centered history of the modern West that pinned the ascent of Hitler and Stalin on the spread of faulty understandings of science, religion, and morality. Likewise, he argued that Westerners were now turning back en masse from an empty faith in science to the established religions because they sensed the world's desperate need of moral guidance. Religion, Gordis explained, filled in the cognitive gaps left by science and overcame the compartmentalized structure of scientific knowledge by offering a comprehensive picture of the world. Scientists cherry-picked the "rational and pleasant" features of reality and ignored "chaos . . . tragedy . . . frustration, suffering, and death," Gordis wrote. Religion, by contrast, highlighted these "negative and painful" elements. Yet it also guaranteed "the triumph of the ideal" by showing "that the human adventure is no accident, destined to be wiped clean from the earth's surface." All genuine religion, wrote Gordis, taught the needed lessons that "man is not alone in a meaningless, unfeeling universe" and that "morality is not an invention of man, but his discovery of a law of life, which is rooted in the cosmos and therefore destined to emerge victorious." Modern science-worship, by contrast, fostered moral relativism, the belief "that right and wrong are merely matters of social convention, that morality is no more than *mores,* the customs and taboos of a particular society." Such beliefs had produced "an age of dissolving standards and vanishing values," characterized by "the collapse of international morality and the wholesale decline of moral standards at home." The "rise of totalitarianism," Gordis asserted, "disclosed the ominous consequences of this approach."[29]

A WIDE ARRAY OF RELIGIOUS CRITICS challenged scientism's apparent reign in the postwar United States and lamented the tragedies of a scientific age. For the most part, they shared a sense of the needed remedy: a framework of transcendent moral laws, inscribed on the world by God and offering guidance today as yesterday. At the same time, these figures defined that moral framework in an immense variety of ways. Some thought the moral law called for unfettered capitalism; others advocated cooperative or collective economic arrangements. Some urged strong political leadership; others sought to decentralize power. A host of other divergences appeared as well: disagreements on foreign policy, aesthetic taste, and much else. Indeed, these religious critics shared little more than the foundational assumption that moral principles had their source in some transcendent, nonempirical domain—and that the failure to recognize such principles had produced the crises of the twentieth century.

As earlier, however, there were numerous other critics of scientism in the postwar era, well beyond the churches and synagogues. Many political conservatives, as well as self-identified "humanists" both inside and outside the humanities disciplines, also saw a need for universal moral principles, even though relatively few in the latter group identified God as the author of those principles. Even social scientists themselves sometimes issued calls for moral guidance. Like religious leaders, these other groups of critics shared key claims about science, scientism, and modernity, despite their many differences. Indeed, those points of overlap fostered a complex web of connections and resonances. As in the World War II era, but on a much greater scale, fears of science's cultural impact led to shifting and often surprising alliances in Cold War America.

6

THE HUMANISTIC OPPOSITION

AS IN RELIGIOUS CIRCLES, many scholars in the humanities disciplines also deplored science's cultural impact after World War II. By 1962, in fact, the chorus had grown so loud that a committee of the Modern Language Association (MLA) sought to dissuade its members from issuing blasts again the cultural sterility of modern society. Attempts "to elevate the humanities by poking fun at or debasing technology" were both impolitic and largely wrong, the committee averred. In fact, it insisted, "the present-day technological culture of the United States is in many ways a stronghold of the humanities and gives great promise of their future development." But to realize that promise, the committee warned, humanists would need not only to refrain from attacking technicians and experts but also to change their own outlook, embracing "the future-oriented view normal to twentieth-century existence."[1]

Despite the MLA's efforts, humanities scholars continued to speak out about science's cultural impact. Since the start of World War II, in fact, such criticism had accompanied what contemporary observers considered a full-blown revival of the humanities in American intellectual life. Its advocates identified the creative works and historical experiences for which they spoke as irreplaceable sources of moral authority, either alongside or in place of religious tenets. Indeed, the humanities revival paralleled the far larger religious revival in several ways. Like its religious counterpart, the humanities revival carried into the 1960s, albeit in modified forms, and powerfully shaped the upheavals of that decade. Also like the religious revival, the upsurge in the humanities took much of its shape from the growing fear that science had cut off moral behavior and human self-understanding at their philosophical roots. Finally, the existence of this revival again threatened to give the lie to the claim of scientific domination, although few observers recognized or addressed that potential inconsistency in the case of the humanities. It seemed painfully self-evident to leading academic humanists that they labored under the yoke of a scientific culture.[2]

Many of these critics argued that the alarming developments of the mid-twentieth century had revealed a sharp rift between two opposing views of humanity, embodied respectively in the sciences and the humanities. Thus, the philosopher Bernard Phillips argued that advocacy for the humanities necessarily took the form of challenging "the objectifying tendencies of modern life and of the sciences of man" in the name of a "non-naturalistic interpretation of man"—and warning citizens about "the tie-up between psychology and an efficient, fascistic society." Writing in 1948, Phillips identified the revival of the humanities in the American universities over the previous decade as part of a "larger reaction against naturalism as a philosophy of life" that spanned Western thought in that era. Perspectives as diverse as vitalism, phenomenology, existentialism, and "Neo-supernaturalism," in his view, highlighted "the unique status of man within nature" and denied that science could explain reality in toto. Thus the crying need for the humanities: Whereas the sciences studied the individual as an object in a world of causal relations, Phillips wrote, the humanities entailed "the study of man *qua* man"—in short, as a subject. Indeed, he asserted, the phrase "science of man" was literally meaningless, because human subjects could never be natural objects. Science could not even see, let alone adequately understand, "man." According to Phillips, the revival of the humanities in the American universities reflected an eleventh-hour recognition that Nazi Germany revealed the inevitable terminus of a scientific culture.[3]

To be sure, the MLA committee's optimistic approach also had vocal advocates. Numerous practitioners of the humanities hitched their wagons to contemporary Western civilization in the 1940s and 1950s. A range of historians, philosophers, and literary scholars offered celebratory portraits of the United States as the West's cultural polestar, upholding its core values of individual freedom and human dignity. But many others set themselves firmly against postwar culture. Arguing that a kind of cultural collectivism had nearly erased the centuries-long heritage of morality and freedom, they identified themselves as the proper cultural guides for unmoored moderns who sensed, dimly but deeply, the pressing need to recover that humanistic tradition. These critics disagreed on how, exactly, the humanities countered what the literary critic Lionel Trilling called the "bland tyranny" of a scientific culture. Some argued that the needed pillars of a rejuvenated self-understanding could be found in the Western cultural tradition: great works of literature and art, or European and American political thought, or both. Others advocated a humanistic ideal, largely unmoored from specific textual sources or investigative methods, that declared human beings spiritually free but subject to reliable, time-tested moral guidelines. Either way, however, most humanistic critics of science's cultural impact argued that Westerners had discovered, through long experience, certain principles of behavior that reliably maintained social order and advanced the cause of human flourishing—even if there was no God to back these principles with a divine fiat.[4]

In the writings of these humanities advocates, the poet, the novelist, and the critic took the place of the minister, the rabbi, and the theologian. Moreover, the threat of determinism typically loomed larger than the loss of moral absolutes. But the enemy of the soulless social engineer, manipulating humanity in pursuit of utopian dreams, remained. Scholars across the humanities agreed with the venerable political commentator Max Lerner that a gulf had opened between the "startling accomplishments of the technicians" and the stunted self-understanding of postwar societies. Americans could no longer focus solely on "specialized knowledge and skills," Lerner urged. "The new frontiers to be opened are no longer in *science* but . . . in *conscience*, no longer in technics but in the arts that give meaning to them and in the wisdoms that give direction to them."[5]

Such rhetoric from Lerner, a fierce advocate of secular progressivism in the 1930s who had written pugnacious works such as *Ideas Are Weapons* (1939) and taught in government departments through most of his long career, also reflected a related development: the fear of some social scientists and many natural scientists that their own disciplines had deepened, or even created, the modern crisis by obscuring considerations of value and seeming to deny the moral freedom of the individual. Between the wars, the campaign for mental modernization, with its twin attacks on laissez-faire politics and religious orthodoxy, had enlisted scholars from the social sciences, philosophy, biology, history, and even some precincts of literary study. It had also dominated the pages of progressive and radical journals such as the *Nation*, the *New Republic*, and *Partisan Review*. By the 1940s, however, an exodus had begun. Across the disciplines, individual thinkers and even whole fields had begun to turn against the modern mind approach, deeming it an excrescence of scientism that was creating a moral vacuum in society. Lerner and other political writers often did the same, fearing that science's cultural influence and social products—technology, bureaucracy, standardization—had nearly extinguished the West's spiritual and ethical flame. In the postwar years, it became a matter of simple common sense for many scholars that the Western world was experiencing an epochal modern crisis and that scientism lay at its root. Such dynamics reveal, once again, the complex networks generated by shared concerns about science as a cultural force.

THE REVIVAL OF THE HUMANITIES

Not all of those who helped build up the humanities in the mid-twentieth century decried science's despotic reign over Western culture. Some, including the well-known literary critic Howard Mumford Jones, agreed that humanistic study bore an important relation to the formation of value judgments but chastised their peers for ignoring the fact that values infused the practices of natural and social scientists as well. Others, such as the Harvard historian Arthur M. Schlesinger Jr., blasted the moral sterility and political recklessness of scientism but insisted that it had not taken hold outside academia—at least, not yet. A card-carrying member of the group that one critic called "Atheists for Niebuhr," Schlesinger charged in 1949 that sociology had "whored after the natural sciences" from its very birth. Since the war, he continued, scientism's champions had convinced administrators and funding agencies that their work promised a magic key to understanding the world. The quantifiers had monopolized research funding and prestige, and their bureaucratic tenor and convoluted jargon squeezed out more critical, humanistic insights on the human condition. Over the long run, Schlesinger warned, scientism might spread to the point of "obscuring from ourselves the ancient truths concerning the vanity of human wishes and the distortions worked by that vanity upon the human performance." Still, Schlesinger saw scientism as a merely academic phenomenon at present, if an extremely powerful one.[6]

More commonly, however, humanities scholars identified scientism as not just an intellectual error but also a driving force in modern culture and society. They took for granted that a scientific age had dawned, wherein "the positivistic and naturalistic spirit has been penetrating ever deeper and deeper into all the literate strata of our population" and causing moderns to lead "schizoid lives." Some of these commentators issued the chronologically sweeping broadsides heard from many religious critics, holding that the scientific bacillus had infected Western culture centuries earlier. For the most part, however, humanities scholars focused on the twentieth-century impact of the behavioral sciences. Indeed, they often contended that natural scientists joined them in honestly seeking toward the truth, whereas social scientists wanted to meddle in human affairs and control people's behavior on the basis of limited, flawed knowledge—or perhaps no knowledge at all. On the whole, humanities scholars distinguished more explicitly and consistently between science and scientism than did religious critics.[7]

These understandings of scientific inquiry, humanistic scholarship, and the cultural drift of the modern world appeared across a wide range of sites after World War II, as scholars and citizens alike grappled with the ramifications of science's postwar ascendance. As commentators thought about questions of intellectual method and authority in this context, they became increasingly likely to divide modern knowledge practices into two categories: "sciences" and "humanities." In 1959, *The Two Cultures*, a short book by the British novelist and science writer C. P. Snow, provided a focal point around which discussions of science and the humanities increasingly revolved. The intensity of concerns about science's cultural meanings in the United States was reflected in the number and variety of American commentators who felt obliged to position themselves in relation to Snow's argument—to endorse it, to reject it, to break it down into pieces, to restate it, and often to misunderstand it.[8]

Before 1959, Snow was known to educated Americans for his novelistic treatments of what he called the "new men": the scientists and bureaucrats who were taking over the task of governance from the traditional British aristocracy in the mid-twentieth century. Having moved from an early career in physical chemistry into various governmental positions after 1940, Snow was intimately familiar with the mind-set of the new men and the training they received in elite British universities. The two cultures of his title were academic cultures. One was rooted in science, by which Snow meant the physical sciences. The other characterized the humanistic fields, led by literature, that still served as the traditional training ground for British leaders. Snow saw great value in both of these cultures, which he straddled in his own person. He lamented the mutual illiteracy that pertained between them—the inability of those trained in each culture to understand the basic concepts of the other. But his ultimate sympathies lay with the scientists.[9]

For one thing, Snow thought that the mutual illiteracy of the two cultures was asymmetrical—that humanists were far less conversant with science than vice versa. More importantly, however, he discerned a deep moral gulf between the sciences and the humanities that played out in their views of world history. Scientists, he averred, had "the future in their bones." Chronically optimistic about humanity's prospects for improvement, they toiled to bring modern industry's material benefits to the people of the world. But humanists had never gotten over the Industrial Revolution, Snow charged. Refusing to come to grips with the modern condition and viewing recent history as an unremitting tragedy, they wallowed in fantasies about the moral superiority of the preindustrial past and offered the global masses nothing but lamentations about their own spiritual alienation. In fact, Snow insinuated that literary thinkers could not even sustain a firm commitment to democracy. He famously quoted an unnamed scientist who charged that Ezra Pound, Wyndham Lewis, and other modernist writers had hastened the coming of Auschwitz.[10]

At the individual level, Snow allowed, the humanists' focus on the tragic dimensions of existence made perfect sense. Individual lives were indeed tragic, full of senseless loss and needless pain. But he denied that this fact bore any relation to the broader social situation. Although individuals continued to suffer psychologically, the modern age featured steady progress in overcoming the material squalor and massive inequality of the preindustrial world. In this context, the scientists' optimism represented the proper basis for political decisions on the material and organizational needs of modern industrial societies. Snow advocated structural changes in British higher education so that future leaders, hitherto fed a diet of literary humanism, would be conversant with the basics of science and steeped in its moral outlook—its hopeful, practical, and morally binding commitment to undoing the evils of colonialism by fostering global progress.[11]

Snow's terminology, if not his specific arguments, spread rapidly in American academic and popular thought. Lerner introduced American audiences to the "two cultures" phrase in 1957, echoing one of Snow's early articles on the subject. When Snow's full essay appeared in 1959, the *Bulletin of the Atomic Scientists* published a lengthy summary and *Science* excerpted a short selection. Henry Luce's *Time* reported on Snow's argument as well. At the 1960 meeting of the American Psychological Association, only a year after Snow's essay appeared, a participant complained that every significant group of American commentators besides psychologists had made their voices heard on the topic. By 1961, a Stanford graduate student had already produced a thesis on Snow's argument.

Interest in the book intensified when the British literary critic F. R. Leavis lashed out at Snow in 1962. Leavis dismissed Snow as a terrible writer, "portentously ignorant" and "intellectually as undistinguished as it is possible to be," lacking even a mind with which to argue. But the deeper problem, for Leavis, was science's pernicious cultural effects: its contributions to the growing massification of modern life; the death of genuine freedom and creativity in a stultifying, bureaucratized society; the spiritual emptiness of a society that continued piling up its material achievements without any sense of meaning. American publications attended closely to the Snow–Leavis clash. The *New York Times* published several lengthy articles about Leavis's attack and Snow's eventual response. Other periodicals, including *Science, Scientific American,* and the *Bulletin of the Atomic Scientists,* also gave blow-by-blow accounts. During the early 1960s, Snow himself was a constant presence in American public culture as well. He published high-profile articles and books, held a fellowship at Wesleyan University, gave talks around the country—including a major address on "The Moral Un-Neutrality of Science" at the 1960 meeting of the American Association for the Advancement of Science—and found himself profiled by *Vogue.* In 1964, the educational publisher Scott Foresman brought out a reader for students on the Snow-Leavis controversy, complete with study questions and excerpts from both British and American contributions to the debate.[12]

Snow's argument had clearly hit a nerve. Yet Americans adapted Snow's catchy phrase to their own distinctive purposes. The cultural and educational configurations that Snow delineated had few parallels in the United States. In Britain, a deeply entrenched cultural split between progressive scientists and conservative literary scholars gave intuitive credibility to the "two cultures" phrase. Americans faced very different patterns of cultural authority. Indeed, many of scientism's critics thought that the situation was reversed in the United States: that scientists held the prestige and authority and influenced public policy, crowding out humanists and their insights. American commentators attached the two-cultures distinction to a welter of views on the nature and relations of the sciences and humanities.[13]

Some of these figures portrayed science and the humanities as harmonious partners in a shared endeavor. Such harmonists included not only humanities scholars but also natural scientists who described science as a creative, essentially artistic enterprise. For example, atomic scientists such as the physicist J. Robert Oppenheimer and Harvard University president James B. Conant portrayed scientific research as a humanistic practice, defined by acts of imagination and infused with an ethos of creativity akin to that of the working artist. In 1953, the well-known science administrator Warren Weaver added this aesthetic argument to his usual description of science as a matter of unfettered curiosity and practical outcomes. By 1960, Weaver portrayed science as "a human enterprise, an enterprise that has at its core the uncertainty, the flexibility, the subjectivity, the sweet unreasonableness, the dependence upon creativity and faith which permit it, when properly understood, to take its place as a friendly and understanding companion to all the rest of life." On the side of the humanities, meanwhile, Jones and others argued that considerations of value shaped all forms of scholarship.[14]

Other commentators in the humanities took a more oppositional approach, identifying science, or at least scientism, as an amoral, materialistic, and dangerous outlook. For the sake of civilization itself, they sought to embed scientific knowledge in a humanistic framework of values and meanings. Many agreed with Snow's critic Leavis, even if they deplored his tone. One Indiana University professor held that Leavis spoke (if rudely) for "the old traditional humanity," whereas "Snow speaks for post-Revolutionary man, for the new majoritarian culture"—in short, for a world that was now "the demagogue's, the scientist's, the technocrat's, the mass man's." Leavis was dead right, such critics believed: it was not literary ideals but Snow's scientism that dominated and endangered the modern world.[15]

Such concerns sometimes appeared among natural scientists as well. To be sure, most ignored, or at least bracketed, Leavis's concern about science's alienating tendency. They followed Snow in calling for more and better science content in the educational system, and added a typically American plug for science popularization as a means of bridging the cultural divide and bringing scientific understanding to the masses. But a significant minority of natural scientists worried about the implications of science's growing presence in what Snow dubbed the "corridors of power" and in American culture at large. The Johns Hopkins biologist Bentley Glass found science's cultural ascendance inevitable but declared that he would rather "perish in a nuclear holocaust" than "live under a scientific tyranny." Only a fusion of the sciences and humanities in a new system of liberal education could forestall the potentially totalitarian implications of science unmoored from human values, said Glass.[16]

This concern was especially common among those atomic scientists who sought to establish a fundamental continuity of method between science and art. After World War II, some of these scientists suddenly feared that they had too much cultural authority rather than too little. They had created nuclear weapons and all kinds of other wonders and horrors. Many of their fellow Americans, they feared, now expected them to remake both the physical and social worlds anew. In this guise, as in a number of others, the widespread postwar concern about scientism appeared at the very heart of the scientific establishment.

At Harvard, for example, Conant worked vigorously to limit the public's sense of what it could expect from science, in politics and culture as well as the development of new devices. Science and its products, he argued, certainly set the tone for the "new age of machines and experts." But that made it all the more important to avoid extravagant visions of science's transformative power. Throughout the postwar years, Conant sought to deflate puffery about science's capacity to solve specific problems. At the same time, he also targeted broader, utopian hopes for scientifically guided social change that he traced back to eighteenth-century rationalism. Science could make nuclear weapons, Conant declared, but it could not eradicate evil from the world. Rather, social progress required a combination of scientific expertise and other sources of moral guidance. The "older humanistic studies," Conant explained, provided knowledge of "liberalism": "how men and women can work together for the maintenance of a nation that is truly free." Religion, meanwhile, upheld "the fundamental premise of our civilization, the premise alike of Catholic, Protestant, and Jew, a belief in the dignity of the individual." According to the Unitarian Conant, American democracy embodied an ancient Judeo-Christian heritage, as filtered through the Reformation, the Renaissance, and frontier conditions. Conant repeatedly sought to rescue Western ideals, including "the spiritual basis of individual freedom," from latter-day outcroppings of the eighteenth-century rationalism and nineteenth-century materialism he deplored.[17]

Darker visions of science's cultural impact sometimes emerged in scientific circles as well. On New Year's Day in 1963, *Bulletin of the Atomic Scientists* editor Eugene Rabinowitch printed a remarkable essay by the émigré historian Theodore H. Von Laue, the son of a famed German physicist. Von Laue wrote that modern society had come under the thrall of "scientism," the persistent hope that science could liberate humanity from "the miserable, irrational world of the old Adam caught in the senseless flux of events—in short, the eternal human mess." Since the early twentieth century, Von Laue wrote bitterly, "science has changed from a philosophic and academic pursuit into a vast social and political effort to manipulate man and nature." Science had "lost its pristine purity conditioned in the laboratory and study" and "entered the fishmarket of psycho-politics, of politics more influenced by the psychic drives of large, rootless populations than by rational considerations." This style of engagement, said Von Laue, could destroy both science and society by turning the former into the handmaiden of the latter: "the agent of forces entirely beyond its insights and even hostile to its nature." Von Laue described the modern citizen as a "most cooperative slave of science and technology," inhabiting the most regimented and conformist society known to history.[18]

INTERNAL CRITICS

The two decades after World War II also brought a growing rebellion within the social science disciplines that had nurtured ambitious plans for mental modernization between the wars. By the early 1950s, high-profile apostates routinely took to professional journals and political magazines to denounce the moral nihilism of modern, secular liberalism. These social scientists joined a host of humanistic scholars, religious leaders, and conservatives in the postwar chorus against mechanistic thinking and social engineering. Challenging what they saw as the endemic scientism of American academia and politics, they sought to restore the distinctively human element to the modern world picture. Such figures argued that no account of reality could be complete—or safe—unless it centered on human subjectivity, moral norms, and other nonobservable but indispensable phenomena. Like their counterparts in the churches and the humanities disciplines, they hoped to protect human self-understandings, moral prescriptions, and social ideals against the intrusions of scientism. In short, they agreed that the quest for a true science of society had badly eroded cultural norms and moral constraints.

By 1960, this ongoing exodus intersected with the two-cultures controversy. Taken literally, the two-cultures frame gave scholars only two possible transdisciplinary identities and caused the social sciences to vanish as a distinctive enterprise. It presented a stark choice: either the social sciences stood on the side of science, and were thus essentially identical to physical science, or they were somehow continuous with the likes of classics, literary criticism, and art history. However, many American commentators saw a third option: the social sciences could serve as a bridge between the sciences and the humanities. Berkeley's M. Brewster Smith made the case from a behavioralist perspective. "In the study of optimal human functioning," he wrote, social scientists could "work toward the clarification of values among which people must choose and of the causal relations that are relevant to value choice." For his part, the anthropologist Lloyd Fallers called the social sciences "a third culture . . . in which humanistic and scientific modes of thought are inextricably intertwined and which thus forms a natural bridge over the chasm." That style of investigation, he noted, demonstrated the centrality of culture as a mediating factor in social change. Because Snow missed the cultural factor, wrote Fallers, he "twists science into a millenarian and potentially totalitarian faith," a worthy descendent of the French Terror. Fallers believed that the third culture of the social sciences could make possible a program of "less totalitarian, more human planning" by revealing "that between man and any rationally worked out image of the future there lie social institutions and cultural values which have worth in themselves—which cannot be swept aside in the interests of the plan without destroying privacy and liberty, qualities without which man becomes, not the free, rational being of the eighteenth century dream, but a lifeless puppet." Harry S. Kantor of the Department of Labor similarly charged Snow with technological determinism, observing that science could be "applied to techniques of subjugation, to the efficient and sanitary mass liquidation of dissenters and the suppression of the individual."[19]

When Snow finally responded to Leavis in the 1963 essay "A Second Look," he too identified a third culture in the social sciences. For the most part, however, Snow had in mind the new social history then emerging in Britain. In his view, social history decisively refuted the humanists' nostalgic view of the preindustrial past, revealing endemic malnutrition, misery, and early death. By contrast, many American social scientists mapped Snow's two cultures onto the familiar distinction between facts and values, while insisting that scholars could never address one at the expense of the other. The image of the social sciences as a third way thus enabled commentators to push for a more value-laden understanding of those fields.[20]

Numerous postwar social scientists saw the behavioralist orientation of their disciplines as a grave and potentially totalitarian danger, not a resource for clarifying ethical decisions. The sociologist Robert M. MacIver, among others, took this line. So did a University of Tennessee political scientist who not only argued that fiction was an important source of insight for the political scientist but added that political science itself was "a fiction . . . and none the worse for that." Another observer praised recent sociologists for resuscitating "the original Aristotelian insistence that ethics and politics go together" and blamed the modern age's intellectual disunity on the "relativism in value theory" that pervaded the social sciences. Numerous other social scientists likewise used Snow's framework to push their colleagues toward a conscious engagement with values, while delineating the grave danger of science unmoored from ethics. Nothing "could be more ominous," wrote the maverick economist Robert Lekachman, "than government by a class that believed that all human problems yielded to the ministrations of science."[21]

In truth, plenty of social scientists were addressing questions of value already. As in religion and the humanities, American social scientists often sought to promote particular understandings of their fields by arguing that those approaches had been relentlessly hounded out of existence by the avatars of scientism. This is not to deny the very real shift of power after World War II toward those social scientists whose research, or at least self-description, resembled the physical sciences. But here, as elsewhere, the saving remnant turned out to be rather substantial—and often highly visible in the public conversation, where the arguments of dissenting social scientists reflected widely shared views of self and society.

Historians of the American social sciences sometimes suggest that a once-vigorous conception of morally engaged scholarship largely vanished in the 1920s or 1930s and was only partially recovered in the 1960s. In fact, however, that vision never really disappeared in the first place. The World War II era saw plenty of changes in the American disciplines, but those changes did not amount to an eclipse of moral engagement per se. Rather, two overlapping shifts took place. First, a narrower definition of the term "science" came to dominate the public conversation, making normative analysis, qualitative methods, and other interpretive approaches seem incongruous. Second, most social scientists and philosophers came to disavow, at least publicly, a goal they had once openly endorsed: that of systematically altering American culture. These figures rejected the ideal, or at least the language, of mental modernization.[22]

Neither change eliminated morally engaged scholarship from the disciplines, however. The list of heterodox social scientists in the 1950s is long and illustrious, especially in sociology but also elsewhere. Moreover, these figures did not simply gain their prestige in retrospect; they were well known and widely respected at the time, even if their work violated the protocols of the foundations that increasingly funded social-scientific work. At Harvard, for example, David Riesman eschewed the behavioralist model in *The Lonely Crowd* and other well-known works of the 1950s. Riesman's Harvard colleague, the sociologist Barrington Moore Jr., levied a more explicit challenge to the search for causal patterns and the concomitant decline of historical and comparative sensibilities in the twentieth-century social sciences. Moore cofounded the Social Studies program, which centered on works of classical theory, in protest against the behavioralist approach of Talcott Parsons and his allies. And Nathan Glazer, one of Riesman's collaborators on *The Lonely Crowd,* repeatedly argued that sociologists should give the public critical insights on social values and practices, not just produce statistical bricks for a scientific edifice.[23]

At the same time, the narrowed postwar definition of science, coupled with concerns about the depersonalizing tendencies of scientism, led some postwar social scientists to jump ship and identify themselves with the humanities. At a 1951 forum discussion, the cultural anthropologist Alfred Kroeber disavowed the term "social science" and called himself "a sort of humanist." In his definition, humanists aimed at insight rather than control: "discerning, describing and understanding values," not "making value judgments for other people or prescribing for other people what they are going to do." By contrast, said Kroeber, social scientists adopted a "clinical" rather than "experimental" orientation, having sought from the very beginning "to improve the world, to remodel other people." Like so many other postwar critics, Kroeber associated the term "social science" with programs of social engineering that ran roughshod over individual freedom. Conversely, he aligned his own approach with that of both humanities scholars and natural scientists, who pursued truth rather than power. For Kroeber and like-minded colleagues, postwar social science represented a sharp and implicitly totalitarian break with the long Western tradition upheld by humanists and natural scientists alike. As the Harvard historian Crane Brinton averred at the same forum, "Shakespeare is nearer a natural scientist than Marx."[24]

Virtually all American historians after World War II agreed with Brinton that they were humanists, not social scientists. This was a significant departure from the interwar period, when figures such as James Harvey Robinson, Charles Beard, and Harry Elmer Barnes led the charge for the modern mind while touting their scientific credentials and often straddling the disciplinary borders between history and fields such as political science. Among a new generation of historians, however, many saw only danger in the application of scientific methods to human affairs. One of the sharpest critiques of scientism came from the Cornell historian Cushing Strout, who charged in 1955 that scientism had thoroughly corrupted American liberalism since the early twentieth century. Strout singled out the utopian "technocratic rationalism" of Robinson, Beard, Barnes, Thorstein Veblen, and John Dewey. In the early twentieth century, he explained, "the triumphs of the machine and the scientists" had "seduced men's imaginations into dreaming of unlimited and perpetual progress through technology," just as pragmatism and its philosophical equivalents had "devalued the importance of principles." Figures such as Dewey and Beard had recaptured "the anti-historical proclivities of the Enlightenment rationalists," creating a science-obsessed progressivism blind to "history's desperation, the grimness of a process which demands absolute decisions from men who live in a relative, finite situation." Such a view, Strout continued, ignored everything except science and technology—most importantly, fields such as politics, philosophy, and history that involved "the creative labors of the human will seeking to organize its values in the world of formal institutions." Indeed, he wrote, the progressives, sure of their correctness, nurtured "an unchecked dream of absolute power" through "the manipulation of society by social 'scientists,' as if individual and social values were mere laboratory data." Strout waxed satirical: "Under the benevolent despotism of this self-elected élite, we could relax in the assurance of social progress. The Guardians of social science would maintain the Republic."[25]

Émigré political theorists often issued equally biting critiques of scientism. These figures inherited European critiques of rationalism and modernity that dated back to the counter-Enlightenment itself. But émigrés such as the political philosophers Eric Voegelin and Leo Strauss reframed long-standing challenges to Enlightenment ideals in new ways. Strauss, though believing only disaster could come from applying the "scientific spirit" to human affairs, traced the Western crisis mainly to the inevitable backlash against scientism: an extreme form of subjectivism that Strauss called "historicism." By contrast, Voegelin focused on the main article itself, painting a dire portrait of scientism's history and cultural effects. Although his fears later eased, Voegelin argued in 1948 that a political system built on scientism stretched around the entire globe. "The damage of scientism is done," he declared mordantly; "the insane have succeeded in locking the sane in the asylum" and "the destructive effects defy repair in any visible future." In the nineteenth century, he specified, the "spiritual eunuchs" drawn to scientism had captured the public mind and fulfilled Helvétius and Bentham's "scientistic-utilitarian dream of transforming society into a prison from which no escape was possible." For Voegelin, the root of the error lay in the magical fantasy that human beings could remake themselves as they remade nature, creating "the superman, the man-made being that will succeed the sorry creature of God's making." But such efforts inevitably failed, Voegelin explained, because the human "realm of substance" did not share "the objective structure of the realm of phenomena." One could not extend the "will to power" into the domain of substance without utterly destroying humanity as such.[26]

More visible in the intellectual commons were the émigré theorists of international relations who came to the fore as commentators on the Cold War. Hans Morgenthau's 1946 book *Scientific Man vs. Power Politics* spelled out a theme shared by virtually all of the émigré political thinkers in one form or another: scientific methods did not apply to politics, a realm of decisions and interests rather than causation and rationality. Yet all of Western civilization, Morgenthau complained, followed the rationalist assumption that history and human beings could be controlled like physical nature, despite the constant failure of that presupposition to produce the desired outcomes. He called the political and military crises of the 1930s and 1940s simply "the outward manifestations of an intellectual, moral, and political disease which has its roots in the basic philosophic assumptions of the age."[27]

As historians seceded from the sciences and political theorists carved out a niche within an otherwise scientifically inclined discipline, American philosophers split on whether to align themselves with science or the humanities. The divide roughly mapped onto—and in some ways helped to produce—the emerging distinction between "analytic" and "Continental" philosophy. In the latter camp, William Barrett, the most vocal American advocate of existentialism, influentially challenged scientism. Of course, many religious critics saw existentialism as of a piece with Freudianism, behaviorism, and other scientific heresies. After all, Jean-Paul Sartre and his ilk flatly denied that stable moral laws guided humanity. Like Bertrand Russell, they portrayed the world as alien and unfeeling, giving no cosmic support to human purposes. But other existentialists, including Barrett, emphasized the other side of the picture: The scientist's mechanistic, deterministic model did not extend to the individual, and it was rank scientism to claim that it did.[28]

In the widely read *Irrational Man* (1958), Barrett identified scientism as part of a broader trend toward rationalism. For five hundred years, under the influence of science, Protestantism, and capitalism, Westerners had steadily denuded nature of the religious content projected onto it by prior thinkers. Although this process was crucial for "man's psychic evolution," said Barrett, it had left Westerners alienated from God, nature, the machine, and finally even themselves. Through the mediums of technology and bureaucracy, rationalism had come to structure all of society, while individuals themselves became less and less capable of thinking reasonably—or perhaps thinking at all. Only by getting back to "the concrete individual" could Westerners escape the modern trap. "Existentialism," Barrett famously declared, "is the counter-Enlightenment come at last to philosophic expression." By this, he did not mean to diminish "everything that goes under the heading of liberalism, intelligence, a decent and reasonable view of life." Quite the opposite: The only way to recapture such values, said Barrett, was to look beyond the "rationalist tradition" of recent centuries to the broader, humanistic conception of reason that had inspired it, and then to dig farther down to the essential fact of "man's existence as a self-transcending self," one that had "forged and formed reason as one of its projects."[29]

In sociology, meanwhile, C. Wright Mills's vituperative dismantling of behavioralism in *The Sociological Imagination* (1959) resonated with a distinguished group of sociologists that rejected postwar pretensions to value-neutrality, including Riesman and Moore. In fact, other American sociologists had penned book-length studies debunking the scientistic pretensions of modern social science before Mills. However, these works have largely disappeared from our histories because their authors stood well to the right of the postwar liberal mainstream. One was Albert H. Hobbs, a conservative sociologist at the University of Pennsylvania. A vigorous critic of liberal approaches to welfare, crime, and the family, Hobbs identified moral discipline rather than government action as the solution to all social ills. He discerned a thoroughgoing liberal bias among his peers, despite their extravagant claims to objectivity, and undertook to document that leftward slant in a pair of books. *The Claims of Sociology* examined introductory textbooks and found them riddled with logical errors, methodological inconsistencies, and meaningless definitions, as well as liberal tenets dressed up as empirical facts. Hobbs extended his criticism to sociology as a whole in *Social Problems and Scientism* (1953). Arguing against the strict value-neutrality that postwar behavioralists claimed, he contended that all empirical work reflected normative commitments and urged his colleagues to replace their technocratic, collectivist ideals with a sounder understanding of the human person. Together, Hobbs's two books portrayed American sociology as a hodgepodge of tendentious, biased claims that worked to eviscerate traditional morality.[30]

At Harvard, Pitirim A. Sorokin espoused a less conventional mode of conservatism, combining his religious faith, traditional social values, and staunch anticommunism—earned through battles with the Bolsheviks during his youth in Russia—with an equally vigorous dislike of modern capitalism's materialism and ruthless competition. Sorokin had fought against value-neutral approaches to sociology since the 1930s, giving him plenty of time to collect the litany of outrages detailed in *Fads and Foibles in Modern Sociology and Related Sciences* (1956). Like Hobbs, whom he cited at the end of the book, Sorokin left his claims about scientism's destructive social effects largely implicit and concentrated on methodological criticism. But Sorokin's final chapter, titled "In the Blind Alley of Hearsay Stuff and Negativism," contained a few of the blasts for which he was already famous. He had long argued that civilizations cycled between three modes of existence: "ideational" (interpreting the spiritual or transcendent as the genuinely real), "sensate" (focusing solely on reality's material side), and "integral" (holding the ideal and the material in balance). In the present day, Sorokin identified centralized state power as a symptom of the disintegration of a sensate period that had defined the West since around 1500. And he scored social scientists for adopting the pessimistic, deterministic tenor of such a mentality, rather than pointing the way beyond to an integral state of culture. The "negativistic notions, nihilistic dogmas, cynical beliefs, and debunking ideologies" of the social scientists, he wrote, had "infected the minds of our intellectual, governmental, business and other leaders, as well as the minds of the masses." This scientism, Sorokin charged, had magnified the other effects of cultural breakdown, fueling "terrible wars, revolutions, and anarchy; the utter atomization of all religious, moral, aesthetic, political, and other values"; and "tornadoes of destruction, bestiality, and inhumanity."[31]

Psychology, which many critics saw as the lion's den of scientism, witnessed an even more significant revolt against a professedly scientific mainstream after World War II. The strict behaviorism of John B. Watson and B. F.

Skinner had never dominated the field, but after 1920 most American psychologists employed experimental methods and sought after strict objectivity. Yet figures such as Gordon Allport, Gardner Murphy, and Lois Murphy sustained an alternative tradition that emphasized the specific individual personality rather than generalized conclusions about "the" human mind. In the 1950s and 1960s, this milieu gave rise to "humanistic psychology." More clinically inclined than Allport and the Murphys, and also friendlier to Freud, humanistic psychologists such as Carl Rogers, Rollo May, and Abraham Maslow defined their field as a means by which individual therapists helped other individuals achieve "self-actualization." Indeed, they portrayed psychology as more closely aligned with religion than with any sort of rationalism or materialism. Maslow called the movement a "resacralizing of science." Rogers, for his part, had studied for the ministry under Paul Tillich and retained much of the framework of Christian existentialism. He spoke of a "divine spark" in all human beings: not a soul in the usual sense, but an innate capacity for psychological growth, for gaining insight and self-knowledge.[32]

The humanistic psychologists were closely attuned to the social and political context for their work, although their political commitments moved in different directions. Maslow, who harbored elitist tendencies, foresaw a libertarian utopia of psychologically healthy citizens, requiring no laws or constitution and featuring no crime or unemployment. Rogers focused more closely on interpersonal relations, seeking to make citizens and experts into equal "co-workers" in the shared task of fostering self-actualization. Widely read outside the field, Rogers debated Skinner on a number of occasions. He repeatedly charged that Skinner's approach, reflecting the orientation of postwar social science more generally, moved toward a kind of "social dictatorship": "control of the many by the few." Elsewhere in the movement, Clark Moustakas argued that the conformist orientation of the postwar social sciences, like their commitment to scientism, facilitated manipulation by corporations and bureaucracies. The board of directors of the *Journal of Humanistic Psychology,* founded in 1961, featured critics of scientism from outside psychology, and even outside academia. In addition to Rogers, May, and their allies, the roster included Riesman, Lewis Mumford, and Aldous Huxley.[33]

WITHIN A FEW YEARS, humanistic psychology would become a crucial vector through which critiques of scientism made their way into the emerging New Left and the counterculture of the 1960s. For the time being, that approach emerged in the context of numerous attempts to turn the social sciences into a bridge between Snow's two cultures: a meeting ground of scientific and humanistic methods. At the same time, it also emblematized the tendency of many heterodox social scientists to reinforce broader perceptions that the social sciences were thoroughly vitiated by their emulation of physics—and that the work of the dissenters did not fall under the heading of "science" at all.

The "two cultures" concept itself produced this effect as well. The debates around Snow's thesis tended to associate science with the kind of bureaucratized, technocratic society that many critics believed had appeared in the United States and Britain since the war. From this point of view, Snow and science's other champions appeared to celebrate an anonymous, heartless, expert-dominated mass society—and to portray it as the translation of science into political terms, a natural outgrowth and extension of science's rise to cultural power. It became increasingly difficult for thinkers on either side of the Atlantic to imagine science taking other social forms, buttressing other political models, or inhabiting other institutional arrangements. In this regard, the two-cultures controversy played a role in helping to conceptually fuse science with a technocratic, high-modernist vision in American political culture.

By 1965, some observers began to step back and to wonder why the two-cultures concept had proven so attractive to Americans. The anthropologist Margaret Mead interpreted the furor as a symbol of widespread fear that American schools and universities failed to produce a truly "educated man; that is, a whole man, to whom nothing in the world is alien, to whom no path of possible exploration is blocked or closed." Mead believed this anxiety was no mere foible. Indeed, she expressed the same fear herself. "Our experience since World War II," she wrote, "has demonstrated with terrible vividness how few men there are—of any age or from any country—who have such an integrated relationship to all that man has been, is, and may become." According to Mead, Snow's popularity bore witness to both the fragmented state of American thought and culture and a deep-seated longing to overcome that state and achieve some kind of integration. As powerful challenges to postwar liberalism emerged across the political spectrum, this longing for integration—intellectual, cultural, personal—and an accompanying critique of scientism's disintegrating tendencies would increasingly define the era.[34]

7

A NEW RIGHT

WHEN C. P. SNOW'S *THE TWO CULTURES* appeared in 1959, writers at the conservative *National Review* immediately took up cudgels. William F. Buckley Jr.'s journal, the primary mouthpiece of the emerging New Right, had not previously taken offense at Snow's writings. Indeed, its reviews of Snow's fiction in the 1950s called him "a novelist of subtle craft and power," though one with certain stylistic flaws. As soon as the two-cultures controversy kicked off, however, Joan Didion bashed Snow's "soporific" writing in a pair of *National Review* pieces. His novels, wrote Didion, "celebrate a world of committees, compromise, decisions and revisions"; they constituted nothing less—and nothing more—than "the literature of the National Health Service."[1]

The criticism did not stop there. The conservative Catholic writer Frederick Wilhelmsen then reviewed Snow's book for the *National Review*. He wrote that Snow was like so many other scientific thinkers who "failed to do their homework in the Western tradition" and thus ignored the essentially moral character of the modern predicament. The real problem was "theological and philosophical," not literary or scientific, said Wilhelmsen; it involved "the relation that ought to exist between science . . . and man himself." Wilhelmsen concluded by calling for the mandatory subordination of science to higher values. Precisely because science had the potential to remake "the very fabric of human personality," he wrote, it needed to be yoked to "the moral and intellectual and even psychic demands of the human substance." There could be no doubt about the primacy of the human person, according to Wilhelmsen: "Where science cannot be disciplined and made to fit the human cloth, it must be suppressed."[2]

Two years later, following F. R. Leavis's attack, Buckley himself penned both an editorial and an essay on the controversy. He was one of the only American commentators to praise rather than deplore Leavis's tone, finding the assault entirely warranted by Snow's outrages against both common sense and basic freedoms. Buckley called Snow the author of "ponderous, middlebrow, pseudo-philosophic novels" who used his "inflated reputation" to push for a scientific plutocracy. The Two Cultures, especially, had been "a major boost for the thesis the atomic scientists have been pushing ever since Hiroshima," namely "that the scientists should . . . rule the world." As for Leavis, Buckley wished the literary critic had actually carried his attack farther. If Snow's novels were not bad enough, said Buckley, he also embraced "that final effrontery of relativism: that in the last analysis, there simply aren't any essential differences between the two contending parties in our great war with Communism." In short, Snow spoke in the amoral "voice of the machine," with its dispassionate relativism. Elsewhere, too, Buckley used Snow to symbolize the totalitarian tendencies of modern liberalism, at one point describing various liberal offenses as "steps on the road to C. P. Snow." In fact, Buckley remained sufficiently exercised that he again heaped criticism on Snow when the novelist died in 1980.[3]

The National Review's campaign against Snow reflected a foundational assumption of postwar conservatives, that of the power of ideas in history. Looking back, historians have tended to highlight two intertwined features of the New Right: its dislike of the New Deal state and its fierce opposition to communism. Yet the new conservatives also focused intently on cultural and intellectual matters, combining their economic and political arguments with claims about science, religion, and the humanities. Viewing institutions and practices as outgrowths of the prevailing intellectual substrate, they traced the collectivist orientation of the New Deal state to shifts in that climate.

Prominent among these shifts were postwar changes in the form and content of the social sciences. New Right leaders argued that both social scientists and government planners had projected a materialistic outlook from the physical sciences onto the world of human affairs, threatening to destroy the social order altogether. Like anticommunism and opposition to the New Deal, distaste for the postwar social sciences represented a crucial point of shared emphasis for the many different groups taking up the label "conservative" in the 1950s. As theorists such as Buckley, Russell Kirk, and Frank S. Meyer worked to revitalize and modernize American conservatism, they constituted it not only as an alliance of individualists against collectivism but also as an alliance of Judeo-Christian believers against secularity and scientism. It is partly coincidental that the New Right and the behavioral sciences arose nearly simultaneously in the 1950s. But as a result of the twinned births, each movement powerfully marked the other. The early development of the New Right reveals how strongly modern American conservatism has been shaped by critiques of scientism and its cultural influence.[4]

THE CONSERVATIVE LANDSCAPE

It is not hard to find instances of postwar conservatives defining their views in direct opposition to the intellectual proclivities of American social scientists. The pages of conservative journals and books are littered with attacks on the social sciences. Even staunch libertarians, who stressed the absolute sanctity of property rights and freedom of contract rather than the authority of divine laws or the accumulated wisdom of mankind, often portrayed social science as the area in which, and through which, liberal dominance had proven most destructive. Reasoning from the tenet that "ideas rule the world," conservatives believed that real power in America rested in the hands of those controlling the means of cultural production and reproduction: namely, professors and journalists. They warned that "academic intellectuals, as well as demagogues, can influence society—that the mild-mannered man with the briefcase and bifocals constitutes a threat to democracy as well as the bullyboys in breeches." Even more galling to many conservatives than the state's adoption of liberal policies was the way that liberal social scientists projected their values onto the fabric of reality itself and defined conservatism as a mental pathology rather than a legitimate political option. All manner of postwar conservatives charged that the dogmatic philosophy of modern liberalism found its most enthusiastic and effective champions in the social sciences. In this manner, a critique of the social sciences became central to the self-definition of the 1950s New Right. As we will see, some conservatives, led by Kirk, even tried to reshape the social science disciplines themselves along friendlier lines.[5]

In the meantime, all of the prevailing varieties of social-scientific thought inspired withering criticism. Conservatives were particularly incensed by value-neutral epistemologies, which they saw as dangerous expressions of moral relativism. Meyer contended that "sociological objectivism as a basis for law and morals" led directly to Skinner's frank rejection of human freedom. And beyond that error lay the inevitable outcome of all "environmentalist" theories of behavior holding that human beings were shaped primarily by their surroundings: totalitarian rule, in which a small elite manipulated the masses to serve its own interests. Yet postwar conservatives, like many religious critics, also rejected morally engaged forms of social science. Social scientists' primary offense, for the New Right, was not their claims about values but rather their use of such claims to turn Americans away from the Western moral tradition, with its roots in religious faith and its fruits in economic and political freedom.[6]

In the final analysis, for these critics, the problem was not one of competing epistemologies or methodologies per se. Rather, they viewed the influence of social science through the lens of cultural authority and representation: Who spoke for the truth? Which cultural leaders could guide a truth-seeking people down the proper path? Conservatives argued that because human beings needed moral guideposts and scientific analyses ignored those guideposts, the task was to establish the proper "relation between a scientific elite and the populace" by carefully hemming in the former's influence. For the New Right, naturalistic but normatively committed varieties of social science were just as dangerous as ostensibly neutral ones, because they substituted for traditional moral principles a new structure of cultural authority centered on social scientists themselves. Whether social scientists spoke in a neutral idiom or issued normative recommendations, conservatives charged, they peddled a thin, attenuated view of man, "stripped of all qualities not accessible to the scientific method" and ultimately derived from social scientists' own desire to consolidate their power over the masses.[7]

Thus, Meyer could attack all the competing schools of social science in a single breath. Modern intellectual and political life, he wrote, was dominated by "a positivism, a pragmatism, an instrumentalism" that produced "a successfully more radical devaluation of values," despite many protestations to the contrary. Americans, he said, faced a stark choice between "the accumulated wisdom of mankind over millennia" and the rapidly growing "tradition of a positivism, scornful of truth and value, the tradition of the collective, the tradition of the overweening state." There could be no middle way: "Either standards derived from the moral order by reason operating within tradition will protect the human rights that stem from that order; or everything is in neutral flux, to be moulded for expediential reasons by the high priests of a 'science of man,' guided by nothing but their itch for control in the testing of successive hypotheses." For Meyer and many other conservatives, any explanation of human behavior in terms of forces other than transcendent moral law led down the path to totalitarianism. Unless a society recognized the importance of self-determination within a framework of permanent moral and political truths, it would instead follow the social scientists and open the door to the control of the masses by an elite minority via the total state. Meyer and his fellow conservatives portrayed any delegation of authority to social scientists as an assault on the higher moral law.[8]

The problem was not simply an abstract one to these figures; they believed that social scientists wielded enormous authority. Buckley declared in the *National Review*'s first issue that social scientists "run just about *everything*. There never was an age of conformity quite like this one." The social sciences figured in the writings of outraged conservatives not only as an expression of the secular, leveling, social engineering outlook of the universities and the liberal establishment but also, and more importantly, as the intellectual source of that outlook. Thus, the social sciences bore much of the blame for the fact that American leaders sought to overthrow inherited traditions in the name of rational planning and technocratic manipulation. "Compared with these fantastic doctrinaires," wrote Richard M. Weaver of John Dewey and his social-scientific followers, "the founders of Utopian communities in the past were realists of the highest order." And these rationalists had snared virtually every social institution in their web, according to figures such as Buckley and Weaver.[9]

In fact, however, conservatives, like other critics of scientism, found important and consequential platforms on which to stand, even as they announced their near-total erasure from the contemporary scene. For the New Right, such opportunities often reflected the collaboration of conservative journalists and their allies in Congress. A particularly striking example from the early 1950s concerned Albert H. Hobbs, the University of Pennsylvania sociologist who was then compiling the litanies of sociological horrors discussed in Chapter 6. The combative Hobbs, who clashed repeatedly with his departmental colleagues, remained at the assistant professor level for more than ten years despite his many publications. A group of conservative journalists—Raymond Moley at *Newsweek*, George E. Sokolsky at the *New York Herald Tribune*, the editor and staff of the *Indianapolis Star*—took up Hobbs's case, arguing that it illustrated the hypocrisies of academic liberalism. Soon afterward, through Moley's intervention, Hobbs found himself explaining to Congress in a series of high-profile hearings how his career struggles demonstrated the ideological character and intellectual bankruptcy of the mainstream social sciences. The context was an investigation to root out liberal bias in the major foundations financing research in those disciplines: Carnegie, Rockefeller, and above all Ford. Not surprisingly, the committee, under Tennessee representative B. Carroll Reece, did find the foundations subversive. Child psychologists and other "social 'scientists,'" it declared, "practiced not science but scientism"; they peddled "a wholly materialistic concept of life and behavior" that led down the Marxist path. In the wake of the investigation, leading foundations substantially altered their funding priorities. Ford, for example, shifted its grants away from basic theoretical work in the social sciences toward less contentious applied research. It also began funding organizations such as the Foundation for Religious Action in the Social and Civil Order (FRASCO), at whose meetings President Eisenhower, Will Herberg, and others expounded on the religious roots of democracy—and the totalitarian tendencies of scientism.[10]

This commitment to moral foundations in the face of scientism united all of the rather disparate schools of American conservatism. One important strand was the economic libertarianism exemplified by the Austrian émigré economists Friedrich Hayek and Ludwig von Mises. Often calling themselves liberals rather than conservatives, these figures carried forward the nineteenth-century liberal program of small government and economic freedom. Yet they had much to say about the character and effects of modern science as well. For example, Hayek's famous assertion that economic planning required authoritarian political control appeared in the British science journal *Nature,* where Hayek criticized scientists for supporting such a policy. Meanwhile, Hayek's case against planning rested on an argument about the limits of human knowledge—especially that of social scientists. Only local actors, he reasoned, could gather sufficient knowledge of their particular situations to reliably advance their interests. Government-employed social scientists, with their abstract, generalized forms of knowledge, simply could not deliver the goods.[11]

As part of his campaign against planning in the early 1940s, Hayek undertook a series of detailed studies of "scientism" that helped to give that phenomenon its familiar name. Scientism was assuredly not science, he insisted. Indeed, it was "decidedly unscientific," a "mechanical and uncritical application of habits of thought to fields different from those in which they have been formed." Hayek further argued that Marxism rested on the error of scientism, which had not only given the system its scientific cachet but also inspired the underlying "demand for universal conscious control" that Hayek identified as the spirit of the modern age. After the war, Hayek focused even more intently on the problem of knowledge and associated psychological questions. He spent several years writing a book that debunked value-neutral approaches in the social sciences. For Hayek, the problems with modern progressivism went well beyond its economic policies. They reflected deeper errors in the realm of knowledge, including the faulty application of natural-science methods to human behavior.[12]

Scientism figured even more prominently in the thought of traditionalist conservatives. These figures emphasized the cultural sphere, arguing that basic philosophical commitments determined economic and political behavior. For traditionalists such as Kirk and the Southern Agrarian Weaver, conservatism represented a comprehensive critique of modern thought-ways, not just an attack on Keynesian economics. Traditionalists identified the spiritual heritage of the West as a framework of fundamental moral laws that could be validated through the careful, open-minded study of societies past and present. Many said this framework should guide the formulation of empirical knowledge in the social sciences as well as personal behavior and public policy. Like many postwar liberals, in fact, traditionalists decried the moral and cultural effects of unfettered capitalism. But they traced those effects to the modern disavowal of traditional Western standards rather than the operations of private property or free markets. Their conservatism centered on foundational moral principles, not absolute economic liberty. This commitment to moral authority cut across other lines of difference between traditionalists, some concerning questions as substantial as whether American political culture was fundamentally liberal or conservative.[13]

The language of moral law also put traditionalists in dialogue with leaders in the churches and synagogues. With its emphasis on moral law and the threat of relativistic social science, traditionalist conservatism overlapped especially closely with Catholic arguments. At the same time, postwar Catholics such as John Courtney Murray downplayed doctrinal claims and asserted the need for a shared framework of "natural law." This opened the door to cooperation, or at least strong sympathy, with Protestant thinkers who discerned a similar structure of moral authority. Indeed, Weaver was among a number of Protestant traditionalists who agreed with Catholics that the Reformation was a major source of the modern era's characteristic malady—its inability to recognize the existence and the primacy of moral law. "For four centuries," Weaver wrote, "every man has been not only his own priest but his own professor of ethics, and the consequence is an anarchy which threatens even that minimum consensus of value necessary to the political state." Other traditionalists moved even closer to the Catholic Church. When Kirk converted to Catholicism in 1964, the shift hardly rippled the surface of his intellectual vision. Many other contributors to the *National Review* and Kirk's *Modern Age* were either cradle Catholics or converts whose religious views reinforced their emphasis on political liberties.[14]

Challenges to social science in the name of the Western moral tradition characterized many other varieties of postwar American conservatism as well. Although the circle around Buckley's *National Review* focused their advocacy primarily on individual freedom and economic regulation, they too believed that freedom rested on a foundation of moral principles, and that the rejection of such principles spelled doom. Indeed, Buckley and his followers defined capitalism itself as the bulwark of an organic society. They argued that economic liberty upheld family, community, and Judeo-Christian values against the atomized individual and total state of modern, scientific liberalism. This "fusionist" perspective, as defined by Meyer, united the traditionalists' religious commitments and the libertarians' limited state by describing these ideals as two sides of an integrated moral tradition that dated back centuries but had recently been overthrown by liberals. In other words, fusionists championed the combination of laissez-faire economics and religious traditionalism that the mental modernizers had targeted in the 1920s and 1930s. Buckley went so far as to read the atheistic libertarian Ayn Rand out of the conservative movement, arguing that her unbelief threatened the spiritual tradition that sustained both property rights and moral commitments in the West.[15]

Linking morality to economics, fusionism virtually guaranteed that postwar conservatives would target social scientists, by far the most powerful and visible group that combined a naturalistic orientation with left-leaning economic views. As we saw above, Buckley and Meyer insisted that social scientists, with their inclinations toward naturalism, quantification, and social engineering, had spearheaded the liberal assault on the organic Western tradition. In this view, social scientists' personal and professional beliefs aligned them with the "Liberal propaganda machine" that Willmoore Kendall accused of "conducting a Terror" in postwar America, "skillfully and relentlessly directed against conservatives in all walks of life." Launching the *National Review,* Buckley described the New Deal as a victory of "the Social Engineers, who seek to adjust mankind to conform with scientific utopias," over "the disciples of Truth, who defend the organic moral order." Elsewhere, he said Yale and other modern universities were drenched in the Dewey-inspired "absolute that there are no absolutes, no intrinsic rights, no ultimate truths." Fusionism did not weld together two fundamentally incompatible viewpoints, economic individualism and social conservatism. Rather, it expressed a dual, integrated commitment to moral law and a limited state that characterized almost all forms of postwar conservatism and set adherents firmly against the social sciences.[16]

Two other groups of conservatives rounded out the New Right and reinforced its critique of scientism. One comprised the conservative political theorists among the European exodus. As we have seen, they joined many other émigrés in lambasting "the dogmatic scientism of our age." For example, Eric Voegelin and Leo Strauss deepened their critiques of science's role in Western history during the 1950s, in successive Walgreen Lectures at Chicago, although Voegelin now thought he discerned a shift away from positivism at the highest levels of American thought. A final group of relative moderates, including the poet and historian Peter Viereck, accepted parts of the New Deal order—as did Hayek, for that matter—but deplored the cultural changes that had turned "welfare laws" into "the welfare super-state." Viereck began with a simple proposition: "History is made by the type of idea known as values." But the twentieth century had discovered the inevitable results of abandoning Christian ideals: namely, a "flood of pagan totalitarianism" rooted in "police-state values." Defining Europe as a "value-system" that was rapidly vanishing in a sea of materialism, Viereck argued that "relativist liberalism" and misapplications of Darwinian images of fitness and competition had fueled an "anti-Western and anti-Christian revolution in ethics" during the 1870s. That revolution had birthed systems of "national ethics" under which Hitler and Stalin turned industrial progress and political turmoil into totalitarian "secular religions." It was little wonder, Viereck concluded, "that Fascist and Communist mass-murder, based on the assumption that every means is permitted to achieve one's ends, followed a century of relativist liberalism and of the most modern 'scientific' enlightenment."[17]

CONTESTING THE UNIVERSITIES

Christian critics of scientism and their counterparts in the humanities worked assiduously to redirect the universities toward their perspectives in the 1940s and 1950s. That impulse did not operate as strongly among conservatives, given their greater institutional marginality. Yet a pair of them, the prolific writer Kirk and his publisher Henry Regnery, did launch a campaign to take back the universities by building ties to cultural critics of other political persuasions and thereby creating an organized humanist opposition. As late as 1968, in fact, Kirk expressed hope that traditionalists and their allies could reverse the universities' leftward drift by seeding the social sciences with conservatives. The efforts of Kirk and Regnery to effect this ideological sea change in the social sciences, and the frustrations they faced, reveal a great deal about the differences among scientism's critics, as well as their similarities. Ultimately, political and religious divisions proved impossible to overcome when the chips were down.[18]

As American conservatives worked to gain cultural influence as well as political power, even the smallest battleground mattered. They protested the liberal tenor of the reference work *Twentieth Century Authors,* just as they targeted Snow's alleged relativism. But above all other cultural institutions stood the universities, with their massive resources and unique forms of intellectual authority. The universities offered an unmatched platform from which to promote moral, social, and political judgments—a platform that conservatives believed had been consciously and systematically denied to them. In seeking to turn the academic ship rightward, conservatives made the postwar university a key site for early attempts to craft an alternative intellectual infrastructure. They mobilized an array of cultural, institutional, and financial resources as they sought to amplify their voices in the academic conversation. Indeed, these developments of the 1950s offer an instructive contrast with the tactics of the 1970s, when conservatives attacked the universities from the outside and created a separate, nonacademic network of knowledge production in think tanks.[19]

Part of the backdrop for conservative initiatives in academia was the tendency of postwar liberals to naturalize their values by projecting them onto reality. Liberal scholars of all varieties partook of the impulse to describe welfare-state liberalism as an objective, empirically grounded system, rather than one ideology among many. But social scientists were especially likely to do so. This naturalizing tendency fueled enormous resentment on the right, not least when academic liberals turned to explaining conservatism itself. The "vital center" orientation of the postwar universities suggested that nonliberal viewpoints were hopelessly unrealistic. Indeed, Richard Hofstadter and a number of other commentators argued that both conservatives and radicals were mentally unstable, incapable of recognizing obvious facts.[20]

In practice, meanwhile, many liberal scholars assumed their counterparts on the right were unprofessional, unscholarly, and unproductive, if not outright dangerous. In this context, conservative scholars felt like members of a persecuted minority in academia, and especially the social sciences. Although Americans routinely heard that "'liberals' slink through the academic corridors in shuddering fear," Hobbs contended in 1954, "they wield the whip over the heads of conservatives" in many universities. This image of liberal domination contained a kernel of truth. Without debating degrees of oppression in the age of McCarthy, one can say that conservative scholars, especially in the social sciences, often faced reactions from liberal colleagues that ranged from dismissiveness to scorn to active opposition. Economics, especially as practiced at the University of Chicago, offered some refuge, and political science departments also tolerated ideological variation among political theorists. But most of the social sciences proved less hospitable to conservatives.[21]

Some of the conflict stemmed from differences of personal style, which overlapped with ideological proclivities. Most liberal scholars valued harmony and tolerance in their interpersonal exchanges as well as their politics. Indeed, the historian Jamie Cohen-Cole has shown that postwar social scientists tended to model their conceptions of both science and politics on their seminar discussions and conversations at cocktail parties. But their conservative counterparts were not interested in intellectual small talk and subtle persuasion. They believed in fighting for permanent truths, against bitter opposition if necessary. Conservatives frequently ran into trouble because they failed to hide their disdain for conventional academic niceties—or their colleagues' substantive commitments. Many were prickly, combative characters who elevated matters of principle above personal comity or institutional reputation. Indeed, they saw the replacement of absolute values with such ephemeral standards as the greatest danger to the republic. The difference in argumentative tactics made for contentious departmental meetings at the very least, and sometimes developed into bitter feuds and public controversies.[22]

Such clashes sometimes resulted in the dismissal of conservatives from university posts. As with McCarthyite depredations for liberals, battles between conservative scholars and hostile departments and universities became causes célèbres for the New Right. Impassioned retellings of these cases circulated through conservative intellectual networks and served as flashpoints for the articulation of a generalized image of the universities as bastions of liberal orthodoxy. Stories of political repression under the cover of scientific objectivity and academic freedom forged a strong sense of persecution among right-wing scholars, strengthening bonds of political identity. Thomas Nixon Carver, Felix Wittmer, William Terry Couch, Frank Richardson, Kenneth W. Colegrove, Father Hugh Halton, Buckley's Yale mentor Kendall— conservatives of all persuasions knew the list of martyrs. When liberals said they were victimized, conservatives felt like they were in a hall of mirrors. "Instead of worrying so much about McCarthy," Regnery fumed, "it might be well for the professors and their friends to worry about the people who prefer Ruth Benedict to Virgil, or to Thomas Aquinas, for that matter."[23]

Conservatives of a more centrist temperament often found stable academic positions: Gordon Keith Chalmers at Kenyon, Robert Nisbet at Berkeley and then Riverside, Eliseo Vivas at Ohio State, Viereck at Mount Holyoke. And there were liberal and radical scholars who treated conservatives of all stripes in a collegial fashion. Still, even the most comfortable of these figures felt the sense of embattlement, of living in the den of an implacable enemy. As conservatives kept up with one another in letters, they shared news about the academic fortunes of their compatriots at other institutions. For these conservatives, the problem of the social sciences was a matter of personal livelihood as well as political principle and cultural authority. Their theoretical critiques of scientism were often closely connected to the career struggles of friends, colleagues, teachers, or students.

Some conservatives responded by seeking to take back the social science disciplines. One key strategy was to write and publish dense, scholarly books, in the hope that these would counter liberal tomes and swing public opinion back toward the moral tradition. Some of these books, such as Kirk's *Academic Freedom* (1955) and E. Merrill Root's *Collectivism on the Campus* (1955), recited instances of liberal oppression in the universities, using cases like that of Hobbs to illustrate their claim that social scientists forcibly silenced those who disagreed with them rather than seeking the truth. Most of them, however, simply offered a straightforward case for one or another variety of conservatism.[24]

Postwar conservatives, like many of their successors today, often portrayed their initiatives as attempts to defend academic freedom and restore intellectual diversity in the universities. They argued that the leading universities and publishers, unlike private foundations, enjoyed the public's trust because they claimed to foster an open, inclusive exchange of ideas. Kirk reported proudly to a journalist that Regnery's top-selling books were *The Communist Manifesto* and Rousseau's *The Social Contract*, illustrating the publisher's intellectual open-mindedness. Other conservatives, however, saw the specters of relativism and socialism behind such calls for academic freedom. Buckley famously wrote a whole book (*God and Man at Yale*, subtitled *The Superstitions of "Academic Freedom"*) demanding that trustees at Yale fire the numerous faculty members who failed to uphold traditional American values.[25]

To figures of Buckley's ilk, open debate was worse than useless when one of the positions was obviously true and the other dangerously false. "There is nothing sacred about colleges," wrote Benjamin H. Namm, a department store owner and trustee of what is now the New York City College of Technology. "If any college permits the students to be inoculated with germs which will ultimately cause the corruption of society, then that college does not merit the support of loyal citizens." Namm found the common claim that children should hear all sides of controversial questions "harmful." On economic questions, for example, "there are not two sides, nor three sides, nor X sides. There is but one side and that is the <u>right</u> side." The ideal of free and open discussion stood in tension with the view that a framework of simple, absolute truths should guide human behavior.[26]

Nonetheless, most conservative scholars in the disciplines embraced the principle of academic freedom and sought to assure that it included their right to call out what they saw as expressions of liberal bias and even blatant violations of university principle among faculty and administrators. Kirk's book *Academic Freedom* gave this view a theoretical foundation. He described academic freedom as a species of natural right, underlying and legitimating the relevant body of positive law. Kirk then argued that modern American society, with its intense pressures toward liberal conformity, "considers university and teacher only so many specks in a tapioca-pudding equalitarian society" and would stop at nothing to bend them to its debased, collective will. The country had forgotten the transcendent importance of academic freedom and other sacred rights.[27]

Traditionalists such as Kirk were particularly likely to seek intellectual change in the universities. To them, the truly important questions did not concern the size of the state or its policies abroad. They focused on moral truth: Who spoke for it? How could it be known? Did science rule it out? What social structures did it imply? Believing that human action flowed from such foundational commitments, traditionalists set themselves against Enlightenment-inspired modes of liberalism rather than liberal policies per se.

This fundamentally moral definition of conservatism created the potential for friction with libertarian champions of market freedoms, even as it opened up lines of alliance with groups beyond the conservative movement. Kirk, for example, viewed the "hopeless doctrinaire" Mises and other economic libertarians as dangerous radicals who shared with Marxists and welfare-state liberals a destructive focus on the material rather than moral side of human affairs. Traditionalists were quick to see economic determinism in the writings of conservatives whose social prescriptions moved exclusively in the realm of policy and who focused on limiting the scope of the state rather than rehabilitating the intellectual and cultural foundations of the West. Conversely, they felt akin to liberals and even radicals who shared their critique of scientism and commitment to moral foundations. They often christened political liberals such as Reinhold Niebuhr honorary conservatives, based on their attention to tradition, sin, and religion's social importance. Some even viewed heterodox social scientists as potential allies.[28]

This was certainly true of Kirk and Regnery. As they worked to move the social sciences toward conservatism, they sought out dissenting figures such as Pitirim A. Sorokin, the sociological critic of value-neutral approaches, and Robert M. Hutchins, the former University of Chicago president who had famously challenged Dewey's educational philosophy in the name of metaphysics in the 1930s. Regnery, especially, felt these figures might be convinced to give the New Right a fair hearing rather than rule it out of court, and perhaps even to come around to its core truths. Believing that journalists and politicians could not win the cultural battle alone, Kirk and Regnery assumed that the struggle between the champions of order and the avatars of relativism and collectivism would also need to be fought on the terrain of the disciplines. They would need allies in order to recapture the social sciences and the universities at large, steering them back toward conservatism and the underlying tradition of moral law.

These overtures by Kirk and Regnery met with varying degrees of success. Before Hutchins left Chicago in 1951, he looked with favor on attempts to increase intellectual diversity and philosophical self-reflection among its faculty. He supported the hiring of Hayek in 1950 and the effort to turn Chicago's Committee on Social Thought into a clearinghouse for the full range of contemporary ideological positions. The committee's John U. Nef, who saw his program as "one of the last strongholds for cultivation of the personal individual talent in a world which tends to divide everything into categories," became a crucial academic interlocutor for Regnery. Hutchins, who had connections to Regnery's father and believed strongly in the importance of foundational philosophical principles, remained a key object of Regnery's attention after he left Chicago. In 1954, Hutchins's ascension to the chairmanship of the fledgling Fund for the Republic—soon to become a thorn in the side of conservatives dismayed by its liberal bent—had Kirk and Regnery salivating at the prospect of a Hutchins-led organization with $15 million and "no idea of what they want to do."[29]

In the case of Sorokin, meanwhile, Regnery published the sociologist's *Fads and Foibles in Modern Sociology and Related Sciences* (1956), which briefly—until the publication of C. Wright Mills's *The Sociological Imagination* in 1959—succeeded Hobbs's study as the most visible and sustained academic critique of social scientists' claims to value-neutrality. Sorokin lauded Hobbs for having "the courage to express his conclusions, regardless as to whether they agree or disagree with the prevalent opinion at a given moment within academic circles." As it turned out, *Fads and Foibles* itself appealed not only to sociological dissenters but also to corporations such as General Electric—the site of Ronald Reagan's conservative turn in the late 1950s. The religion-oriented Lilly Endowment also helped to fund its publication. It says much about Regnery's understanding of conservatism that he was eager to publish a book offering detailed methodological criticism of a body of academic work rather than attacking the New Deal, McCarthy's critics, or other typical targets of right-wing ire.[30]

Kirk likewise hoped to remake the disciplines from within, despite his famous disdain for postwar academia. Ironically, Kirk's belief that genuine universities—unlike foundations—stood above the political fray and thus carried the authority of truth led this famously sour apostate from academia to repeatedly seek a path back into it, though on rather extravagant terms. Kirk had seemingly left the academic life for good in 1953. Profoundly antipopulist, he had hoped to bring the message of conservatism to a saving remnant of right-thinking Americans, who would then lead the degenerate masses. Few institutions could have been less hospitable than Michigan State College (later University), where football and vocational skills were the order of the day. Hired in 1946 to teach the history of civilization, Kirk quickly grew disillusioned with the academic boosterism and "growthmanship" of the post-GI Bill era. He resigned seven years later, "in protest against a deliberate lowering of standards, calculated to attract more students and pay for more dormitories," and retreated to an inherited home in upstate Mecosta. There, he sought to revitalize the great tradition of humane letters that had flourished in an age before social-scientific empiricism. Among the results of his labors were a series of widely read books, starting with the path-breaking *The Conservative Mind* in 1953.[31]

Yet Kirk's desire to turn the universities rightward contradicted his loudly announced rejection of academia—a stance that he continued to elaborate in his writings even as he considered a variety of academic posts behind the scenes. As early as 1953, Regnery informed Kirk that the Volker Fund would bankroll him, as it had Hayek and Mises, if he attached himself to an academic program such as Chicago's Committee on Social Thought. At Chicago, Nef wanted to create for Kirk a position parallel to that of Hayek. In response, Kirk originally told Regnery that he hoped to stay out of academia for two years. He then suggested that he would join the Committee on Social Thought only if he could serve in absentia, as had the theologian Jacques Maritain and the poet-critic T. S. Eliot. Perhaps equally problematic, both Hayek and the conservative economist Frank Knight opposed Kirk's appointment. But there were other options to be considered. In 1954, Kirk fielded offers from Queen's University and the University of Detroit, and Regnery raised the prospect of Notre Dame in 1955.[32]

Kirk's search for an academic post was closely tied to his vision for the journal that became *Modern Age*. In Kirk's mind and those of many supporters, the new journal required an academic platform; only academic credibility would allow him to counter the depredations of scientism and liberalism. At the same time, Kirk insisted that he should be free to focus almost exclusively on the journal rather than teaching. "Under no circumstances shall I become a real professor all over again," he declared to Regnery, after turning down an Earhart Foundation grant that would have required him to teach at Long Island University (LIU) in exchange for publishing *Modern Age* there. Kirk and Regnery undertook a complex, multiyear set of negotiations with several foundations and academic institutions—including Kenyon College, Wabash College, and Ripon College, as well as LIU and Chicago—as he sought an academic home for his journal without committing himself to teaching. In the end, Kirk did land at LIU, which had earlier hired Colegrove to teach political science after Kirk reported that it was "talking about setting up a conservative department" in that field. The Marquette Charitable Organization of Illinois provided the lion's share of the funds for Kirk's fledgling journal.[33]

Throughout these endeavors, Kirk worked closely with his indefatigable publisher Regnery, whose boundless energy, optimism, and networking skills made up for his financial shortfalls. At times, Regnery also moved in influential circles. He attended at least one meeting of the fabled Mont Pelerin Society, and he joined Kirk, Buckley, Moley, and several others in a failed attempt to create a conservative counterpart to the liberal Americans for Democratic Action, with behind-the-scenes financial support from General Motors. Most often, however, Regnery toiled in relative obscurity, seeking the funding that selling books by renegade conservatives did not provide. Regnery courted foundation officials and Texas oil barons alike in his attempt to drum up financial backing for the conservative intellectual revolution. Like any good publisher, he also sought out sources of mass sales for his books. Regnery joined hands with outfits as varied as the United States Information Agency and the Joanna Mills Company to promote Kirk's *The American Cause* (1957) and other titles.[34]

Reflecting their constant financial shortfalls, the letter exchanges between Kirk and Regnery in the 1950s are filled with reports of cooperation with corporate allies, conservative foundations, and lobbying groups, along with proposals for future endeavors. In 1953, for example, Kirk was paid to consult for the National Tax Equality League, lunched with the president of United States Sanitary Supplies, and solicited funds from companies ranging from General Foods to Perfect Circle Piston Rings. Meanwhile, Regnery reported proudly that Marshall Field, owner of the *Chicago Sun,* had read Kirk's book *The Conservative Mind.* The two men also watched with great interest—if some suspicion of the project's ambitions—the expansion of the Smith Richardson Foundation by Robert R. Richardson and his father, Vick Chemical Company president H. Smith Richardson. Indeed, when conservatives imagined new intellectual initiatives, they thought simultaneously about funding sources: "people like Pierre Goodrich and E. F. Hutton, who might be persuaded to help underwrite this venture." Such money financed a wide variety of initiatives, large and small, as when J. Howard Pew of Sun Oil forwarded $1,500 to Regnery to facilitate the publication of Hobbs's *The Claims of Sociology*.[35]

Kirk became particularly adept at moving between the journalistic and corporate worlds, not least through his attempts to secure financial and institutional support for *Modern Age*. A February 1959 lunch at Chicago's Union League Club found Kirk, Regnery, the literary scholar Weaver, and the publisher David Collier—Hayek had been invited as well but could not attend—dining with executives from Great Lakes Solvents, United Electric Coal Company, First National Bank of Chicago, Nationwide Food Service, Kemper Insurance, and several other area businesses. A few months later, Kirk reported to Regnery that he had lined up $1,000 from the Smith Richardson Foundation, was waiting to hear back from the Lilly Endowment and 3M, and was about to take a trip that would include visits to the General Foods Fund and the A. O. Smith Company of Milwaukee. Two years earlier, a local washing-machine manufacturer had ended up footing the bill for Kirk's failed attempt to establish *Modern Age* and an accompanying institute at Ripon College in Wisconsin. Despite such calamities, Kirk and Regnery remained hopeful that their understanding of social inquiry, rooted as it was in timeless human truths, would one day prevail.[36]

AS THESE INITIATIVES SHOW, Kirk and Regnery, like many other conservatives, worked at the intersection of two rather different intellectual alliances. One was the emerging New Right, defined by a set of prescriptions for social and economic policy. The other was a vaguely delineated cultural opposition, comprising a broad range of figures who favored humanistic and religious alternatives to scientism. When faced with a choice between these options, postwar conservatives took different paths. Buckley concentrated on political journalism, which he saw as a means of giving voice to the powerless conservative majority. He helped to spearhead a political coalition encompassing almost all of the groups—minus secularists such as Rand—who wanted to minimize the state's role in economic affairs and fight communism around the globe. By contrast, Kirk and Regnery stressed axes of intellectual agreement and downplayed lines of political connection as they sought to build a big-tent movement against social-scientific liberalism.

Still, the pair continued to operate far outside the intellectual mainstream. Their search for allies proved arduous and only occasionally rewarding, as political differences between scientism's critics repeatedly fostered mutual suspicion. Setting aside local contingencies, the obstacles that Kirk and Reg-

nery encountered suggest that institutional bridges across political and religious barriers could hold only so much weight. There remained crucial,
irreconcilable differences between the various critics of scientism, on matters
of policy as well as assumptions about society and the nature of the human
person. To be sure, traditionalist conservatives shared much with humanistic liberals and radicals: an emphasis on individual moral responsibility,
a critique of materialistic interpretations of human behavior, a sense that
the Western literary and spiritual traditions operated alongside political experience as potent sources of ethical guidance, and a belief that it was illegitimate to use scientific techniques to manipulate individuals, no matter
how socially advantageous the resulting behavior might be. Yet many of
their progressive counterparts—the humanistic psychologists, for example—
worried that a society dominated by scientism stifled individual creativity
and freedom of expression. They wanted more, rather than less, social experimentation. By contrast, traditionalist conservatives identified relativism as the core problem with a scientifically guided society: scientific
techniques obscured the existence of a stable moral order behind the flux
of experience. Such differences ultimately set limits on the degree to which
traditionalists could find common ground with critics of scientism on the
other side of the political aisle.

A number of more conventionally political disagreements reared their
heads as well. For even the most centrist liberals, the circle of potential allies usually omitted the rightward fringe of American conservatism. By contrast, figures such as Kirk and Regnery tended to accept as compatriots all
who spoke of individual liberty and moral order, even if those individuals
also harbored outrageous views on other questions. Regnery himself had a
penchant for highly unpopular positions. For example, he vigorously, if not
always consistently, supported World War II revisionism—the view that
the Western allies, led by Roosevelt, had been the aggressors in a pointless
and avoidable conflict. This claim, seemingly grounded in the belief that
Roosevelt could do no right, drew Regnery closer to a few maverick social
scientists, especially the revisionist historian Harry Elmer Barnes. But it
placed a high wall between him and most other scholars, whom he occasionally judged on the basis of their past enthusiasm for American intervention
rather than on their current views of economic regulation and related matters.
Here, as elsewhere, the philosophical threads connecting scientism's hetero-
geneous postwar critics coexisted with sharp political distinctions that divided them.[37]

8

CROSS-FERTILIZATION

RUSSELL KIRK'S ASSAULT on the modern social sciences reached a new peak in a 1961 article for the *New York Times Magazine*. There, Kirk called human beings "the least controllable, verifiable, law-abiding and predictable of subjects," entirely unfit for scientific analysis. Numbering himself among the "professors of social disciplines," Kirk urged his fellow investigators to turn away from the quixotic, logically incoherent goal of a "social science" and to view their study as a frankly normative "social art," of the kind undertaken by "poets, theologians, political theorists, moralists, jurists and men of imagination generally."[1]

The content of Kirk's lengthy article deserves analysis. So, too, do the remarkable line drawings that accompanied it, each poking fun at the quantified, jargon-filled work of social scientists or the highly programmed and constantly monitored lives of modern citizens. The vigorous rebuttal from the Columbia University sociologist Robert K. Merton, who was profoundly concerned with social science's public image, is important as well. But there is another consequential feature of this episode, one that is subtler but equally revealing: Kirk's pattern of citation. Though Kirk was writing in a genre that required no references, he chose to draw on numerous critics of postwar scientism, pulling examples, terminology, arguments, and quotations from their work. And importantly, he did not restrict himself to conservative commentators.[2]

Instead, as in his other initiatives, Kirk reached across lines of political division to enlist a wide range of scientism's critics. Given Kirk's familiarity with Pitirim A. Sorokin, "perhaps the best-known of American sociologists," it is unsurprising that he drew on Sorokin for several observations: the crying need for social theory in the disciplines, the "sterility" of the behavioral sciences, the "quantophrenia" besetting those fields, and Sorokin's definition of sociology as a science that nevertheless incorporates nonempirical evidence. (Kirk disagreed with Sorokin on the last point, holding fast to a narrower understanding of science as a purely materialistic enterprise and favoring the term "art.") Among more mainstream critics, Kirk looked to Jacques Barzun's writings for a telling example of social scientists proving the obvious at a hefty cost, as well as Barzun's suggestion that the term "behavioral science" reflected a "desperate conviction that man does not behave and should be made to with the help of science." But Kirk also reached farther afield, to critical scholars who would soon influence the New Left. Here, he cited C. Wright Mills, "a radical gadfly among sociologists," on the mediocrity of the researchers in large behavioral science institutes and quoted the maverick political theorist Andrew Hacker's dire warning about "the spectre of predictable man."[3]

Kirk's article represents one of the innumerable instances of cross-fertilization, or at least strategic borrowing, between mid-twentieth-century commentators who shared a humanistic critique of scientism—and often little else. At times, this cross-fertilization involved minds truly meeting, as figures with otherwise divergent views found substantive points of commonality. Often, however, it reflected the extraordinarily flexible nature of this mode of criticism itself. Virtually anyone could identify their own values as quintessentially human and describe the alternatives as impositions by a scientific elite. This is not to say that postwar critiques of scientism were purely arbitrary. But the critics' shared use of a highly capacious language facilitated borrowing, even as it raised the likelihood that they would talk past one another without realizing it.

When these critics were contributing essays to a collective volume, or sending appreciative letters to an author, or giving a series of talks at an institute, or—as in Kirk's case—citing isolated phrases and observations from others, they did not necessarily have to face the hidden points of disagreement. In such instances, commentators with an enormous range of views could converge on a single institutional or cultural site. The results ranged from mildly noteworthy—Barzun's presence at a 1964 conference of humanistic psychologists—to quite surprising: In his 1956 best seller *The Organization Man*, William H. Whyte invoked the conservatives Friedrich

Hayek and Eric Voegelin when asserting that the so-called social sciences remained merely "social studies" because their areas of study lacked the "objectiveness" of physical phenomena. This chapter illustrates the possibilities for cross-fertilization by examining how a remarkably diverse set of reviewers and ordinary readers responded to the literary critic and nature writer Joseph Wood Krutch's acerbic attack on social science in the 1954 book *The Measure of Man*, which won the National Book Award and drew acclaim from many quarters over the years.[4]

Not all contexts facilitated the easy overlapping of voices in different registers, however. When critics of scientism tried to develop collective plans of action, the differences suddenly came into sight. Superficially like-minded figures found themselves struggling to reconcile competing views of exactly what scientism meant, how it worked in the world, and how best to combat it. Even differences of writing style could become roadblocks. As we have seen, a range of disagreements cropped up in Kirk and Henry Regnery's search for a humanistic alliance across political lines. Such problems also emerged elsewhere in the late 1950s, as when the New York–based editor Ruth Nanda Anshen sought to harmonize a range of humanistic perspectives by creating an ambitious, interdisciplinary research center, the Institute for the Study of Man. Like the efforts of Kirk and Regnery, Anshen's failed initiative revealed not only the deep substrate of concern with science's cultural impact after World War II but also the obstacles, both profound and prosaic, to building stable institutions on that broadly shared but extremely abstract ground.

Still, the conceptual ferment of the 1950s and early 1960s left its mark. Remarkably variegated groups of figures circled around Krutch's book and Anshen's proposed research center, just as they moved through a host of sympathetic institutions and published in congenial journals. These dynamics illustrate the immense range of sensibilities contained within postwar American liberalism. At the same time, they also reveal the porousness of the boundaries between conservatism, liberalism, and radicalism in the mid-twentieth-century United States. It is easy in retrospect to sort historical actors into those three categories on the basis of their economic policies. Yet it is remarkably hard to find such neat lines of demarcation on questions of science, morality, and cultural authority. There, ideas and individuals flowed freely across apparent boundaries, fueling round upon round of cross-fertilization in the charged ideological climate of early Cold War America. Indeed, such exchanges sometimes inspired the hope that an organized opposition to scientism and technocratic liberalism had begun to take shape.

INSTITUTIONAL RESOURCES

Institutions matter. Even a philosophical movement needs spaces, both cultural and physical, in which to flourish: publications, lecture halls, conference rooms, university departments. By the early 1960s, critics of scientism from many political and religious backgrounds operated in a network of loosely linked institutional nodes in the United States, calling for humanistic dialogue and championing the whole person against the planners and social engineers. These sites included colleges and universities, including the University of Chicago, the New School for Social Research, Brandeis University, Kenyon College, Long Island University, the University of California at Riverside, and progressive colleges such as Bard and Bennington. Critiques of scientism also appeared in journals such as *Commentary, politics, America, The Christian Scholar, The American Scholar, The Kenyon Review, Modern Age,* and *The Review of Politics.* The Hazen Foundation and other philanthropies offered resources as well.

For critics of various persuasions, the University of Chicago offered a kind of mecca after World War II. A bastion of quantitative social science in the 1920s, it had been turned in a very different direction by new president Robert M. Hutchins and his close ally, the philosopher Mortimer Adler, in the 1930s. Emphasizing the metaphysical underpinnings of all thought, including scientific inquiry, Hutchins and Adler challenged what they considered the dangerously relativistic stance of Chicago's influential sociologists and political scientists. Their solution centered on the "Great Books," a set of classic texts said to contain the crystallized moral heritage of the West. But Hutchins also made the university hospitable to other oppositional perspectives as well.[5]

Chicago proved especially important as a haven for conservatives. The political theorist Leo Strauss was among the many right-leaning figures who found a home there. The university also featured a bevy of economists who championed free markets: figures such as Frank H. Knight, Jacob Viner, Henry Simons, and later Milton Friedman. Although we now associate Hayek with the Chicago School of economics as well, he did not teach in that department because he considered many of Chicago's economists— themselves already critical of the application of mechanistic, deterministic views to human affairs—too positivistic and rationalistic. To Hayek's eyes, they were overly concerned with mathematics, insufficiently critical of modern theories of knowledge, and naive about human capacities and traits. Arriving from the London School of Economics in 1950, he designated himself a professor of "social and moral science" and joined the Committee on Social Thought.[6]

That program itself became a particularly important site of resistance to scientism at Chicago. The Committee on Social Thought was shaped by the views of the economic historian John U. Nef, an important institutional actor who joined forces with Knight and the cultural anthropologist Robert Redfield to launch an initiative that would steer between the poles of John Dewey and Adler in Chicago's curricular battles of the 1930s. Earlier attracted to the pragmatism of his wife's uncle, the philosopher George Herbert Mead, Nef eventually turned toward Adler and other sharp critics of scientism, such as T. S. Eliot and Jacques Maritain. The program he created with Knight and Redfield would eventually house such luminaries as Hannah Arendt and Eliot and Maritain themselves, along with Hayek and many other critics of scientism. In 1950, Nef and Hutchins joined forces with the conservative publisher Regnery and a few colleagues from the Committee on Social Thought to publish the short-lived journal *Measure,* dedicated to "the power of the mind and spirit to resist material processes." The journal included pieces by Continental thinkers such as Martin Heidegger alongside works by Kirk, Maritain, Eliot, and various Committee on Social Thought members.[7]

Important centers of resistance to scientism appeared at other universities and colleges as well. Of course, Catholic institutions and their conservative Protestant counterparts featured entire departments filled with scholarly critics of naturalistic philosophies, many of whom challenged positivistic conceptions of social science as well. Other schools housed notable émigré thinkers. The New School for Social Research's University in Exile brought nearly 200 European social scientists to Greenwich Village, including ardent critics of scientism such as the Gestalt psychologist Max Wertheimer and the economist and sociologist Eduard Heimann. Strauss also taught at the New School from 1938 to 1948 before decamping for Chicago. Brandeis University likewise hired numerous émigré scholars after its 1948 founding, and around one hundred émigrés found positions at Berkeley as well. At North Carolina's experimental Black Mountain College, meanwhile, leaders such as the émigré artist Josef Albers and the American-born poet Charles Olson carried on running polemics against the social sciences, despite the interest of many of their students in those fields. Other progressive colleges, including Bard and Bennington, housed critics of the academic mainstream as well. Another vocal critic of scientism's effects on modern education, the literary scholar Gordon Keith Chalmers, led Kenyon College from 1937 to 1956. At the new University of California campus in Riverside, the sociologist Robert Nisbet, a progenitor of today's communitarian sensibilities, became the founding dean in 1953 and fought to keep Riverside a liberal arts college rather than a research university.[8]

Not coincidentally, the renascent field of political theory flourished in these institutions, enlisting many émigré critics of technocratic liberalism and sustaining a dialogue that included radicals and conservatives as well as heterodox liberals. An innately normative pursuit, political theory took shape in direct opposition to scientism. Arendt herself, though no friend to scientism, rejected the widespread attribution of totalitarianism to scientific and rationalistic tendencies in Western thought. But other émigré political theorists, led by Voegelin, offered sophisticated critiques of scientism's effects. At Duke, John H. Hallowell targeted scientism in his textbook, *Main Currents in Modern Political Thought* (1950), and in *The Moral Foundation of Democracy* (1954). Hallowell traced modern ills to the illusion of "the autonomy of human reason" and urged a return to the liberalism of old, which had assumed "a transcendental order of truth accessible to man's natural reason."[9]

Small but important philanthropies launched by conservative business leaders, including the Volker Fund and the Smith Richardson Foundation, helped to finance some of these attacks on scientism, as did an expanding network of personal and corporate donors. Such donors paid the salaries of figures such as Hayek and Mises, as we have seen. They also sponsored various research initiatives and intellectual exchanges. Other entrepreneurs and executives concerned with religious or philosophical issues occasionally did the same, as when the oilman H. A. W. Myrin founded the Myrin Institute for Adult Education at Adelphi College—the collegiate offshoot of the Waldorf movement—to meet what he considered the modern world's deepest need by reconciling "the modern scientific attitude with a spiritual world concept."[10]

The Myrin Institute exemplifies how a low-stakes pattern such as a series of individual talks could facilitate the convergence of vastly different thinkers. Speakers at Myrin in the 1950s included many middle-of-the-road liberals and the occasional international figure. But Myrin also hosted conservatives such as Kirk, *Freeman* editor Frank Chodorov, and Foundation for Economic Education president Leonard E. Read. Also invited in the 1950s were Chicago's Nef, the Jesuit scholar John E. Wise, the Great Books advocate Stringfellow Barr, the humanistic psychologist Rollo May, and the ecologically minded anthropologist and nature writer Loren C. Eiseley. Although the institute's budget grew increasingly tight by the end of the decade, its leaders went forward in the 1960s, believing that "countless tragedies today, such as unmotivated crime, drug addiction, alcoholism and a general sense of rebelliousness and frustration prevailing among our youth" derived from "doubt about the meaning of existence" and "the individual's loss of a true connection with life itself." Myrin leaders consistently traced such calamities to "the 'scientific' skepticism that poisons devout feeling and stifles idealistic enthusiasm."[11]

Other business leaders likewise encouraged philosophical and religious opposition to technocratic liberalism. Ironically, executives from high-tech industries were especially likely to associate genuine science with individual ingenuity, free enterprise, and nature appreciation rather than economic planning or large, bureaucratic projects. Indeed, W. C. Mullendore of Southern California Edison identified such outcroppings of collectivism as the fruit of a "psychic epidemic": a materialistic mind-set that only the moral traditionalism and authentic individualism of a Kirk or a Richard Weaver could heal. Mullendore cautioned fellow executives that continuing down the existing scientific and technological paths would lead more Americans to adopt the materialistic mind-set of anticorporate radicals. They should instead address the sphere of intellect directly and attack "the domination of American philosophy by the radical liberals." Similar discontents, though not always framed in conservative terms, inspired the Aspen Institute in Colorado, where a group of business leaders drawn to Adler's version of the Great Books built a center for the commingling of philosophers, artists, and other cultural leaders.[12]

Other kinds of educational programs for corporate executives also brought philosophically inclined business leaders into dialogue with scholars alarmed by prevailing cultural and political tendencies. Without great fanfare, and sometimes with a conscious commitment to avoiding publicity, a number of universities established programs of humanistic education for corporate leaders after World War II. In 1953, for example, the University of Pennsylvania philosopher Elizabeth Flower asked her New York University colleague Sidney Hook—known for his fierce anti-communism and his defense of individual agency in the 1943 book *The Hero in History*—to contribute a lecture to "a year's intensive program for a small group of the junior executives of the A. T. & T." The aim, said Flower, was to "educate and humanize these potential men of industry." The following year, Hook joined the sociologist Marvin Bressler and others in a repeat performance. The history of these corporate programs remains largely unexcavated, and not all of the participants challenged scientism—Bressler, for example, advocated the behavioral science approach—but they constituted another space wherein critics of scientism, coming from various points on the political spectrum, could meet and exchange views while mingling with sympathetic business leaders. Here, as elsewhere, institutional sites, often financed with corporate money, provided important bases of organization and sources of legitimacy for scientism's critics.[13]

CONVERGENCE

"Science is wonderful but sometimes I wish it would stay in the laboratory where it belongs instead of coming out to bother me." Krutch's lament, from an unpublished essay of the 1950s, neatly captured the ambivalence of many postwar critiques of scientism. In a common pattern, Krutch embraced the natural sciences but found the social sciences terrifying. Indeed, he worked diligently to limit their cultural influence, publishing innumerable blasts against scientism and its disastrous effects between 1950 and his death in 1970. As letter exchanges, reviews, and other invocations of Krutch's *The Measure of Man* (1954) and related writings show, these publications served as points of convergence for myriad critics of scientism who shared a common vocabulary but pushed that language in wildly different directions. Politically, Krutch was a centrist Democrat; religiously, he called himself a pantheist. Yet readers of virtually every stripe could embrace his main argument in *The Measure of Man:* that moral freedom defined humanity as such, and yet the humanistic tradition of the West had been virtually lost, drowned out by calls for social engineering and assertions of cultural determinism.[14]

Back in 1929, Krutch's *The Modern Temper* had symbolized for many readers the corrosive potential of a scientific worldview. Like Bertrand Russell, Krutch had lamented the total erasure of meaning under the sign of modern science but insisted that it was entirely, if tragically, true. The Krutch of 1929 painted a stark picture of a world that was utterly indifferent to human purposes and psychologically tolerable only through what he later called an "aesthetic Existentialism": behaving as though there were meaning in the world, even though there was not. The book cemented Krutch in the public mind as a leading religious skeptic, alongside such notorious atheists as Russell and the philosopher Max Carl Otto.[15]

Shortly thereafter, however, Krutch took the key step in his transformation from au courant New Yorker to antiurban crusader against all things modern and progressive. A populist critic of intellectual orthodoxies by temperament, he had embraced Herbert Spencer's agnostic naturalism as a youth in theologically conservative Knoxville, Tennessee. But he reversed his perspective in the early 1930s, concluding that the truly dominant, oppressive orthodoxy of the modern age was the technocratic progressivism on display all around him in Manhattan, not the pinched Christianity of his childhood. What moderns needed to resist at all costs, for the sake of their essential individuality, was the secular radicalism of his colleagues at the *Nation*, not conservative religion. Krutch began to seek a humanistic middle way between scientism and traditional faith, viewing each as a form of determinism. He also ruminated on the real-world impact of ideas, writing in an unpublished 1941 article that Nazism and other global catastrophes stemmed from the philosophical errors of Western liberals—specifically, their belief that human behavior was fully determined and moral freedom was a sham. He continued to develop these views through the 1940s and 1950s, coming to be seen as the leader of "a band of militant humanists who charge that science is costing man his humanity" and giving back little of value.[16]

As Krutch was turning against philosophical naturalism, he began to find a source of meaning in nature itself. In 1950, he left the city for a small desert plot outside Tucson, where he spent his last two decades ruminating on the natural world and lambasting modern civilization. A steady stream of nature writings, followed by television programs on the desert and other wilderness areas, made him once again a minor celebrity for a new generation. In fact, Krutch is most often remembered today as a leading architect of the ecological sensibility that coalesced in the 1960s. He was an early champion of ecology's holistic, nonmechanistic approach and a hero to figures such as the radical environmentalist Edward Abbey.[17]

As with Abbey, however, Krutch's reverence for nature mirrored his disdain for modern civilization. He deplored phenomena as varied as supersonic transport and the mainstreaming of garlic, previously associated with immigrants. Krutch's long-running column in The American Scholar also skewered the suburbs, vivisection, hunting, the space race, and overpopulation, as well as a host of technocratic and relativistic tendencies in politics, law, and education. Behind such phenomena, Krutch saw the determinism of the behavioral sciences. When applied to everyday affairs, he contended, that mind-set threatened all that made people human: their individuality, their autonomy, their moral freedom. The species could not long survive in a world controlled by the behavioral scientists. Krutch insisted that what made human beings human was their capacity for moral self-determination, in the face of group pressures if necessary. Adopting the opposite approach—the materialistic, deterministic outlook fostered by modern science—led to a host of abominations: permissive parenting, the life adjustment curriculum, Kinsey's studies, and developments in criminology, among innumerable others. In each case, Krutch contended, the social sciences eradicated traditional morality and created a totalitarian pattern wherein experts sidestepped democracy by conditioning individuals to conform to social imperatives. In short, the United States was rapidly sliding toward B. F. Skinner's Walden Two.[18]

Indeed, Skinner's book provided the central exhibit for Krutch's indictment of social science in The Measure of Man. Like Walter Lippmann before him, Krutch ruled out religion as a solution to the modern crisis. Science had won the war, he declared, and rightly so. But now, said Krutch, Skinner's book demonstrated that an even deeper battle was underway: the battle over human freedom itself. This time, the combatants were science and humanism, or "literature and sociology." It was "the poet and the prophet versus the social scientist . . . insight and imagination versus the questionnaire, the poll, and the 'study.'" Krutch no longer equated science itself with determinism, as he had in 1929. Indeed, he now argued that natural scientists had long since scrapped the mechanistic sensibility of nineteenth-century physics. Yet the social scientists still clung to the old, discredited approach, Krutch lamented. Dogmatically equating science with materialism and determinism, they ruled out by fiat the moral freedom of the individual, which Krutch said was clearly evident from immediate, internal experience, though not external, sensory evidence. The spread of this reductive mind-set had created "a pragmatic, utilitarian, instrumentalist, materialistic and norm-worshiping civilization" that threatened humanism, and thus humanity, at its core. The issue was simple, Krutch summarized: "Can we really learn anything about human nature from a Shakespeare or must we, if we really want to know something, consult a Kinsey instead?"[19]

The Measure of Man and Krutch's many other writings on the topic resonated with an extraordinary array of readers. Particularly striking, given Krutch's open disavowal of theistic frameworks, were the responses of religious critics. After trading his youthful Episcopalianism for Spencer's naturalism, he never again considered religion a live option. But he recognized the obvious points of overlap with theistic frameworks. As he told an associate, he rejected "the traditional Christian creeds" but was "most definitely not anti-religious." In fact, he was "firmly convinced that there are such things as the eternal verities," and that "our tendency to disbelieve in them" was "responsible at the bottom for the most fundamental ills of our time." In making such claims, Krutch noted, "I have always tried to defend them in such general terms that they would be acceptable on the one hand to the most orthodox believers and the other to those who like myself have no such comfortable detailed certainties but are trying to be as certain as we can about what is most fundamental." Krutch sometimes framed the target of his scorn in religious language as well. By throwing out moral self-determination along with traditional theology, he wrote, "little men and little intellects, numerous, voluble, and strategically placed, have fixed upon the popular mind a sort of secular Calvinism which substitutes 'proper conditioning' for 'grace' and takes even the possibility of salvation out of every individual's hands." Perhaps the Catholic writer G. K.

Chesterton was right, Krutch mused, when he said that "the doctrine of original sin provides the only really cheerful view of the world."[20]

A broad range of readers argued that Krutch, deep down, actually shared their own religious commitments. They found Krutch's reinforcement of core religious tenets more convincing than his disavowal of Christian doctrine: How could a self-professed pantheist endorse moral absolutes and criticize scientism so sharply? Although a few respondents either welcomed or deplored Krutch's nontheistic approach, many simply refused to believe it. Some declared that Krutch's arguments pointed directly toward belief in God. John LaFarge, writing for the Catholic journal *America,* stated that Krutch's dissection of materialism cleared the way "for an equally logical mind to move on to the idea of a purely spiritual soul and to a Creator." Other respondents claimed that Krutch had already adopted an implicit religious position, while disagreeing on the nature of that position. A former student of Krutch's called him essentially Catholic, and a Catholic Worker pacifist likewise expressed his admiration of Krutch's work. But a *Jewish Forum* editor opined that "your philosophical viewpoint is Hebraic—The Prophets—Amos Ezekiel etc." Meanwhile, Krutch's close friend, the Columbia literary scholar Emery Neff, wrote in 1956 that the pair of them, along with the heterodox biologists N. J. Berrill and Paul B. Sears, were "converging upon a new religion, which resembles Lao-Tse's Taoism, illuminated by later scientific knowledge." Another reviewer called Krutch a "religious Humanist—religious in the sense that he admits mystification, Humanist in that he finds man the chief mystery." Readers recommended to Krutch a diverse array of religious thinkers—the radical Jesuit Daniel Berrigan, the Episcopalian thinker J. V. Langmead Casserley, the anthroposophist Rudolf Steiner—that they expected to meet his approval and comport with his views.[21]

Krutch also found a receptive audience among postwar conservatives. As we have seen, many conservative thinkers were deeply engaged with questions about science, religion, and the humanities. And while Krutch was a committed, lifelong Democrat, he worried about the psychological effects of welfare provisions and decried the loss of individual responsibility they entailed. His writings reinforced many conservatives' belief that modern applications of science undermined moral truth and political rectitude. He went far beyond the standard postwar complaints about mass conformity, attacking many liberals' deepest assumptions about human nature and the role of governance. In a 1960 address, for example, Krutch deplored the loss of personal morality in the modern world—not a vague "social consciousness," of which there was plenty, but rather "the supreme importance of purely personal honor, honesty, and integrity," and behind it "the strong clear sense of good and evil, the most important realities of the human being when we contrast him with the animal or the machine." Americans, he contended, had adopted the relativistic view that "morality means mores or manners and usual conduct is the only standard." Krutch painted the welfare state, in particular, as a staunch enemy of personal freedom and morality. In published writings and especially unpublished pieces, he distinguished policies that "fill the world with fools" by serving to "protect the weak from the consequences of their folly" from legitimate public aid—the provision of services to the blind, for example. "Welfare and state-subsidized culture are uncomfortably suggestive of the bread and circuses" that pacified Rome, Krutch declared.[22]

Numerous conservatives welcomed Krutch's portrait of a dominant liberal elite that would trample individual freedom and morality in its quest to remake everyone else in its own image through psychological conditioning. The economist Henry Hazlitt, a colleague during Krutch's time at the *Nation,* quoted *The Measure of Man* extensively and compared it to Hayek's *The Counter-Revolution of Science. Freeman* coeditor John Chamberlain considered Krutch firmly "anti-New Deal," because his insistence on moral responsibility and the limits of conditioning ruled out "the whole idea of a 'planned society,' or of making State-dictated 'welfare' the measure of all things." Another reviewer summarized Krutch's argument as follows: "If you believe in man you will not pamper, protect, guide, coddle and social-welfare him into a state of moronic second-childhood." He welcomed Krutch's assault on "the fatal security of the hive" offered by "social planners, welfare staters, totalitarians and brave-new-worldlings." A conservative political theorist argued that Krutch, "the best-known living American exponent of skepticism," utterly denied "that men can have the ability to transcend their own subjective limitations and secure reliable standards of truth." Such skepticism, he wrote, buttressed "a full-fledged Tory viewpoint, characterized by an aristocratic social philosophy and an emphasis upon authority and tradition." If it flourished, it would drive "a revival of traditional conservatism with a force comparable to that which led to the spread of intellectual radicalism under the stimulation of the Great Depression." Claremont's Institute on Freedom and Competitive Enterprise reached out to Krutch for assistance with a workshop, and ordinary readers signaled their approval by equating Krutch with the conservatives Raymond Moley and Albert H. Hobbs.[23]

Although Krutch did not participate in the Claremont event, he often returned the praise of conservatives. He recognized right-wing individualists such as Albert Jay Nock, Frank S. Meyer, and Friedman as compatriots. In *The Measure of Man,* he hailed Voegelin's critique of scientism as well. In

1956, Krutch joined a group of conservatives, including Hayek, Friedman, Weaver, and John Dos Passos, in Princeton to discuss the fate of individualism in a massified world. Meanwhile, Krutch's critique of the "educationists" followed that of Bernard Iddings Bell so closely that he worried Bell might charge him with plagiarism. Other correspondents and reviewers likewise compared Krutch to Bell. Krutch published in the libertarian *Freeman* in the early 1950s, reviewed a manuscript by Meyer for Regnery's press in 1963, and was reading Kirk's journal *Modern Age* by the late 1960s. His attacks on scientism also drew attention from many business leaders, for whom his moral individualism outweighed his ferociously anti-industry, antidevelopment stance.[24]

Perhaps more surprisingly, Krutch's analysis also resonated with natural scientists, especially after 1960. Some rejected his argument out of hand and circled the wagons, defending their expertise. But many agreed that scientism threatened society and had little to do with genuine science. A dental surgeon predicted that Krutch's book would come to be seen as "a turning point in man's struggle toward self evaluation." Brookhaven National Laboratory invited him to give its Pegram Lectures, and enthusiastic letters flowed in: a Stanford geneticist, a chemist in Ohio, a physician from Albuquerque, the chair of the Caltech biology department.[25]

As the 1960s progressed, natural scientists increasingly welcomed Krutch himself into the fold. Krutch was indeed an amateur researcher as well as an ecological theorist; many letters find him exchanging specimens and detailed observations of native species, including his beloved boojum tree. By the mid-1960s, Krutch was closely aligned with ecologically minded thinkers such as the anthropologist Eiseley and the biologists Berrill, Sears, and Marston Bates. Of course, some of those who drew on his expertise in natural history, science education, and popular science writing stood outside the scientific mainstream. Thus, Notre Dame's Great Books program asked Krutch's advice on how to teach biology in a humanistic manner, after the fashion of his popular book *The Great Chain of Being*. But he moved in more mainstream circles as well. In 1965, an official of the New York Academy of Sciences invited Krutch to speak to its environmental science division, noting that scientists as well as humanists were now pondering the normative as well as technical dimensions of their work. As early as 1959, in fact, the Biological Sciences Curriculum Study (BSCS) of the National Science Foundation consulted Krutch as an expert on science education and then added him to the group's steering committee after BSCS chair Arnold B. Grobman embraced Krutch's conception of biology as a humanistic field. A few years later, he was invited to join Isaac Asimov and others on the editorial board of a major new venture in popular science proposed by the publishers of the *World Book Year Book*.[26]

Krutch's attack on scientism also earned him admirers in the social sciences themselves, where a growing number of practitioners chafed at the value-neutral ideal and sought to push their fields in more humanistic directions. Kurt H. Wolff, an émigré sociologist at Ohio State, applauded *The Measure of Man* and proposed an anthology of "transpositivistic" readings for students that would feature Krutch alongside Max Horkheimer, Strauss, and other heterodox thinkers. In spring 1960, Berkeley's political scientists gave Krutch a visiting position so that he could help them develop a "humane social science." Krutch also received numerous invitations to speak on "The Old Fashioned Science of Man," as one put it. Psychiatrists and psychologists were particularly receptive to Krutch's sharp criticism of their fields. He was invited to address the American Psychiatric Association in 1955 and the American Psychological Association in 1959. The psychiatrist Karl Menninger, a longtime admirer of Krutch's columns, called him a prophet and read his book "as a devout religionist might read a book of meditations."[27]

Overall, Krutch's forthright defense of individualism and morality facilitated the convergence of virtually all of the streams of postwar criticism on a shared concern with scientism and determinism. Humanistic scholars and natural scientists alike could find much to appreciate in *The Measure of Man*. So, too, could political conservatives and religious critics, despite Krutch's Democratic affiliation and staunch rejection of theism. Only theologically conservative Protestants kept their distance from Krutch's work.

Still, what could be done about scientism besides writing outraged books? It was easy enough for these thinkers to pen reviews of Krutch's works, or send him letters, or invite him to speak. But did the various groups of critics have enough in common to constitute themselves as a coherent oppositional force? Could they take some form of practical action?

SYNTHESIS

The fruitless effort of Anshen, a philanthropist, philosopher, and religious syncretist, to establish an Institute for the Study of Man in the late 1950s suggested that scientism's critics could hardly even agree on a platform of broad principles, let alone apply those principles to the real world and then put them into motion. Standing much closer to the liberal mainstream than Krutch, Anshen adopted a rather different critique of scientism—more messianic than apocalyptic—that enlisted the support of prestigious natural scientists as well as humanities scholars, religious leaders, and social scientists. Whereas many postwar critics took a frankly oppositional stance toward the academic establishment, insisting on the need for an alternative, nonscientific method or body of knowledge, Anshen and her collaborators operated from a position of great prestige and spoke in the language of harmony, integration, and synthesis. They sought a full-fledged cosmology that would transcend science's interpretive limitations by embedding its findings in an expansive vision of human unity and creativity. Still, these figures shared many assumptions with scientism's more combative critics. Anshen and her many eminent allies agreed that science said nothing about the human condition and blamed a fundamentally inhuman, mechanistic view of the world for the total crisis that had gripped humanity in the twentieth century.

Historians have ignored Anshen's initiatives, and she failed not only to launch the Institute for the Study of Man in the 1950s but also to create a journal of international thought in 1944. Nonetheless, Anshen left a significant mark on American culture through another mode of cross-fertilization that required little negotiation of interpersonal differences. Anshen developed and edited no fewer than seven different book series, running from 1940 through the 1980s. An indefatigable cultivator of intellectual relationships with a deep sense of mission and a strong mystical streak, Anshen began her editing career with eight star-studded volumes in the Science of Culture series, stretching from 1940 to 1957. Each volume featured up to forty luminaries, from Albert Einstein to Thomas Mann to Henri Bergson. In 1954, Anshen also began issuing books—sometimes several a year—in the World Perspectives series, which explored "the new organicity of man with nature, new yet very old, but only now beginning to achieve concretion morally and politically." Additional series followed: Credo Perspectives, Perspectives in Humanism, Religious Perspectives, Tree of Life, and finally, in the 1980s, Convergence.[28]

Overall, Anshen solicited more than 130 original or translated books by a galaxy of mid-twentieth-century figures, including Konrad Adenauer, W. H. Auden, Martin Buber, Martin D'Arcy, Christopher Dawson, Theodosius Dobzhansky, René Dubos, Mircea Eliade, Erich Fromm, Walter Gropius, Martin Heidegger, Werner Heisenberg, Karl Jaspers, Pope John XXIII, Jacques Maritain, Margaret Mead, Lewis Mumford, Gunnar Myrdal, Swami Nikhilananda, Jaroslav Pelikan, I. I. Rabi, Sarvepalli Radhakrishnan, Paul Ricoeur, D. T. Suzuki, Paul Tillich, and many more. Anshen's relentless insistence that these works constituted a single, unified body of thought is open to debate. But she certainly facilitated a great deal of influential writing. To take just two examples, Anshen convinced Rabi to put his thoughts on paper and also published Fromm's 1956 blockbuster *The Art of Loving* (as well as two of Fromm's other books). Throughout her career, she encouraged authors to spell out, in a broadly accessible fashion, their views of knowledge, value, and faith in the modern world. As it turned out, Anshen's frequent collaborators—figures such as Tillich, Fromm, Rabi, the biologist Edmund W. Sinnott, the émigré political scientist Ernst Jäckh, and the sociologist Robert M. MacIver—and many of her other authors agreed with her that advocates of scientism had wrongly projected a restrictive philosophy onto the world of experience. (Sinnott characteristically lamented the influence of "scientific materialism," which left the human person "in danger of degenerating into a selfish and soulless mechanism.") Yet these figures insisted that science, if properly understood, buttressed an expansive view of moral and spiritual freedom.[29]

Overall, Anshen's series helped give expression to a mode of scientific humanism that strongly inflected American thought from the 1950s through the 1970s. They appealed to liberal and left-liberal readers who shared a humanistic bent and mystical, universalistic spirituality with many 1960s rebels but eschewed totalizing critiques of capitalism, modernity, or science writ large. Scientific humanists of this variety sought to unify reason and faith or body and soul, not to drop out or smash the state. This inclusive sensibility was largely continuous with humanistic psychology but emphasized that movement's constructive visions while moderating its polemic against behaviorism. Meanwhile, it took for granted that the physical sciences, and perhaps biology, were now thoroughly antimaterialistic and could anchor a synthetic worldview. Such a framework would appeal to the world at large and even unify the species, fostering global peace and allowing humanity to advance to unimagined heights by unlocking its innate creativity.[30]

In certain regards, Anshen's initiatives overlapped with a series of other integrative endeavors dating back to the late 1930s. The Conservative Jewish leader Louis Finkelstein spearheaded several of these from his post as chancellor of the Jewish Theological Seminary, including the wide-ranging Conference on Science, Philosophy and Religion in their Relation to the Democratic Way of Life (CSPR) and the more scholarly Institute for Religious and Social Studies (IRSS). Groups of prominent scholars also met under various other auspices after World War II, attempting to hammer out a reconciliation of science and religion and give the modern world a unified, morally robust, spiritually satisfying culture, thereby reversing the process of social disintegration. After 1954, for example, the annual conferences of the Institute on Religion in an Age of Science brought together scientists and religious leaders. At Harvard, Sorokin's Research Center for Creative Altruism fostered dialogue as well. Other initiatives sought to include the arts alongside science and religion in a broad cultural synthesis. The psychologist A. A. Roback issued his own books and those of congenial colleagues, including Tillich's *Can Religion Survive?* (1962), through his Sci-Art Publishers. Many individuals participated in several such enterprises. MacIver, for example, was a stalwart in the CSPR and IRSS, a frequent contributor to Anshen's edited volumes, and a founding member of her ill-fated Institute for the Study of Man. Such endeavors, like so many other postwar projects, rested on the assumption that social institutions and practices took their shape from a shared, underlying philosophy. In short, social unity required intellectual unity. Humanity would not be at peace, either individually or collectively, until science, religion, philosophy, and perhaps the arts had been thoroughly reconciled and the resulting synthesis distributed to the peoples of the world.[31]

The religious valences of these integrative projects varied. Most of their leaders sought to include representatives of non-Western faiths such as Buddhism and Hinduism. Indeed, organizers and participants often argued that shared ethical claims underlay all of the world's "high" religions. They discerned a fundamental moral law for humanity, available through non-scientific means such as intuition, mythological symbolism, or mystical insight. Most framed this law in terms inherited from Immanuel Kant: Human beings should never treat one another as merely instrumental means to their own individual or collective ends. In American culture more widely, as Jennifer Ratner-Rosenhagen has shown, "wisdom literature" circulated alongside more conventional expressions of theism in forums such as *Wisdom* magazine (1956–1962). Theorists of religion, including Joseph Campbell, author of *The Hero with a Thousand Faces* (1949), and the Romanian émigré Eliade, emphasized the commonalities in religious experience, as captured in mythological symbols. This expansive, universalistic vision often came packaged with criticism of social scientists for emphasizing practical needs or cultural differences rather than spiritual and moral unity.[32]

Although Anshen's formulation of this universalistic sentiment leaned in a strongly mystical direction, she downplayed the religious side of her project when she set out to build a new research enterprise. By 1957 she had identified a site—the Alexander Hamilton Estate in Tarrytown, New York—and a potential funding source for her grand new endeavor: an Institute for the Study of Man that would promote the integration of modern thought through coordinated exchanges between leading scholars. Adopting her usual mode of sending a barrage of postal entreaties, Anshen lined up the physicist Niels Bohr as president and organized an illustrious group of founding members for the proposed Institute: Barzun, Fromm, Heisenberg, MacIver, Maritain, Rabi, Sinnott, the management theorist Chester Barnard, the biophysicist and Rockefeller University president Detlev Bronk, the mathematician Richard Courant, the psychologist Kurt Goldstein, the historian Allan Nevins, the physicist J. Robert Oppenheimer, and the philosophers Susanne K. Langer and F. S. C. Northrop. Anshen also had a lavish starting grant in hand—or so she thought. Associates of H. Smith Richardson, the conservative Vick Chemical magnate, promised that he would donate $9 million to the institute project, and perhaps even his entire fortune of $40 million in the end.[33]

Planning for the institute proceeded, although tensions arose between key actors. One minor issue was stylistic: Barzun, who spent much of his career advocating simplicity and clarity in writing, repeatedly warned Anshen that the florid, abstruse phrases in her proposal would confuse and alienate the Richardson officials. Still, Barzun remained fully on board with the institute idea. He had been drafting key sections of the proposal and rewriting Anshen's contributions since at least June 1957, and he later served on a small organizing committee that met frequently with the Richardson staff. Despite Barzun's irritation with Anshen's prose, he pushed forward with the organizing work for the institute, patiently reminding Anshen to write for nonspecialists and temper her flights of fancy.[34]

The thorny matter of what questions the participants should pursue first also arose. The astronomer Fred Hoyle wanted to begin by exploring the evidence for divine creation. Courant, a close confidant of Anshen's, proposed that they should focus instead on whether complementarity and other principles drawn from the natural sciences could explain human experience. Barzun thought both of these inquiries were too cosmic and overly ambitious. His own plan, which he considered more modest, was for institute fellows to establish art's cognitive value; specify its relations to philosophy, religion, and the sciences; produce authoritative canons of historical writing; determine the proper training for teachers at every grade level; gather biologists, psychiatrists, and anthropologists to rule conclusively on the existence of a universal human nature; divide the useful material in the nonscientific fields from the "hangovers" and thereby set the direction of future research in those areas; and examine all of the operations of language, especially its function as "the indispensable complement of Number for storing our knowledge."[35]

It was not the absurdly ambitious character of these goals that ultimately sank Anshen's institute project, however. Quite the contrary: It was the proliferation of proposals for such integrative, global endeavors, which came from many others as well as Anshen. She managed to put off Finkelstein, who wanted to fold his proposal for such an academy into hers. But the Institute for the Study of Man collapsed in the summer of 1958, after the Richardson representatives revealed that they had been cultivating proposals for rival "World Institutes" and accused Anshen of stealing her plan from one of the competitors.[36]

Practical matters aside, it seems likely that deep political differences between Anshen's associates and the Richardson faction also helped to scotch the institute. One former Richardson trustee highlighted the political mismatch, even as he insisted that Anshen's proposal would be judged on its intellectual merits. It is difficult to imagine a conservative foundation looking favorably on the liberal bent of Anshen's institute—nor, perhaps, its religious syncretism either, given the Richardsons' deep Presbyterian faith. Under pressure from both Barzun and the Richardson officials, Anshen did finally rewrite her proposal to stress the institute's practical outputs, although the document's mystical framing material remained largely in place. Yet the expansive global changes she expected to flow from the project could hardly have warmed the elder Richardson's heart. She asserted that the "proved genius of the participants" would not only produce theoretical results as powerful as Einstein's theory of relativity but also ensure both reliable knowledge and effective control of a vast array of postwar social dynamics, from automation and technological unemployment to "the interplay of social, economic, political and cultural forces in an age of science and technology." Beyond that, Anshen promised her funders authoritative findings on a host of issues: "The new implications of education for the atomic age. . . . Population explosions and the future of eugenics. . . . The problem of leisure. . . . Ways to combat economic depressions. . . . Meaning and use of mass media of communication . . . The implications of the centralization and decentralization of government. . . . The development of undeveloped areas. . . . How the individual may be secured against the threatening anonymity and collectivity of mankind." Closing on a technocratic note, Anshen wrote that "new and deliberate effort will be needed to control processes of social change" under modern conditions. Even Barzun was put off by the new proposal, scribbling at the top of his copy, "Abysmal nonsense, vulgarity and fallacy."[37]

In the end, Anshen's vision was tailor made to excite spiritually minded scientists and scientifically inclined theists and humanists. Even Barzun's supposedly more restrained version, however, was wildly impractical. Moreover, Anshen's associates diverged significantly from the staid, business-minded Richardson crew in their understanding of the postwar political scene. Still, Anshen's efforts were hardly for naught, even if this particular product of her fertile mind died stillborn. Her book series provided an outlet for a substantial cadre of spiritually inclined writers and readers within the broad liberal mainstream to argue that the world could not realize the promises of either industrial production or economic planning without re-discovering the core values of the great religious and humanistic tradi-tions. This style of antiscientistic liberalism, operating alongside the New Left and intersecting at key points with the more oppositional countercul-ture, would significantly shape American thought and culture in the 1960s and 1970s.

NOT EVERYONE SHARED the deep fear of modern science's cultural ef-fects that drove Krutch, Anshen, and many others in the 1950s. By the time Krutch's book appeared, one reviewer thought its main message was thoroughly cliché. Scores of recent works had ingrained the litany of charges: "that democratic man is on the way to becoming like totalitarian man, a conditioned thing; that it is a false idea of science which is respon-sible; that, moreover, this idea has so dominated the social sciences that they are now nothing but pseudo-sciences threatening to gain a complete domi-nation over us." But if this reviewer met *The Measure of Man* with a yawn, others saw it as a clarion call to action in a world teetering on the abyss. Scores of readers came out of the woodwork to explain to Krutch why his book had spoken to them so powerfully—and often why he should, or im-plicitly did, agree with them on religious or political matters as well.[38]

In turn, *The Measure of Man* was only one of many sites at which ideas about science and scientism circulated through postwar American culture, in a complex process of cross-fertilization that fostered everything from mis-understanding to inspiration—and much in between. In the 1950s and early 1960s, a series of institutional niches gave critics of scientism the opportunity to articulate their perspectives and hammer out the points on which they overlapped and differed. Against a mainstream discourse that, in their view, took scientism for granted and readily drew its manipulative conclusions, these critics sought to create spaces in which they could build, through both personal and intellectual connections, an oppositional force: one that took moral freedom for granted and sought to clarify its mean-ings and applications.

Yet it proved much easier for the various critics to attend the same con-ferences or publish collective broadsides against scientism than to find common ground. Above all, political agreement eluded postwar critics of scientism. A few of their endeavors incorporated figures from across the political spectrum, but most covered only part of it. Some stretched from the right to the center-left, leaving out left-liberals and radicals. Others spanned from left to center-right, omitting many groups of conservatives. Anshen's efforts, like those of Kirk and Regnery described in Chapter 7, highlight the power and variety of critiques of science in the postwar United States, as well as the obstacles to organizing this energy and channeling it against technocratic liberalism. Among other things, they reveal the limits, for the purpose of concrete action, of a critical framework that left such capacious terms as "human" and "moral" undefined. These expansive categories facilitated conceptual alignments among critics of many persuasions, but they did not say much about what to actually do.

Political differences would reassert themselves with a vengeance in the 1960s and 1970s, even as the networks of thinkers and publications built by figures such as Anshen bequeathed significant interpretive resources to a new generation. On the left, many radical critics would weave earlier strands of criticism—radical, left-liberal, humanistic, theistic—into a potent assault on a technoscientific elite. Disgusted with postwar liberalism, though often borrowing from it as well, many activists and scholars would renew the project begun by Mumford, Reinhold Niebuhr, and others between the wars: creating a more organic, authentic American radicalism, purged of scientism, statism, and other incursions of bureaucratic rationality. Meanwhile, a host of conservative theorists, activists, and politicians would reframe New Right arguments in a more populist manner that pitted ordinary citizens and their moral traditions against manipulative experts. Both radical and conservative critics, in fact, would specifically target a "new class" of managers, bureaucrats, and technicians. Despite such persistent points of overlap, however, ideological divisions would deepen further among the disparate advocates of a nontechnocratic, morally committed form of politics.

9

A NEW LEFT

"NO ONE HAS TO WORRY that the colleges are producing radicals," wrote the historian Irwin Abrams in 1958. Within a few years, however, American higher education would do just that. The 1960s brought a dramatic leftward turn among college students, as well as some of their professors. Yet while the upheaval of the 1960s is often seen as a comprehensive break with the past, it also featured important lines of continuity. One was the tendency to contrast the morally desiccated, science-dominated society of the post–World War II United States to a humanistic alternative. Belying the conservative charge that they threw off morality altogether, many 1960s radicals embraced moral universals, even as they detached them from social conventions past and present. And like so many liberals and conservatives of the early Cold War years, these young critics insisted that a culture disfigured by science made it impossible to live authentically human lives.[1]

Of course, the activists and theorists of the 1960s had new views of what humanism meant and what morality demanded. But they often slotted those conceptions into a well-established pattern of cultural criticism that identified science and its outgrowths as the leading threats to human flourishing. They, too, lamented the atomized individuals, the bureaucratized routines, and the insidious consumer culture of a scientific age. They contrasted universal human values with the technical, rationalistic orientation of the society around them, and they portrayed postwar America as a domineering, technocratic society, left bereft of value considerations by its adoption of a scientific mind-set and protected from challenges by the cultural dominance of scientism. In short, they radicalized the critique of postwar liberalism as a science-obsessed framework that systematically eradicated moral autonomy and human dignity alike. As with conservatives, in fact, a concern with science's social effects provided a key point of overlap between the divergent strands of radicalism that emerged in the 1960s. Countercultural dropouts, social-democratic "politicos," Maoist revolutionaries, and post-Marxist social theorists all shared an instinctive distrust of bureaucratic organizations and instrumental rationality—and often traced these to science's cultural impact.

The 1960s generation grew up surrounded by such portraits of their society. Casey Bohlen has suggested that many young people came to radical activism through the "social action" groups organized by liberal churches and synagogues in the 1950s, which urged the youth to put their faith into action by closing the gap between democratic ideals and realities. As we have seen, many leaders of those institutions also told their parishioners and readers that handing over the cultural reins to scientists had brought the world to a state of total crisis. Some secular thinkers issued startlingly harsh rhetoric in the postwar years as well. "Under industrialism," one high-profile commentator charged, modern society had become "the waste land": "Who can live without desperation in a social system which represents organized frustration instead of organized fulfillment?" Industrialism, the writer continued, "has inflicted savage wounds on the human sensibility; the cuts and gashes are to be healed only by a conviction of trust and solidarity with other human beings." The communitarian tenor and language of alienation would not have been out of place in the young activists' Port Huron Statement. But these words had been penned in 1948 by Arthur M. Schlesinger Jr., that hated pillar of the liberal establishment, in his anti-ideology tract *The Vital Center*. Such laments were part of the air that the 1960s radicals breathed as they came to maturity.[2]

At the same time, the new dissidents were far less likely than their postwar predecessors to find moral resources in the social and economic patterns of the recent past. Whereas other groups of critics often looked back to the middle-class societies of the nineteenth and early twentieth centuries, these radicals rejected bourgeois norms as well as the capitalist dynamics that supported them. Instead, they sought agents of change and reservoirs of humanistic concern beyond the Western middle classes. For the most part, moreover, their ideal society lay in the future, beyond structures and beliefs that had shaped the world for centuries. When they indicted science, these new critics were less likely to insist that it destroyed familiar moral conventions and more likely to argue that it upheld deeply entrenched and profoundly amoral patterns. Here, the fruits of modern science included not just the familiar enemies of totalitarianism and social engineering but also a vast array of formal and informal mechanisms for the domination of marginalized social groups and the repression of individual subjectivity. To such observers, liberation and morality went hand in hand against the interlocking forces of scientific rationality and social control. Many enlisted religion and the humanities as well, though not necessarily in their traditional Western forms.

Thus, the young radicals amplified minor themes from postwar critiques of scientism and added new strains, even as they continued to sound familiar notes as well. As before, many lamented the divide between religion and the natural sciences and identified the putative value-neutrality of the social sciences as an ideology that legitimated either pervasive moral relativism or social engineering by experts. Yet the 1960s critics were more likely than their predecessors to target societal rationalization or Western rationality as a whole, employing concepts such as alienation, authenticity, and responsibility. They also placed new emphasis on the danger of technics out of control, as well as the capacity of the social sciences—especially psychology—to ensure conformity by equating repressive social norms with mental health. Finally, these critics targeted the natural sciences as well as the social sciences. They decried weapons-related research in both areas and often argued that modern science as a whole provided ideological cover for a totalizing system of social domination. In all of these areas, the new radicals of the 1960s developed and extended arguments about science's cultural impact that continued to resonate, albeit in different forms, among many liberals and conservatives as well.

This was the context in which activist students and professors deemed postwar American society, quite literally, a machine. ("Moloch whose mind is pure machinery," Allen Ginsberg called it in his iconic poem "Howl.") They decried the military-industrial complex, with its enormous investment in death and destruction; the dominance of large, bureaucratic organizations and rationalized practices across all domains of life; a government that failed to provide basic rights to large groups of its citizens; and an increasingly suburban, consumption-driven culture that emphasized conformity and keeping up with the neighbors. They blasted the militantly anti-ideological, realist stance of postwar liberal thinkers, including even religious leaders. And they targeted the behavioral sciences, which were spinning off rational choice theory, operations research, systems theory, and other highly depersonalized approaches to the study—and control—of human behavior.[3]

For many 1960s critics, as for many of their predecessors, these features of a soulless, alienating, inhuman society stemmed from a philosophical outlook, rooted in the sciences, that elevated the material above the moral—or even denied that the latter existed. It was no accident that the adjective "human" carried immense rhetorical weight among these activists and theorists. That term signified an ideal of individual liberation that extended far beyond material comfort into the emotional and spiritual realms. It involved freeing individuals from the grip of "positivistic" thinking as well as conventional habits. When young radicals rejected "faceless and terrible bureaucracies" and called for the replacement of "material values" by "human values," they targeted a public culture that many believed had been nurtured in the cradle of science. Some of these critics focused solely on economics, politics, and international relations. But for others, the United States was a capitalist monolith and an imperial power in part because it was, at a deeper level, a soulless technocracy in which scientific reason and technical expertise crushed out moral impulses and eradicated the possibility of even imagining an alternative to prevailing patterns.[4]

UNDERSTANDING THE MACHINE

Many forces shaped the cultural associations of science for the massive generation of baby boomers that entered college in the 1960s. The most important was the energetic mobilization of science by a particular kind of state, one committed to rolling back communism, managing capitalism, and asserting the universal validity of its principles as it sought to spread them around the globe. The military-industrial complex, which had grown steadily since the late 1940s, was one highly visible product of that nexus of science and state. More often than not in the 1960s, physical scientists and mathematicians worked for the defense industry, either on campus or at an independent research center. At the same time, most of them employed the postwar rhetoric of scientific neutrality and insisted that they bore no responsibility for the applications of their research. "What I'm designing may one day be used to kill millions of people," noted a graduate student during a protest against military-sponsored research at MIT in 1969. "I don't care. That's not my responsibility. I'm given an interesting technological problem and I get enjoyment out of solving it." Many of the social scientists working on defense-related projects at the RAND Corporation and elsewhere adopted a similar stance of detachment.[5]

As natural and social scientists committed themselves to pursuing national goals, the worlds of elite academia and policy formation converged in a manner not seen since the 1930s. During the Kennedy years, the steady stream of scholars from Harvard and other leading universities into the executive branch shaped decisions on both foreign and domestic policy, as well as organizational practices throughout the federal government. Internationally, the Kennedy administration hewed to an emerging framework of modernization theory that aimed to analyze and guide the industrialization process across the newly christened "Third World." Modernization theorists assumed that the various "traditional" societies, including new nations in the decolonizing world as well as older ones, would universally seek the material comfort promised by industrialization. Policymakers, they asserted, could steer that process into democratic channels by providing forms of aid that fostered certain social and cultural preconditions in each developing country. Above all, they should aim at creating centrist, realist, scientific, and secular political cultures, in which citizens avoided the fatal mistake of viewing politics as a means toward personal, spiritual, or ideological ends and recognized that it was simply a matter of practical negotiations over the distribution of material goods and the expert delivery of services such as health care and education. As secretary of defense under both Kennedy and Lyndon Johnson, meanwhile, Robert McNamara not only oversaw the continued expansion of the military-industrial complex but also brought insights from systems theory and operations research into the federal bureaucracy, streamlining organizational structures across the government. At the National Science Foundation and elsewhere, a cadre of nuclear physicists and other scientist-administrators enjoyed considerable leeway over the distribution of federal research funds, leading observers to dub them "the scientific estate," "the new Brahmins," or even "the new priesthood."[6]

American intervention in Vietnam knit the universities even more firmly to the military and defense contractors, even as it sowed the seeds for an eventual split between elite scientists and the White House. Both McNamara and the modernization theorists helped shape the early phases of the conflict in Vietnam. The Kennedy administration scrapped its predecessor's largely religious interpretation of Asian events and viewed the situation in Southeast Asia through the twin lenses of the Cold War and its recent announcement of a "Decade of Development." During 1962–1963, American military forces, heeding the advice of modernization theorists, sought to make proper modern citizens out of Vietnamese peasants by forcibly relocating them from traditional villages to planned urban communities. When that plan backfired and the war escalated, the military-industrial complex swung into action more fully. Alongside the nuclear warheads, the country's equally vast arsenal of conventional munitions soon included devastating new weapons such as napalm and cluster bombs.[7]

On the domestic front, legions of social scientists helped design and imple-ment a significant expansion of the American welfare state under Johnson's Great Society programs. The president's commitment to a "war on poverty" and subsequent development or augmentation of programs such as Medi-care, Medicaid, and Aid to Families with Dependent Children brought rafts of additional experts into the federal bureaucracy. Critics on the left, as well as the right, found much to dislike in these Great Society initiatives. Many radicals deemed the implementation of American welfare programs deeply paternalistic. They decried the operation of stereotypes about race, gender, and class among social workers and bureaucrats, especially when coupled with the assertion of scientific neutrality. Particularly galling was the wide-spread belief that "cultural deprivation" and non-normative family struc-tures explained the persistence of urban poverty. These critics pushed to deepen the liberal establishment's commitment to cultural diversity and apply its emphasis on material security in a more inclusive and egalitarian fashion. Of course, conservatives saw paternalism in the Great Society as well, al-though they advocated individual initiative and moral character and rejected the radicals' call to empower local communities. So, too, did the small but influential group of apostate Democrats later dubbed "neoconservatives." To all of these groups, the Great Society represented a dramatic acceleration of the ongoing substitution of technocracy for democracy.[8]

Meanwhile, a host of mainstream scholars still contended that postwar liberalism was the only form of politics compatible with mental health. Not surprisingly, psychologists figured especially prominently in this discourse. Building on the insights of *The Authoritarian Personality* and similar studies, a growing body of research differentiated "open" and "closed" minds in a manner that mapped neatly onto American liberalism. As had Theodor Adorno and his collaborators, psychologists such as Milton Rokeach de-clared that individuals with open minds embraced religious and racial diver-sity, along with a moderate welfare state and a mixed economy. A range of other social scientists and historians offered similar judgments about the pathological character of both radicalism and conservatism through the 1950s and early 1960s. Widely read books such as the 1955 volume *The New American Right* averred that individual psychological imbalances ex-plained dissent from postwar liberal realism. Although the sociologist Daniel Bell's *The End of Ideology* (1960) analyzed the contemporary scene in a more nuanced fashion, disputants on all sides ignored its contents and em-ployed its title as shorthand for the complacent, dismissive character of much postwar liberalism.[9]

Yet the radical critics of the 1960s could also draw on key resources from the postwar period. The humanistic psychology of Carl Rogers and his allies and Paul Tillich's Christian existentialism were only the most visible of a host of earlier approaches that bequeathed potent understandings of interpersonal relations, ethical obligations, and genuine selfhood to the rising generation. Figures such as Pitirim A. Sorokin, whose rejection of materialism and individualism rang louder than his social conservatism to young radicals, also found receptive audiences. Newer interventions such as *The Broken Image* (1964), an extended analysis of machine metaphors in the social sciences by the humanistic liberal Floyd W. Matson, bridged generational and political differences as the current of criticism swelled. To be sure, the older figures had surprisingly little to say about the questions of political economy that loomed so large in the 1960s. But young activists, no longer feeling the need to look over their shoulders for McCarthyite attacks, often radicalized the prevailing critiques of science's cultural impact by turning them against capitalism as well as bureaucracy and conformity.[10]

There were a few earlier models that the critics of the 1960s could follow in rejecting the managed capitalism of the postwar era as well as its laissez-faire predecessor and linking both to science's pernicious influence. Even in the depths of the McCarthy years, scattered groups of radicals continued to theorize the nature and sources of modern domination. The turmoil of the 1940s led a number of American Marxists to identify the scientific bent their comrades shared with many postwar liberals as the cause, not the cure, of modern ills. Some responded by embracing traditional forms of religious faith, as when the Jewish socialist Will Herberg turned to the writings of Reinhold Niebuhr or the communist leaders Louis Budenz and Bella Dodd converted to Catholicism under Fulton Sheen's watch. For the most part, however, these newly minted theists also moved rightward politically in the 1950s. But a number of more radical figures, including the "renegade Marxist" Dwight Macdonald, the sociologist C. Wright Mills, and the psychoanalytic thinker Erich Fromm, connected scientism squarely to the managed capitalism of their day. Picking up themes from Max Weber as well as Marx, such figures argued that the modern era had witnessed the thoroughgoing rationalization of society as well as knowledge.[11]

Macdonald, far more than Mills or Fromm, embraced traditional understandings of the human person even as he continued to identify as a socialist. After the war, he dismayed many former allies by turning his new journal *politics* into a forum for criticism of scientism's deleterious effects on the radical cause. His own writings, such as "The Root Is Man," called for a new, humanistic form of radicalism: "We must emphasize the emotions, the imagination, the moral feelings, the primacy of the individual human being, must restore the balance that has been broken by the hypertrophy of science in the last two centuries." Macdonald identified all mainstream political thinkers as "Progressives," celebrating humanity's control over nature and viewing history as moving inexorably toward that end. That outlook, said Macdonald, redefined the negative outcomes of science and technology as mere products of human error or malice. More dangerously, it could justify any harm to present-day individuals in the name of a better future. Whereas "the Progressive makes History the center of his ideology," Macdonald summarized, "the Radical puts Man there." He described radicals as "anarchists, conscientious objectors, and renegade Marxists like myself . . . who reject the concept of Progress, who judge things by their present meaning and effect, who think the ability of science to guide us in human affairs has been overrated, and who therefore redress the balance by emphasizing the ethical aspect of politics." Such figures, he continued, preferred to court "technological regression" in "adjusting technology to man" rather than vice versa. Genuine radicals, according to Macdonald, rested their hopes on "those non-historical values (truth, justice, love, etc.) which Marx has made unfashionable among socialists."[12]

Indeed, Macdonald contended that the radicals' "tragic, ethical and nonscientific" outlook "corresponds partly with the old Right attitude." It rested on a dualistic understanding, he explained: There was the world as science viewed it, but also a separate "sphere of human, personal interests." This "traditional sphere of art and morality" rested on nonempirical judgments and portended a very different kind of future. The continued advance of science and technology at all costs, whether in Marxist and Deweyan form, produced the "Bureaucratic Collectivism" of Hitler and Stalin, Macdonald explained. But judgments of value could anchor a "libertarian socialism" that judged historical change in ethical rather than technical terms. The atomic bomb, Macdonald wrote hopefully, might convince Westerners "that they don't want electric iceboxes if the industrial system required to produce it also produces World War III, or that they would prefer fewer and worse or even no automobiles if the price for more or better is the regimentation of people." One way or another, Macdonald insisted, something needed to change or the human enterprise would come to a crashing halt. Like so many other postwar critics, Macdonald feared that Western scholars and politicians had left out the one thing that truly mattered: humanity. In his laments for the lost individual, he sounded more like the cultural critic Paul Goodman, whose anarchist sensibility and biting commentary on "the dominance of science and applied science" and the "superstition of scientism" also resonated with 1960s radicals, than the Marxists with whom Macdonald retained filial ties.[13]

Whereas Macdonald focused primarily on conceptions of the human person, Mills emphasized institutions and structures in writings such as *The*

Sociological Imagination (1959), where he flayed behavioral science approaches. As we have seen, Mills was not only a precursor to 1960s radicals but also a successor to the motley company of Sorokin, Albert H. Hobbs, Joseph Wood Krutch, and others who dissected the flaws of value-neutral social science in the 1950s. To be sure, he worked independently of their influence, and each of them stood well to his right politically. Yet Mills shared their intense dislike of a style of investigation centered on highly specialized research projects that aimed at empirical certainty about carefully delineated features of present-day society. These critics of value-neutral approaches, despite their many differences, agreed that social scientists should instead sensitize the public to broad, structural changes in modern societies that threatened both freedom and morality.[14]

Mills, however, came at the problem of science's cultural influence from a new angle and took the fight to the heart of the social science establishment. He worked alongside Robert K. Merton, Paul Lazarsfeld, and other architects of the postwar behavioral sciences at Columbia University. Mills's detailed studies of class and power in modern America also gave his methodological criticism far more credibility among social scientists than did Hobbs's cranky moralizing about permissive parenting and sexual promiscuity or Sorokin's calls for Christian love and "creative altruism" in the face of a social apocalypse. But Mills had a nightmare vision of his own, one that was shared by a growing number of Americans after 1960: the fearful prospect that the critical theorists of the Frankfurt School termed a "totally administered society." Indeed, Mills sounded many of the same notes as Herbert Marcuse and other Frankfurt School critics, though without the Continental terminology, mythological discursions, antipopulist aesthetics, and bleak pessimism that made their writings such difficult nettles to grasp. Like the Frankfurt theorists, Mills drew out the implications of Weber's assertion that modern societies were undergoing sweeping, inexorable forms of rationalization that could leave them with enormously sophisticated means but no moral ends.[15]

Mills described scientism as merely one symptom of a deeper process of societal rationalization that had reshaped virtually all forms of thought and action in the modern world. In the social sciences, he identified scientism ("abstracted empiricism") as one of the two prevailing ways—along with the grand, abstruse theorizing of the Harvard sociologist Talcott Parsons and his ilk—that social scientists fit themselves into "rationally organized social arrangements" and shirked the true vocation of human reason: namely, creating a society of ethically engaged and politically efficacious citizens. Mills wanted social scientists to help replace the "Cheerful Robot" of modern, rationalized societies with "The Renaissance Man": "the self-educating, self-cultivating man and woman; in short, the free and rational individual." To do so, he said, scholars would need to join a wider cultural impulse that was already pushing beyond modern forms toward a "post-modern" outlook. Such an outlook would center on the sociological imagination of Mills's title: a mode of rationality that grasped large-scale changes in social structure and connected personal experiences to those broader forces.[16]

Fromm, who combined Mills's sociological orientation with resources from Freudian psychiatry and humanistic psychology, offered a slightly different but equally terrifying vision of an administered society that reflected the early Marx more strongly than Weber. Although Fromm had broken with Max Horkheimer and Theodor Adorno in the 1930s, he had been an important figure in the early development of the Frankfurt School and shared its other members' distaste for positivistic scholarship and technocratic politics. Unlike theirs, however, his books were highly readable and sold in the millions from the late 1940s through the 1960s. In *The Sane Society* (1955), Fromm offered the ubiquitous observation that moderns needed "a human renaissance" far more than "airplanes and televisions." Unlike the vast majority of postwar critics, however, he traced the crisis back to capitalism. That system, Fromm argued, had now begun to exert its routinizing and depersonalizing force through science and technology as well as the logic of the market. "Man has been thrown out from any definite place whence he can overlook and manage his life and the life of society," Fromm wrote. "He is driven faster and faster by the forces which originally were created for him," through a mode of capitalist alienation that Fromm called "abstractification" and saw behind modern science as well as political economy. The result was a total loss of autonomy, of even minimal control over one's fate, in "a system which has no purpose and goal transcending it, and which makes man its appendix." In the 1960s, Fromm's vision, like that of Mills, would find a receptive audience among disaffected youth and critical scholars alike.[17]

Like Alfred Kroeber and other methodological dissidents who called themselves humanists, many of these radical theorists chafed at the term "science." They called for sustained, empirical inquiry but wanted nothing to do with technocratic nightmares such as *Walden Two*—let alone *Brave New World*. Mills, for example, reluctantly used "social sciences" for the sake of readability but preferred the broader phrase "social studies." A few years later, the latter term became the title of a Harvard undergraduate program created by critics of the behavioral science model. Even as many postwar social scientists rushed to associate themselves with the natural sciences, then, a surprising number of their colleagues kept their distance.

Many of these dissenters believed that scientism had badly weakened or even eliminated modern citizens' capacity for moral reasoning. Not all of them looked to moral absolutes; some argued that values were as personal as tastes. But they insisted that moral commitments were substantive and important—and that scientific perspectives ruled them out of court, with dire effects. These methodological dissidents, whether radical, liberal, or conservative, remained important interlocutors for the behavioralist mainstream and also exerted a significant influence among the general population. As card-carrying members of the social-scientific guild, in fact, their claims about the views and ambitions of their disciplinary colleagues held special weight in the clashes of the 1960s.[18]

RESISTING THE MACHINE

Those clashes centered on the universities, which expanded rapidly in the 1960s and became ever more closely tied to government. In the highest tiers of academia, weapons research and policy advising reflected a broader conception of the research university as what radical critics—and many humanistic liberals and conservatives—dismissively called a "service station" for the surrounding society. University of California head Clark Kerr famously portrayed the modern "multiversity" as a sprawling, decentralized collection of units, each addressing a specific area of social need by producing instrumental knowledge to guide economic activity and public policy. Writing in 1963, Kerr acknowledged that the multiversity shortchanged undergraduate teaching, and indeed predicted that it would sow dissatisfaction among students. Undergraduate education was hardly a minor issue, as became apparent when the Free Speech Movement erupted at Berkeley the following year and Kerr found himself the target of students' ire. But for Kerr and other liberals in the early 1960s, the postwar university simply could not justify its existence unless it addressed pressing, practical concerns through specialized research while training an "elite of talent" to administer social institutions. Kerr thus identified the multiversity as an irreplaceable component of a technologically sophisticated society while also calling it a fundamentally apolitical institution with no partisan agenda.[19]

A labor economist by training, Kerr adopted an emerging "human capital" approach that outraged many radical critics by equating education with industrial production. The human capital framework defined knowledge and students as the universities' twin products, in a quite literal sense. Like many university presidents and other liberal commentators, in fact, ·

Kerr believed a "knowledge economy" was emerging in which technical expertise replaced raw materials as the key economic resource. The key driver of the knowledge economy, in this view, was an ever-expanding fund of intellectual capital—including scientific and engineering knowledge—rather than financial capital. As usual, Kerr identified significant drawbacks to the resulting conception of the university as a hub of the knowledge economy. He noted that the "great machine" of modern scholarship "turns out its countless new pieces of knowledge . . . with little thought for their consequences" and offered a modest call for social engagement, urging humanists and social scientists to "add wisdom to truth" and natural scientists to study population growth, the nuclear threat, and environmental degradation. Yet Kerr's recommendations appeared in the middle of a passage announcing the inevitability of changes in higher education and the broader economy: "The process cannot be stopped. The results cannot be foreseen. It remains to adapt." In the knowledge economy, Kerr wrote, the "university and segments of industry are becoming more alike. The two worlds are merging physically and psychologically."[20]

Among the massive new cohort entering college in the 1960s, many students recoiled at such visions and the social norms and institutional patterns they justified. Rates of college enrollment rose significantly in the 1960s, and the baby boom meant that the increase was far greater in absolute terms; in 1970, there were more than twice as many college students in the United States as there had been in 1960. In their postwar childhoods, these students had heard the constant, celebratory refrains about their country's uniquely free and moral character. They were also sensitive to deviations from the stated ideals, however, and they tended to hold existing institutions up to those ideals rather than emphasizing the superiority of the postwar United States to other historical societies. Racism, poverty, war, and a pervasive sense of meaninglessness: none of these comported with the dominant portrait of a United States that now had the technical tools to solve age-old human problems. Nor did these students share their liberal elders' fear of a resurgent McCarthyite conservatism—perhaps in part because those elders had described conservatives as a backward-looking, rapidly vanishing fringe group with views unmoored from reality, let alone policy.[21]

By the middle of the 1960s, a growing number of students rejected the bureaucratic, professional roles for which the knowledge economy model slated them. Those who turned leftward rather than rightward—the conservative movement also took root on American campuses in the 1960s—attacked the nexus of university, industry, and government that Kerr celebrated. Experiencing their society as impersonal, inauthentic, and immoral, they refused their appointed roles as managerial elites for a knowledge economy. These students crafted new identities and sought new vocations outside "the system" or "the establishment." Some radical activists identified themselves as the core of a new working class of highly educated but deeply oppressed white-collar workers who were being forcibly trained to run the soul-destroying, technocratic machine that was modern America. Inverting plaudits for the knowledge economy, Mario Savio and other Free Speech Movement leaders, as well as Students for a Democratic Society (SDS) head Greg Calvert, joined the influential cultural critic Goodman in calling undergraduates at top universities "the major exploited class" in the postwar United States.[22]

Radical students and their faculty counterparts decried the "grotesquely distorted" character of higher education in their day. They urged a comprehensive return to considerations of value in the universities and disciplines. "It is ours to demand meaning; we must insist upon meaning!" declared the leaders of the Free Speech Movement in 1965. Student activists sought to render their university "a loving community" by applying "the relentless hammer-blows of conscience" to the "ramparts of rationalization which our society's conditioning ha[s] erected about our professors' souls." Marked as the future technicians, the students hoped they could take the whole system down from within. They sought to begin by recovering the language of values and using it to tear down the presumptions of objectivity and universality that surrounded American liberalism and the associated forms of social science in the early 1960s.[23]

Of course, many postwar liberals, even at the highest ranks of the establishment, also emphasized the normative dimension of politics. Kerr, for example, deplored Americans' "embarrassment about the expression of moral seriousness" and urged them to embrace "love, courage, self-development, commitment, meaning," and similar concepts. Yet Kerr also used the vocabulary of psychological health to rule radical criticism out of bounds, along with traditional Christianity and other worldviews that defined morality for so many Americans. Young radicals (and conservatives) saw hypocrisy in liberal scholars' tendency to dismiss ideology while advocating extensive programs of social engineering on behalf of New Frontier and Great Society liberalism.[24]

Not surprisingly, the perspectives of 1960s radicals often reflected the welter of complaints about science's cultural impact that had circulated since World War II. What had squeezed values and meanings out of the universities, and American life more generally? Many 1960s critics, like their 1950s predecessors, concluded that a scientific mentality had remade the public conversation and left its stamp on social institutions of all kinds.

Behind the "idolatrous worship of things" that Tom Hayden excoriated in the Port Huron Statement such radicals saw the desiccated liberalism of the postwar era, with its insistence on the purely material goals of politics. And behind that framework, in turn, many saw the cold hand of science.[25]

Such an analysis informed many of the core tenets that distinguished New Leftists from earlier radicals. For example, the New Left turned away from the labor movement as an agent of historical change because they believed it had been co-opted, both institutionally and culturally, into a bureaucratic establishment that many critics traced to science's elimination of meaning and value. These figures looked instead to groups that had escaped the grip of technocratic rationality: the impoverished underclass, racialized minorities, "Third World" peoples—and, they hoped, themselves. Similarly, perceptions of science's cultural meanings both shaped and reflected the New Left's emphasis on the internal, subjective dimensions of oppression rather than objective, external conditions and forces. The young radicals refused to follow liberals in separating public from private, the social from the personal, basic security from emotional fulfillment—in short, political change from cultural and even spiritual change. Many knew from personal experience that material comfort offered no protection against inner emptiness and misery, and they believed that social change would require inner transformation on the part of individuals as well as collective organizing and other social practices. And many of them identified the stifling liberal mindset they needed to overthrow with modern science.[26]

Many leading radicals, in both the student and faculty ranks, combined this analysis of science's impact with a firm commitment to transcendent, universal moral principles that often reflected their religious upbringing (or the values of secular Jewish parents). In the New Left, inheritances from Catholicism (Hayden, Savio) operated alongside expressions of Judaism, both religious and secular, and strands of Protestantism that variously combined a Social Gospel emphasis on creating a beloved commonwealth with the Christian realists' recognition of the ubiquity of power in human relations, a personalist belief in the infinite value of the human person, and an existentialist commitment to forms of immediate, total engagement and transformation that broke down the barriers between I and thou, self and other. Young activists spoke of becoming human, finding their authentic selves, and establishing morally responsible relations to one another and to the world—at significant personal risk, if necessary. Psychological and spiritual languages intertwined, each addressing potent questions of mutual obligation and the meanings of personal authenticity and responsibility. Radical students and professors alike believed that transformative change was possible, that it lay within their collective reach, and that they would bear a grave responsibility if they shirked the task.

But where, exactly, did the problem lie? Both radical and left-liberal critics within the academic disciplines often framed it in terms of professional responsibility and personal choice. These dissidents argued that their establishment counterparts had either consciously or unconsciously traded their integrity for status and comfort by supporting a corrupt political system. In their view, the many natural and social scientists enmeshed in the military-industrial complex had actively chosen to help build the war machine, while other social scientists had voluntarily devoted themselves to legitimating policies that harmed Americans in the short or long run. A range of Marxists, Maoists, and left-liberals challenged the presumption of scholarly neutrality, at least regarding which research problems to tackle, and urged their colleagues to devote themselves to the needs of the people, not the wealthy and the powerful. They argued that science and technology had the potential—and the duty—to serve everyone, if control over them could be wrested from elites. The radical disciplinary caucuses and interdisciplinary programs (especially Black studies, women's studies, and ethnic studies) that emerged in the late 1960s typically reflected this voluntaristic emphasis, as activist scholars urged colleagues to dedicate their expertise to the common good and abandon the pretense that their choice of research problems had no moral implications. Rejecting the rhetoric of value-neutrality, they aimed to consciously align their work with the needs of the people.[27]

Many of these activists insisted not only that value commitments shaped research directions but also that scholars could discern moral truths, achieving a robust consensus in what might appear to be the purely subjective realm of value judgments. In the Port Huron Statement, for example, Hayden explained that empirical studies implicitly took their form from "orienting theories" and "basic principles," especially "conceptions of human beings, human relationships, and social systems." By denying the relevance of such concerns, modern views of knowledge impoverished public life. Whereas earlier progressives possessed "vision without program," Hayden wrote, "our own generation is plagued by program without vision." He added, in the face of what he saw as a prevailing moral relativism, that value judgments were hardly "beyond discussion and tentative determination," despite the lack of pat "formulas" or "closed theories" in that realm. Older critics, such as the historians Eugene Genovese and Christopher Lasch, likewise sought "objective ethical standards" to guide both technical research and social practice. These figures echoed the lament of the political theorists John H. Schaar and Sheldon Wolin that "values are no longer shareable as knowledge" in the modern age. But in truly free processes of intersubjective reasoning, they believed, moral universals would shine through.[28]

1960s radicals often targeted what they considered the abuse of genuine science through its application to immoral ends. Much antiwar protest, for example, focused on the universities' entanglements with the military-industrial complex and the war in Vietnam. Dow Chemical, which manufactured napalm, came in for sustained criticism. MIT and other universities saw huge protests against contract research for the military, eventually leading MIT to shut down all classified work in 1969. Campus activists often emphasized the personal motives and commitments of individual researchers. "The scientists who are called upon to construct the ABM [anti-ballistic missile] need not do so; the social scientists who are invited to preside over the management of some helpless society—perhaps our own—can refuse," MIT's Noam Chomsky declared. Yet the prevailing conceptions of knowledge ruled out such moral choices. The "social scientists pride themselves on being value-free," lamented the physics graduate student Joel Feigenbaum at MIT, while "natural scientists disclaim responsibility for the uses to which their work is put."[29]

Still, the members of organizations such as the Union of Concerned Scientists (UCS) and Science for the People (SftP) emphasized the immense potential value of science even as they decried its typical applications. "Misuse of scientific and technical knowledge presents a major threat to the existence of mankind," ran a UCS statement of 1969. That group aimed to redirect scientific research "away from the present emphasis on military technology toward the solution of pressing environmental and social problems." SftP, though more radical overall than the UCS, took a similar approach as it challenged the ideal of value-neutrality. SftP activists insisted that "scientific activity in a technological society is not, and cannot be, politically neutral or value-free," given the intimate connection between research advances and policy options. For the most part, however, these critics did believe that neutrality was possible at the level of theory formation, especially in the natural sciences. They mainly focused on the choice of research problems and applications. For UCS and SftP members, it was the abuse of scientific knowledge by a corrupt society, not the methods or worldview of science itself, that accounted for napalm and other evils. Ironically, that position aligned radical science groups with many elite scientists and science advisors, who increasingly turned against Johnson's Vietnam policy by the late 1960s.[30]

While the UCS and SftP deplored natural scientists who worked with the establishment, the radical linguist Chomsky and a host of other critics charged social scientists with the same sin. "It is the responsibility of intellectuals to speak the truth and to expose lies," Chomsky wrote forthrightly in his 1967 article "The Responsibility of Intellectuals." He pursued an Enlightenment-inspired strategy of removing the public's ideological blinders, tearing away the veil to reveal the naked workings of power that lay behind the prevailing conceptions of knowledge. Radical scholars in this vein assumed that the truth would inevitably resonate in the public mind, if only it could be extricated from the fog of conventional deceptions. Most practitioners of this politics of truth also insisted that there were powerful moral truths as well as empirical ones, and that the social scientists' faulty pretense of value-neutrality represented one of the establishment's primary ideological weapons.[31]

Chomsky thus identified "the desperate attempt of the social and behavioral sciences to imitate the surface features" of the natural sciences as a major source of the public's reluctance to recognize America's imperialist ambitions. Scholars, he contended, must go beyond "technical" and "tactical" problems to debate matters of "principle," including "moral issues and human rights," and even "the traditional problems of man and society, concerning which 'social and behavioral science' has nothing to offer beyond trivialities." Chomsky and like-minded critics believed there were clear moral principles that applied in all domains, from foreign policy on down. Unlike many of their academic successors, the radicals of the 1960s often assumed that moral commitments were powerful forces in history (many also considered them objectively true), that moral tenets should shape all social institutions and habits, and that meaningful social change depended on individuals hewing to moral principles and altering their culture and institutions accordingly.[32]

CONFORMITY, RATIONALITY, AND MODERNITY

Calls for morally committed scholarship and criticism of mainstream scholars' political complicity meshed easily with a nascent tendency among radical activists to identify the root of contemporary ills as a hegemonic culture rather than an institutional configuration or a dominant class. This cultural turn in 1960s radicalism could be seen in the claims of many activists that students had all the power they needed to take full control of their lives; they had actually forged their own chains by passively embracing spoon-fed beliefs. Some analysts of the ideological functions of modern knowledge, such as Chomsky, argued that Americans simply needed to rip off the blinders of scientism in order to see the evils of domination by the capitalist class. In this view, scientism served to legitimate modern capitalism, forestalling movements for economic and political change by conceptually eliminating the possibility of morally grounded criticism. Other critics viewed the class structure differently, identifying scientism as the characteristic ideology of a "new class" of technicians and bureaucrats that had supplanted the old bourgeoisie as the true ruling power in advanced capitalist societies. But many radicals credited science with more than a merely ideological role. Instead, they viewed modern social institutions as direct outgrowths of scientific thinking or technical rationality.[33]

At MIT, for example, Feigenbaum moved immediately from his lament about the stance of scholarly neutrality to the claim that contemporary ills reflected the substantive content of modern science. Across the universities, he asserted, one found "a prevalent feeling of materialistic determinism, which tends to reduce people to automata" and to foster a thoroughgoing "moral relativism." In the antiwar movement and elsewhere, this sensibility aligned radical critics with humanistic and religious liberals such as Martin Luther King Jr., who spoke out against Vietnam policy by charging that "the within of our lives," the "realm of spiritual ends expressed in our literature, morals, and religion," had "become absorbed in the without," the "complex of devices, techniques, mechanisms, and instrumentalities by means of which we live." For many younger critics, the soul-crushing weight of scientific thinking helped to explain not only American policy in Vietnam but also the moral bankruptcy of the cultural norms and social structures that American leaders sought to impose on the Vietnamese.[34]

Given the prevalence of such arguments about the cultural influence of science and technical rationality in the postwar United States, it is no surprise that they continued to reverberate in the 1960s. Even older Marxists such as Paul A. Baran and Paul M. Sweezy sounded such themes. In their iconic *Monopoly Capital* (1966), which sought to adapt Marxist theory to an age of massive, sector-spanning corporations, Baran and Sweezy laid out their structural argument—that an economy centered on monopolistic firms would struggle to find investment outlets for its accumulated capital—but also rooted the historical development of science and technology in a rationalization process that had undermined "society's psychic police force" because it "fatally undermined faith in many of the basic moral principles guiding men's conduct" and offered no substitute for traditional values. The pair argued that the forms of behavior made possible by the decline of moral strictures had lost their "emotional content" and thus their "meaning and power to gratify."[35]

For some critics of the 1960s, however, the loss of morality was the main story. A deep tension regarding science's relationship to contemporary forms of oppression ran through the radical movements of the era. Many activists and theorists argued that science served to legitimate a system of structural domination by a privileged class. Others, however, transposed that claim into a different register, sidelining structural dynamics and highlighting instead the forms of philosophical domination that generations of American critics had discerned since the 1920s. The arguments about science's cultural influence that these 1960s radicals inherited often inclined them to see scientific thinking as the historical root of modern social structures and not simply an ideological cloak for them. Echoing the familiar laments about secularization, relativism, social engineering, and the modern crisis, they assumed that twentieth-century citizens suffered mainly under the yoke of a misguided philosophy and its academic and political champions, not an economic elite. And like their predecessors, they sometimes traced that philosophy back centuries rather than decades. That approach offered an explanation for the intense alienation felt by many well-off undergraduates and facilitated the mixture of post-Marxist and libertarian sensibilities that defined much of the era's radicalism. It also suggested that the characteristic forms of modern oppression could be thrown off voluntarily, through freely willed acts of intellectual and cultural resistance. In terms of institutions, the real struggle centered on the means of socialization, not the means of production.

This tension between charges of ideological obfuscation and philosophical domination could be seen in the era's innumerable blasts against "positivistic" social science. Radicals in the 1960s issued innumerable challenges to the detached, technocratic, and implicitly totalitarian approach of the postwar social sciences. They chafed at the scientific tenor of Marxist thought as well, leading some to reject the common view of Marxism as a deterministic science of history and reframe it in more humanistic terms. Such critics turned toward the tradition of "Western Marxism," which developed in opposition to Soviet models between the wars and emphasized the independent role of cultural forces in determining the structure of class relations. Above all, young dissidents took inspiration from the Frankfurt School of critical theorists, who combined Marx's early insights with themes from many other Continental thinkers, including Weber's portrait of a world dominated by a soulless, instrumental rationality. In their 1944 opus *Dialectic of Enlightenment*, Frankfurt icons Horkheimer and Adorno had expanded the school's long-standing criticism of putatively neutral social science into a nightmare vision of the modern age.[36]

The 1960s generation encountered these ideas mainly through Marcuse's writings and personal example. Building on Heidegger's critique of the technological orientation of modernity, Marcuse connected the impersonal, deadening character of contemporary life to views of science in a manner that resonated powerfully with American activists. His *One-Dimensional Man* (1964), which introduced many young radicals to the theories of the Frankfurt School, argued that the concept of neutral knowledge reflected a culture-wide elevation of means over ends that prevented the inhabitants of the modern world from recognizing their complicity in their own oppression. The scientific philosophies of behaviorism and operationalism and the modes of technological manipulation they made possible set the terms of "discourse and action, intellectual and material culture," in the modern world. "Technological rationality has become political rationality," Marcuse declared, with the result that the relentless system of social controls "appear to be the very embodiment of Reason for the benefit of all social groups and interests." Technical rationality produced essentially totalitarian results by establishing a "comfortable, smooth, reasonable, democratic unfreedom."[37]

Marcuse's emphasis on "society's domination over the individual" rather than specific class dynamics, coupled with his emphasis on personal (including sexual) liberation from collective repression, found a receptive audience in the campus left. His writings reinforced the cultural turn among young radicals, many of whom located modern domination in various forms of cultural expression rather than tracing it to class dynamics per se. In the academic disciplines, for example, Marcuse's work suggested that dissidents could advance the cause of liberation by recovering a frankly normative conception of knowledge and applying it to the cause of human freedom. Marcuse himself hardly waxed optimistic about the prospect of such a renewal, given the entrenchment of the rationalization process that had spawned positivism. *One-Dimensional Man* countered the hopeful, Deweyan tones of the Port Huron Statement with a stark, apocalyptic image of the modern world as a bureaucratic monolith that featured only the tiniest cracks for dissent. Still, it gave critical students, activists, and scholars a concrete task that fit well with the prevailing understandings of science's cultural impact: namely, tearing away the façade of putatively neutral scholarship and reasserting the primacy of human values. Marcuse himself came to see the student movement as a key expression of the form of resistance he had dubbed "the Great Refusal." He eagerly aided young activists and became an important presence in the protest culture of the 1960s, whose leaders often shared his bleak vision of "an advanced society which makes scientific and technical progress into an instrument of domination."[38]

Other combinations of Marxist themes with humanistic sensibilities also flourished in the 1960s. The radical historians Genovese and Lasch decried the "vulgar instrumentalism underlying bourgeois ideology and practice" and urged scholars to take on this "instrumental conception of culture" wherever they found it. In the universities, they insisted, the "myth of scholarly neutrality, which has now spread from the 'hard' to the 'soft' sciences, must be challenged at every opportunity" and the "ethical and philosophical concerns" characteristic of the arts and humanities restored to their properly central place in scientific and technical pursuits. Only when the prevailing "technological anti-culture" gave way to a "new cultural synthesis" would the workers be able to throw off the shackles of bourgeois rule, with the help of working-class intellectuals who would raise up the masses by challenging the foundational philosophical tenets of their society. Genovese and Lasch thus connected the class struggle to a traditional model of liberal education, equating the cause of radicalism with the "restoration of academic standards" and a return to the humanist's "quest for meaning, order, and intellectual synthesis."[39]

A remarkably broad array of radicals argued in the 1960s that an instrumental, technical orientation associated with modern science anchored the characteristically modern system of domination. Yet a range of important questions remained open. Was this scientism a mere tool of middle-class domination or a ubiquitous, faceless force that also—perhaps especially—controlled middle-class individuals themselves? And what concrete forms did this scientism take? How, exactly, did it do its work of obfuscation and pacification? Which institutions and practices pointed beyond a rationalized, technocratic world, and which were complicit in it? What kinds of tactics might be effective in challenging the cultural monolith? Which social groups would need to undergo an inner revolution in order to foster the needed cultural and social change? Was that inner revolution already in motion or just a distant vision? What sort of culture would emerge on the other side? A multitude of answers to these questions coexisted with the general claim that scientism and instrumental rationality anchored social repression and should be overthrown.

Critics sometimes identified specific institutions or practices as the breeding grounds of conformity. One account of how a repressive culture retained its grip on individuals accompanied the burgeoning antipsychiatry movement. A libertarian tendency that could lean in various directions politically, antipsychiatry echoed the sporadic complaint of dissident practitioners over the years that the surrounding society, not individuals, needed adjusting. Antipsychiatry cut deeper than jabs at the palliative use of psychotropic drugs such as Valium, which the Rolling Stones famously satirized in the 1966 song "Mother's Little Helper." Antipsychiatry advocates such as Thomas Szasz and the Scottish critic R. D. Laing challenged the cognitive and cultural authority of the entire psychiatric enterprise. Szasz, a Hungarian-born psychoanalyst, contended that the "biopsychological world view" of mainstream psychiatry, inherited from the nineteenth century by way of Freud, obscured the centrality of ethical choices and moral freedom in human behavior. By reinterpreting voluntary action as the result of biologically based conditions that paralleled physical maladies, psychiatrists were "obscuring and disguising moral and political conflicts as mere personal problems." Szasz called psychiatry a "pseudo science" rooted in "scientism," which he defined as the habit of deferring to "science as an institutional force, akin to organized theology in past ages." Scientism had authorized a "therapeutic state" that based its laws on the "morally judgmental and socially manipulative" expertise of psychiatrists and claimed an absolute right to institutionalize citizens it dubbed mentally ill, just as earlier states imprisoned religious heretics. Szasz's critique intersected at key points with less totalizing rejections of psychiatric authority by feminists and other critics. By the 1970s, films such as *One Flew over the Cuckoo's Nest* (1975) registered—and reinforced—a growing popular suspicion of psychiatry.[40]

Radical pedagogues, meanwhile, described the mainstream schools as incubators of conformity. Iconic works from abroad, especially Paulo Freire's *Pedagogy of the Oppressed* (1970) and Ivan Illich's *Deschooling Society* (1971), reflected themes that also circulated among American advocates of "free schools." Illich's volume, published as volume 44 of Ruth Nanda Anshen's *World Perspectives* series, described contemporary schooling as a means of ensuring blind obedience to scientifically informed bureaucracies. He championed "the poetic surprise of the unplanned" against a society whose every single feature had been "scientifically developed, engineered, planned, and sold to someone," producing a "pandemic inflation of dysfunctions." From the vantage point of the dawning 1970s, Illich discerned a broad-based cultural reaction against the "naive reliance on magical technologies" and belief in "a scientific millennium" that had long anchored modern societies. "Only ten years ago conventional wisdom anticipated a better life based on an increase in scientific discovery," Illich wrote. "Now scientists frighten children." Yet that widespread frustration with science's social applications had not—and perhaps could not—touch an educational system that inculcated mindless rule-following and bureaucratic rationality. Illich thus advocated a comprehensive deinstitutionalization of education.[41]

Suspicion of such institutions also permeated the counterculture of the late 1960s, whose members sought to detach themselves from modern society and create new spaces in which a more natural, genuinely human way of life could flourish. Embracing authenticity and liberation from the rigid social constraints of an empty, artificial society, they often targeted "Western rationality" in particular: a pinched, repressive form of consciousness that they saw behind mainstream forms of Christianity and Judaism as well as science and technology. From this perspective, even the radical activists in SDS and other organizations appeared hopelessly complicit in the technocratic, managerial outlook that defined the modern West. Through a variety of alternative spiritual practices, as well as the use of psychedelic drugs, members of the counterculture worked to undo centuries or even millennia of faulty cultural programming. They aimed to escape the narrow blinkers of "the Western version of consciousness" and recapture the deeper, holistic insights cultivated by all other peoples and cultures. Breaking down "the walls fashioned by Western science and religion," Greenpeace cofounder Robert Hunter explained, would reveal that its absurd constraints, and not the hippies and dropouts, represented the truly "unnatural and freaky" force in the world. Whereas the political activists of the era tended to focus on twentieth-century developments in politics and the sciences, countercultural dissidents often had much older and deeper cultural tendencies in mind when they described modern society as a vast, repressive machine fueled by technical rationality. "The story of the mind exiled from Nature," Hunter asserted, "is the story of Western Man."[42]

Other sympathetic elders likewise identified the counterculture as a solvent to the centuries-old sway of scientific rationalism and argued that it portended an entirely new stage of human consciousness—perhaps what the ecologically minded law professor Charles A. Reich famously dubbed "Consciousness III." At its core, Reich explained, this emerging mode of consciousness embodied "freedom from the domination of technology." This did not mean freedom from technology itself, he specified; Consciousness III was hardly "anti-technological." It simply believed that machines, rather than "run man," should "do the bidding of man, of man who knows and respects his own nature and the natural order." At the same time, Reich suggested that putting machines in their proper place would involve freeing human consciousness from the machinelike tendencies that had come to dominate it, in part due to the influence of scientific reasoning. Consciousness III, he explained, privileged "'non-linear,' spontaneous, disconnected" forms of communication over "logic, rationality, analysis." In fact, Reich identified the scientific stance of detachment as "an ideal cloak for the personal human will," such that those who claimed the mantle of science could manipulate their fellow human beings as they pleased. Rather than using technology to "further rationalize society" by facilitating such manipulation, Reich contended, Consciousness III would use machines to ensure material comfort and allow people to spend their time pursuing "aesthetic and spiritual" goals.[43]

The historian Theodore Roszak, in the 1969 book that introduced the term "counterculture," indicted science even more broadly. Behind the prevailing "technocracy," hated by activists and hippies alike, Roszak saw the unquestioned authority of experts. And that authority, in turn, reflected the ideological keynote of modern science: a "myth of objective consciousness" that deemed human subjectivity antithetical to the discovery of genuine knowledge. Citing the likes of Laing, Lewis Mumford, Abraham Maslow, and the philosopher of science Thomas Kuhn, Roszak argued that modern life reflected "the psychic style which follows from an intensive cultivation of objective consciousness": "Objectivity as a state of being fills the very air we breathe in a scientific culture; it grips us subliminally in all we say, feel, and do. The mentality of the ideal scientist becomes the very soul of the society." Such a scientific culture waged "open warfare upon joy," he elaborated, leading people to become like computers as they worked to "degrade, disenchant, level down" all outcroppings of their own subjectivity. But in the counterculture, Roszak saw the possibility of a comprehensive turn from theoretical abstractions to immediate experience and from knowledge to ethics, "subordinating the question 'how shall we know?' to the more existentially vital question 'how shall we live?'" (Ironically, he noted, Dr. Spock's permissive recommendations to parents had opened the door to technocracy's destruction by freeing postwar youth from the traditional strictures of adulthood.) He hoped the rising generation would replace science's "egocentric and cerebral mode of consciousness" with a culture based on "the non-intellective capacities of the personality." The counterculture, Roszak believed, might finally throw off the deadening hand of modern science.[44]

In truth, few members of the counterculture rejected either science or technology whole cloth. As Reich's analysis suggested, they embraced alternative, oppositional versions that aimed to replace an artificial or mechanical orientation with human values. Above all, humanistic psychology informed the counterculture, as it did feminism and many other movements of the 1960s and 1970s. The work of Rogers and his counterparts Maslow and Rollo May had already fed into Esalen, which took shape at the nexus of Western and Eastern faiths in the late 1950s. Esalen leaders pioneered many iconic countercultural practices, including yoga, meditation, encounter groups, and the use of psychedelic drugs, as well as soaking nude in hot tubs. Humanistic psychology also made its way into the counterculture more directly. Maslow, though hardly radical in his politics, found himself besieged by militants and hippies—including his former student Abbie

Hoffman—who were drawn to his concepts of "self-actualization" and "peak experiences." As the 1960s progressed, another alternative, oppositional science, ecology, also proved highly influential among countercultural critics young and old. Psychedelics provided a further point of contact to heterodox forms of science, as did midwifery and other feminist approaches to medicine.[45]

Nor did the counterculture eschew technology, even as it decried the massive systems and instruments of the modern age. To be sure, technology vied with bureaucracy, managerial capitalism, and positivistic social science for the mantle of chief villain in the radical thought of the 1960s. Politicos and hippies alike welcomed Marcuse's Heideggerian blasts at technologism, along with those of humanistic elders such as Mumford and the French philosopher Jacques Ellul. During the 1960s, Mumford joined Ellul in arguing that modern history took its shape from the emergence of large-scale technological systems and the distinctive forms of unfreedom they created. Technology seemed to have a life of its own in their writings, pushing inexorably toward greater standardization and larger aggregations of power. Ellul, for example, had declared that the machine "pursues its own course," seeking "the elimination of all human variability and elasticity." Yet few critics in the 1960s believed that this process exhausted the possible meanings or uses of technical know-how. Although critics charged that the hippies would return humanity to the Stone Age, small, human-scale devices and ecologically sound techniques proliferated in countercultural spaces. After 1968, Stewart Brand's *Whole Earth Catalog* served as a clearinghouse for such grassroots innovations. Selling a wide range of seeds, tools, and books by the likes of the maverick architect Buckminster Fuller, the *Whole Earth Catalog* promoted a countercultural vision of technological development based on cooperative living and the self-realization of creative, liberated individuals. Indeed, Brand and his countercultural followers latched onto the first personal computers as potent tools of self-empowerment and contributed to the famously libertarian sensibility of Silicon Valley.[46]

In the universities, the countercultural phenomenon that historians have labeled "groovy science" shaped developments in physics as well. A San Francisco–area subculture that eventually took shape as the Fundamental Fysiks Group brought theoretical insights from quantum mechanics, and even broad metaphysical speculation, back into a physics establishment that had become largely instrumental and results-oriented since the Manhattan Project. The group built ties to organizations as diverse as the CIA and the Esalen Institute while making important contributions to quantum theory and its applications. One member, Fritjof Capra, would later introduce the idea of an Eastern-inflected science to wider audiences with his popular 1975 book *The Tao of Physics*. Similar attempts to integrate science with mystical sources of wisdom, both Western and Eastern, permeated new modes of science fiction that appeared in the 1960s. Walter Miller's *A Canticle for Leibowitz* (1960), works by Kurt Vonnegut and Philip K. Dick, and a host of "New Wave" writings explored the dangers of prevailing modes of science and technology. Yet they often left open the possibility of genuinely human, morally engaged, and even spiritually enlightened variants. All of these critics, though diverging considerably in their political views, saw an abstract, reductive, technical form of rationality at the heart of postwar American society. Like so many other countercultural expressions of the 1960s, groovy science and its fictional counterparts sought to tear down the edifice of bureaucratic rationality and remake science and technology from within.[47]

IN THE 1960S, arguments about science's cultural effects took more institutionally specific forms than they had in the postwar years. Even when they attacked a diffuse cultural sensibility—modern thought, modern liberalism—radical critics often identified particular targets, if still large ones: the military-industrial complex, industrial corporations, the bureaucratic state, and so forth. The student movement and the counterculture also provided specific points of focus for those who feared the cultural domination of science. Overall, the 1960s radicals were less likely than their predecessors to champion traditional pillars of Western cultural authority—religion, philosophy, politics, the humanities, the arts—and more likely to look ahead to cultural configurations as yet unimagined, or at least to embrace non-Western approaches.

For many other Americans, however, the antidote to science's cultural sway lay closer at hand, in the churches and synagogues, humanistic learning, and other familiar institutions and practices. Above all, religion served as a point of contrast for those who deplored the scientific age. Of course, nontraditional faiths and spiritualities proliferated in the 1960s. Members of the counterculture, especially, explored an array of alternative beliefs and practices, from Asian traditions to peyote rituals to Esalen. From this cauldron emerged the hybrid known as the New Age movement, among other innovations. Mysticism and mythology of all kinds also appealed to the new generation, as they sought alternatives to mainstream sensibilities. But many observers in the churches and synagogues saw signs that the spiritual hunger of the 1960s youth would lead them back to the fold. The Catholic sociologist Andrew M. Greeley and his émigré Lutheran contemporary Peter L. Berger were only two of the many critics of science's sway who now discerned a "return to the supernatural" among young people, and perhaps even what Greeley called a "post-secular" mentality emerging. Farther to the theological right, of course, many conservative evangelicals viewed the radicalism of the 1960s as the quintessential product of "secular humanism." They traced the drug culture, the sexual revolution, and other revolts against social authority to the secularizing influence of a scientific worldview. But there were also evangelicals among the hippies: the "Jesus people," who combined biblical faith with communal lifestyles and countercultural aesthetics. And the political activism of the era was hardly as secular as conservative critics suggested. By the 1970s, in fact, critical accounts of science's influence would increasingly take the form of laments about secularization. As the evangelical ranks swelled, and as many on the left also called for a renewal of spiritual values, public debates around science's cultural influence would come to focus once again on its relation to the established religious traditions.[48]

10

SKEPTICISM INSTANTIATED

"**THERE IS AN IMPENDING SENSE OF CHANGE** in the world of ideas," *Time* announced in a remarkable series of feature articles in April 1973. The dominant framework—"liberalism, rationalism, scientism"—was deeply embattled: "Man's confidence in his power to control his world is at a low ebb. Technology is seen as a dangerous ally, and progress is suspect." Critics in every corner of the intellectual world shared a "deep, even humble" sense of the complexity of the world and "the recalcitrance and perversity of man." At the heart of this Copernican revolution, the final article concluded, lay a denial that science held "a stranglehold on truth" or that its "cold, narrow rationality" offered sufficient guidance for human existence.[1]

At first glance, the *Time* articles seem to reinforce the familiar view that the 1970s witnessed a society-wide explosion of irrationality or "antiscience." Yet they surveyed a remarkably variegated set of impulses, many within the sciences themselves: not just the expansive antirationalism of Theodore Roszak, the asceticism of the Hare Krishnas, Ivan Illich's deschooling proposal, and the British writer Arthur Koestler's belief that quantum physics accommodated extrasensory perception, but also Rollo May's humanistic psychology, Lionel Tiger and Robin Fox's assertion that evolution explained human behavior, and Arthur Jensen's theory of fixed racial differences in IQ, among dozens of others. The target was not science per se, although mechanistic worldviews were heavily implicated. Rather, it was a particular style of hopeful, technocratic liberalism, combining liberal religion with mechanistic science to produce boundless optimism in experts' capacity to drive both technological and social progress. *Time*'s articles identified four pillars of the postwar mentality: "the ruling doctrines of the Freudians and behaviorists," with their determinism and commitment to "human engineering"; liberal Protestantism and Judaism, with their lack of fixed guideposts and attenuated sense of the sacred; a belief in the capacity of educational reform to produce social equality; and an abiding faith that science and technology could solve social problems.[2]

There is much to be said for *Time*'s analysis. By the mid-1970s, science's declining status had become a common theme among journalists as well as scientists and their advocates. Magazines and newspapers exhaustively detailed the range and extent of public skepticism, often highlighting creationism and various countercultural and New Age movements, such as ufology, parapsychology, cryptozoology, and crystal healing. But as *Time* suggested, the target was not so much science as particular understandings of its nature, social potential, and role in governance. Those understandings echoed the laments of myriad critics since the 1920s: Science is relentlessly mechanistic, deterministic, reductive, quantitative. It is culturally dominant, defining every facet of our lives. And its advocates worship it with unparalleled devotion and naïveté. Expecting utopian outcomes from science's application, they will stop at nothing to impose its harsh discipline on humanity.

The groundswell of resistance to what Roszak called simply "the technocracy" was both political and philosophical. Since World War II, the American state had pursued three broad goals in funding and applying research: rolling back communism, sustaining economic growth, and fighting disease. Each goal reflected a moral choice: namely, to prioritize civil liberties, economic security, and bodily health over other valued ends. But liberal politicians, commentators, and scholars typically took these moral judgments to be universally true, reflecting the latent preferences of human beings everywhere. Meanwhile, they also assumed that the state could reliably deliver the goods in question. There were always dissenters, of course. An energized conservative movement challenged most aspects of that technocratic project. And a significant subset of liberals, as we have seen, denied that scientists and technicians could be trusted to promote the needed values. But by the 1970s, every facet of postwar liberal governance—the ends, the means, and the relations between them—had come under sustained attack, from multiple angles and in varying combinations.

Many factors accounted for postwar liberalism's travails. One was the maturation of the baby boomers, many of whom harbored an acute moral sensitivity and a deep faith in the possibility—and necessity—of aligning everyday practices with human values. Another was a series of scandals that seemed to reveal the corruption of both science and the American state, especially at the points where they intersected. The 1965 revelation of Project Camelot, a secret counterinsurgency project, suggested the willingness of both the military and leading social scientists to subvert democratic procedures, while the 1975 exposure of MKUltra, a decades-long CIA project that tested mind control drugs on unwitting human subjects, suggested that government scientists endangered the lives as well as the autonomy of American citizens. The dangers posed by common chemicals registered even earlier, as the anti–morning sickness compound thalidomide was shown in 1961 to have caused numerous birth defects and Rachel Carson's *Silent Spring* (1962) revealed the deadly effects of the popular insecticide DDT. In 1966, Barry Commoner's *Science and Survival* connected environmentalism to governmental negligence by noting the unacknowledged health threats from widespread nuclear testing as well as chemical pollution and other sources. A few years earlier, the psychologist Stanley Milgram's controversial experiments, aiming to illuminate the Holocaust by showing that ordinary people would harm one another if ordered to do so, had also raised ethical questions about the experimental use of human subjects. Philip Zimbardo's famed "prison experiment" at Stanford in 1971 heightened the controversy. In 1972, the revelation of the four-decade Tuskegee study, in which the Public Health Service denied poor Black men with syphilis both diagnoses and treatment, brought the issue of social inequality to the fore and highlighted the willingness of government researchers to risk the lives of what one classified document from the nuclear testing program called "low use segment[s] of the population."[3]

Technological innovations also stoked controversy and raised questions about the allegedly progressive effects of modern science. Job loss due to automation was back on the table in the 1960 election, as employment lagged behind productivity growth. It remained contentious as talk of a new "knowledge economy" heated up in the 1960s and automation proceeded in industries such as steelmaking and shipping. The introduction of automated teller machines at banks caused great consternation among Gallup's respondents in the late 1960s and early 1970s. Widespread public resistance to the government-funded development of supersonic aircraft, which caused disruptive sonic booms on the ground, led to the cancellation of federal funding for that project. Some groups also decried the fluoridation of water supplies, calling it a prime example of researchers experimenting on citizens without their consent and in ways that threatened their health. By the 1970s, bipartisan resistance to the unregulated development of biotechnology had also emerged. As early as 1963, the *New York Times* had asked, "Is mankind ready for such powers?" The British writer Gordon Rattray Taylor's 1968 book *The Biological Time Bomb* proposed that genetic engineering might "destroy Western civilization, perhaps even world culture, from within." As the pace of innovation accelerated in the 1970s and the biotechnology industry began to take shape, criticism arose from many quarters, including Congress. In 1977, the city council of Cambridge, Massachusetts, established an oversight system for recombinant DNA laboratories within its borders, which encompassed Harvard and MIT as well as numerous commercial firms.[4]

It was little wonder that the champions of science and technology felt besieged in the 1970s. Indeed, that decade birthed the trope of a comprehensive, epochal crisis of science—the idea that a "world-wide anti-science movement" had taken hold since the late 1960s. In fact, a good deal of the resistance was actually shallower and more narrowly targeted than earlier criticism, though it was more widespread. "The new skepticism, at bottom, is not antiscience at all," observed the writer Frank Trippett in 1977. "If the layman on the street has discovered that science is fallible, that hardly makes him its permanent enemy." Trippett found the new mood "essentially political in its aspirations"; it merely sought public oversight of decisions previously left to scientists themselves, about the orientation and application of research. Although more sweeping forms of skepticism persisted, and even multiplied in some circles, many commentators focused solely on the risk to human lives and values from specific practices, not any inherent danger in science as such. As Trippett put it, the public had simply learned that "the promising fruits of science and technology often come with hidden worms."[5]

Yet the political climate was changing in much larger ways as well. Since the early 1960s, the Democratic Party had worked to deepen its commitments to freedom, security, and health by interpreting those goods in new ways and extending them to additional groups. At a time when conservatives and radicals already felt profoundly unfree, the Kennedy and Johnson administrations sought to extend putatively universal forms of freedom to the Vietnamese. At a time of growing skepticism toward both bureaucracy and psychology, federal officials expanded their efforts to eradicate poverty and mental illness under the Great Society rubric. And in the Nixon years, Democrats in Congress advocated the legalization of abortion and the regulation of business activity to protect workers, consumers, and the environment. All of these projects proved massively controversial. Even the intended recipients often rejected the underlying assumptions about how individuals, families, and communities should live—assumptions that were deeply embedded in the mainstream social sciences and shared by many natural scientists. The cost of these new initiatives also fueled consternation, especially where they were caught up in fierce contests over race, gender, class, and religion as well as clashing economic visions. Such concerns accelerated when the oil crisis of 1973 led to an unprecedented combination of stagnation and inflation, calling into question the government's capacity to manage the economy at all.[6]

In this context, new understandings of American history and politics proliferated. Although liberals and progressives celebrated the growing inclusion of previously marginalized groups, most of the other accounts posited a steep decline in the nation's fortunes—and often blamed the prevailing mode of technocratic liberalism. Each of these portraits identified a turning point at which the country had lost its moral bearings and begun a rapid downhill slide, due in part to its embrace of a scientifically inspired form of liberalism. Of course, these narratives of decline differed considerably in their details. Economic conservatives targeted the 1930s, while the 1960s radicals decried postwar changes. New groups emerging by the 1970s, especially the Christian Right and neoconservatives, located the decline squarely in the 1960s. Despite their many disagreements, however, each camp sought to reverse the slide by bringing American culture and institutions back into line with moral values. By the 1970s, these rapidly multiplying challenges began to significantly alter the country's governing system as a whole. The New Deal order gave way to an era of endemic suspicion toward Washington—and often science as well, given its intimate association with the liberal mainstream. Forms of skepticism that had circulated widely in the previous decades now began to remake American politics in fundamental ways.

THE CONSERVATIVE COALITION

The 1970s brought a dramatic rightward shift in American electoral politics. Socially conservative evangelicals mobilized to reshape an increasingly secular public sphere and shifted their allegiance to the Republican Party when Ronald Reagan ran against Jimmy Carter in 1980. Many blue-collar and "white ethnic" voters turned their backs on the party of the New Deal as well. By the early years of Reagan's presidency, the current Republican coalition of economic and social conservatives was firmly in place. Historians of that conservative turn sometimes argue that it reflected an "anti-statist" sensibility that had anchored the nation's political culture since its founding. In this view, only the unprecedented economic crisis of the

1930s allowed an exception to the otherwise unbroken reign of antistatism: namely, the postwar New Deal order, which then dissolved with remarkable rapidity as the system snapped back into its usual pattern in the 1970s. It is important, however, to recognize a key difference between pre–New Deal antistatism and its current iteration: Today, many Americans see the state they fear as secular, impersonal, and technocratic. Since the 1930s, critics have been more likely to associate Washington with unelected, unaccountable bureaucrats than with venal would-be tyrants. It is no coincidence that many of those bureaucrats are also experts who administer, perform, or apply scientific research of some kind.[7]

As the conservative movement began to take its current shape in the 1970s, fears of science's cultural effects played a significant, if hardly determining, role. Observers often marvel at the emergence and longevity of the Republican coalition of religious traditionalists and economic libertarians. Of course, shared opposition to the New Deal state and associated liberal tenets helped to midwife the convergence of fiscal and social conservatives that characterizes today's Republican Party. It also matters, however, that the governing institutions and outlooks of the postwar era were widely associated with science. Since the 1970s, all manner of conservative critics have decried the influence of a scientific elite whose worldview rules out genuine human personhood. Secularism, moral permissiveness, abortion, Keynesian economics, environmental regulations, weakness in the face of communism—each of these was intimately connected to science in the minds of many 1970s conservatives. Despite their differences, they could find common ground in their shared opposition to the technocratic liberalism of the postwar decades. Caricatures of the amoral social engineer and the soulless technician circulated widely as the modern Republican coalition took shape.

The centrality of arguments about science and expertise in modern conservatism represented something of a departure from the 1960s, when such themes had often receded into the background. The generation of conservative students that entered the universities in the 1960s tended to be more libertarian, and often more secular, than fusionists such as William F. Buckley Jr. and traditionalists such as Russell Kirk. Some of these students followed postwar patterns, but organizations such as Young Americans for Freedom (YAF) also cultivated a more populist, subversive style. Many conservative students gravitated toward the libertarian atheist Ayn Rand, whose novels, "objectivist" philosophy, and frank celebration of self-seeking individualism gained numerous adherents even as Buckley sought to build a firewall between her movement and the *National Review*'s religiously committed approach. Like their radical counterparts in Students for a

Democratic Society, YAF activists developed highly politicized personal identities and sought to close the gap between American ideals and postwar realities through direct, personal action. Indeed, some YAF members considered the draft an infringement on personal freedom, favored decriminalizing drug use, and, especially on the West Coast, joined the drug culture themselves. Of course, these young conservatives targeted the nexus of social science and government, challenging top-down decision making and planning by experts. But the secularization of American culture did not generate as much outrage in YAF circles as it had among their elders.[8]

On a national scale, especially, anticommunism was the glue that bound together the conservative movement in the 1960s. Conservatives did argue that secular liberals were especially prone to communist sympathizing, failing to recognize the Soviet Union and its satellites as both moral and military threats. And as before, many assumed that the atheism of the Soviets explained the totalitarian character of their regime. But the discourse of conservative anticommunism was largely secular on its face. When movement conservatives targeted social scientists in the 1960s, they focused mainly on political positions, highlighting the liberal policy preferences of scientific thinkers rather than their secularism or relativism. Although the YAF's 1960 Sharon Statement grounded the principles of limited government, free enterprise, and anticommunism in the "God-given free will" of the individual, the principles themselves came to the fore in that manifesto, as in the organization it launched.[9]

Similarly, Barry Goldwater, the Arizona senator and conservative darling who headed the Republican presidential ticket in 1964, foregrounded the perils of state power and the evils of communism. To be sure, the main argument of his 1960 book *The Conscience of a Conservative*—ghostwritten by L. Brent Bozell, Buckley's brother-in-law and the *National Review*'s senior editor—comported well with attacks on the scientific outlook. Bozell listed the individual's transformation "from a dignified, industrious, self-reliant *spiritual* being into a dependent animal creature" among "the great evils of Welfarism." Indeed, *The Conscience of a Conservative* argued that a flawed theory of human nature anchored modern liberalism. "Conservatives," it stated, "take account of the *whole* man, while the Liberals tend to look only at the material side of human nature." The book also excoriated progressive educators for sowing mediocrity, ignoring the intellect, and failing to transmit an all-important "cultural heritage" to the young. The question of scientism still percolated below the surface of the conservative movement, especially when postwar fusionists such as Bozell shaped the discourse. But such cultural concerns became increasingly muted after Goldwater's book appeared in 1960. Even Phyllis Schlafly, whose defense of the

Christian woman would figure so prominently in the cultural politics of the 1970s, said little about religion when she penned *A Choice Not an Echo* for the Goldwater campaign.[10]

By the late 1960s, the swelling evangelical churches began to reveal the outlines of the nascent Republican coalition. Yet many of the theological conservatives who would soon become the backbone of that coalition still favored New Deal economics and voted Democratic, especially in poorer and more rural areas. At the same time, these figures found the cultural changes of the 1960s nothing short of apocalyptic. They discerned an all-out assault on Christian values, beginning with Supreme Court decisions in 1962 and 1963 that sharpened the church-state line by banning school prayer and Bible reading in public schools. The participation of noted atheist Madalyn Murray O'Hair in one of the associated cases seemed to prove that a tyrannical, secular state now aimed to stamp out religious belief. Theological conservatives decried the secularization of the American public sphere, especially during the broad turn away from prevailing forms of Judeo-Christian faith toward Eastern religions, "secular theology," and outright skepticism that followed the Court's decisions. In the schools, meanwhile, the National Science Foundation (NSF) funded a new set of biology textbooks that reintroduced the Darwinian theory of natural selection for the first time since its widespread removal from curricular materials in the wake of the 1925 Scopes Trial. Conservative Protestant fears about secularization and scientific naturalism would profoundly shape national politics after 1973, when *Roe v. Wade* inspired a sustained political challenge to "secular humanism."[11]

As critics of the liberal state began to find common ground in the 1970s, in fact, concerns about science came to the fore once again. For religious conservatives, the naturalistic tenor of modern science remained the main issue. Secular humanism had thoroughly infected modern societies, according to the evangelical critics of the 1970s. What else could explain the Supreme Court's portentous decision to legalize abortion? *Roe v. Wade* played a critical role in forging the new conservative majority by aligning Protestant evangelicals with conservative Catholic opponents of abortion and secularization. Both groups now argued that the country had gone over to a relativistic, materialistic worldview that denied the God-given dignity of the human person. "Today the view that man is a product of chance in an impersonal universe dominates both sides of the Iron Curtain," wrote Francis Schaeffer and C. Everett Koop in 1979. "Having rejected God," they explained, "humanistic scientists began to teach that only what can be mathematically measured is real and that all reality is like a machine." Now, the rising generation had "taken these theories out of the lab and classroom and into the streets," fueling a radical devaluation of humanity. A year later, Moral Majority leader Jerry Falwell charged that B. F. Skinner and his naturalistic allies controlled the American schools. From this perspective, *Roe* was just one symptom of a deeper cultural disease: the total loss of moral standards by citizens and leaders alike. Indeed, a number of religious conservatives contended in the 1970s that the American government had established secular humanism as an official faith.[12]

A raft of widely read books by prominent evangelicals pressed the case against secular humanism. The theologian Schaeffer traced the moral rot back centuries, while singling out the Enlightenment-era transition from explicitly Christian forms of science to what Schaeffer called "modern modern science." In the early modern conception, he explained, "God and man were outside the cause-and-effect machine of the cosmos." But with the Enlightenment came the image of an all-inclusive "total cosmic machine." In such a view, Schaeffer wrote, "the mechanical cause-and-effect perspective is applied equally to psychology and sociology" as well as the natural sciences. Tim LaHaye drew on Schaeffer's analysis in his best seller *The Battle for the Mind* (1980). Like Schaeffer, he identified the secularization of science itself as the hinge in Western history. LaHaye asserted that Enlightenment thinkers had imposed on science a materialistic filter that revealed only "the baser appetites of mankind" while obscuring the reality of absolute moral principles and much else. Yet scientific humanism was not just a philosophical error; it was also an organized conspiracy against the family. "Amoral humanists are really after the young," LaHaye wrote. "That is why—in the name of 'health care,' 'child's rights,' 'child abuse,' and 'the Year of the Child'—they are pressuring political leaders to pass legislation taking the control of children away from their parents and giving it to the state."[13]

Christian conservatives saw secular humanism behind a host of cultural changes in the 1960s and 1970s. LaHaye was typical in deploring "permissiveness in child raising." Religious critics had blasted the advice of Dr. Benjamin Spock and other experts since the 1960s. The phrase "permissive parenting," coined in 1966, became an all-purpose pejorative for what critics perceived as a relativistic, "anything goes" approach to raising children. It did not help that Spock vigorously opposed the war in Vietnam and allied himself with the Black Panthers and other radicals. Many religious conservatives feared that scientifically informed educators and parenting experts had cut the moral guide ropes of Western culture in the 1960s, abandoning even the pretense of disciplining children or exerting adult authority. James Dobson and other Christian leaders penned alternative childrearing manuals that portrayed children as inherently sinful, in need of strict discipline rather than freedom and encouragement. Such concerns aligned them with the "law and order" faction of the Republican Party, as Vice President Spiro Agnew was known for his "anti-Spockmanship" and Nixon decried the "fog of permissiveness" in contemporary society. Theological conservatives saw a total loss of moral strictures behind the sexual revolution of the 1960s as well, and the push to legalize abortion— by physicians, among others—further reinforced their belief that a scientific outlook denied the very concept of morality. In the schools, they deplored the teaching of evolution and the sympathetic presentation of non-Western social practices in *Man: A Course of Study,* another curricular package from the NSF.[14]

Meanwhile, a growing number of free-market thinkers targeted science in the form of expertise related to domestic economic and social policies. The Great Society reforms of the 1960s alarmed many Americans, including some who embraced the economic regulations and social safety net that Franklin Roosevelt had established. Lyndon Johnson's creation of new programs and consolidation of numerous initiatives in the Office of Economic Opportunity faced resistance on multiple fronts. Fiscal conservatives and libertarians charged that Johnson had completed the revolution begun by Roosevelt, creating a massive, centralized federal bureaucracy that offered cradle-to-grave welfare provisions. As with the New Deal, they often identified the Great Society as the brainchild of a technocratic cadre that assured its employment and professional status by claiming the ability to predict and control human behavior. Decrying liberal social engineering, conservatives called for a return to moral character, community self-help, and free enterprise.[15]

As the conservative movement found its footing in the 1970s, figures such as the libertarian economist Milton Friedman reiterated Friedrich Hayek's argument from the World War II era: that no citizenry could cede economic control to the state and expect to remain free of totalitarian rule. New Deal and Great Society liberals were fools to believe that economic and political liberties could be separated. Even moderate, Keynesian interventions in the economy portended dictatorship, in this view. At bottom there were only two choices: "private initiative operating in a free market" or "central direction by government." On matters of economic causation, these conservatives saw faulty liberal principles behind the putatively neutral, scientific theories that had guided presidents since the 1930s—above all, the belief that federal policy could drive economic growth more effectively than the self-interest and practical knowhow of entrepreneurs. The unprecedented "stagflation" of the mid-1970s gave conservatives a golden opportunity to argue that mainstream economists, steeped in the faulty and implicitly authoritarian assumptions of technocratic liberalism, did not actually understand how the economy worked.[16]

The champions of free markets also decried a potent new mobilization of scientific expertise in the early 1970s. The Nixon administration, in tandem with a Democratic Congress, undertook a series of initiatives to protect consumers, workers, and the environment, often by creating whole new agencies to regulate various forms of economic activity. These acts embedded scientific research deeply in federal policymaking, especially at the new Environmental Protection Agency (EPA). Indeed, the early 1970s brought an entirely new type of scientific research: regulatory science, tasked with assessing levels of risk to human health and the environment so that policymakers could set legal thresholds. The emergence of regulatory science, like the policies and agencies that produced it, fueled consternation among economic conservatives. "Whatever the announced objectives," Friedman declared, the new initiatives used the heavy hand of government to throttle the economy and halt economic growth: "They have been opposed to new developments, to industrial innovation, to the increased use of natural resources." Evangelicals began to speak out against these regulatory endeavors as well. "Only a perverted society," wrote Falwell, "would make laws protecting wolves and eagles' eggs, and yet have no protection for precious unborn life."[17]

Whereas predecessors such as Kirk and Henry Regnery had sought to take back the universities, few conservatives in the 1970s held out hope that academia could be redeemed. Instead, conservative business leaders funded an array of think tanks, providing experts to counter the messages coming out of the universities. New groups such as the Heritage Foundation, the Cato Institute, and the Manhattan Institute joined older organizations such as the American Enterprise Institute in the 1970s, giving rise to a full-blown intellectual counterestablishment—a parallel, conservative system of knowledge production designed to serve as an alternative source of expertise. There were think tanks on the left as well, but their conservative counterparts proved far more influential and helped to facilitate a rightward turn in American politics. These groups provided statistics, causal explanations, and scholars to make the case for conservatism in the media, at a time when trends such as stagflation and the persistence of poverty and crime called the prevailing social-scientific theories into question.[18]

Not all conservatives had given up on the universities, however. In 1971, Lewis F. Powell Jr., who would soon join the Supreme Court, laid out a plan to steer the academy away from liberalism. A corporate lawyer known for defending the tobacco industry, Powell lamented that "respectable liberals and social reformers" had joined their radical counterparts in undermining

American economic principles. Powell's memo to the US Chamber of Commerce, titled "Attack on American Free Enterprise System," argued that a concerted campaign to establish a socialist society was under way in the United States. He urged business leaders to use their financial resources to fight back by creating a powerful counterforce in the realm of ideas, and not just by lobbying politicians. Although Powell identified the media and the courts as additional problem areas, he called higher education "the single most dynamic source" of the anticapitalist sentiment threatening the country. After all, the universities had trained the legion of critics that had taken over the establishment and convinced even corporate executives to follow the path of Keynes rather than the American founders. Within the universities, Powell singled out the social sciences, home to Marxists such as Herbert Marcuse as well as "the ambivalent liberal critic who finds more to condemn than to commend." In those departments, he lamented, "freedom of speech has been denied to all who express moderate or conservative viewpoints." Powell called on business leaders to restore balance by multiplying friendly voices in the universities, through the reapportionment of speaking opportunities and other means. He also proposed a "faculty of scholars" at the Chamber of Commerce and a group of experts to "evaluate social science textbooks, especially in economics, political science and sociology," and craft substitutes that emphasized the strengths of the free enterprise system. Although Powell held out more hope for academia than the leaders of the conservative think tanks, he joined them in arguing that the dominant framework of technocratic liberalism found its main source of strength in the pervasive cultural influence of the social sciences.[19]

APOSTATES

The disaffected Democrats who came to be called neoconservatives likewise worked to turn the social sciences against what they considered technocratic, manipulative, and naïvely utopian projects. Many of the neoconservatives were unreconstructed postwar liberals who supported moderate forms of welfare provision and economic management but felt that both the Democratic Party and the culture at large had abandoned them in the 1960s, moving sharply to the left. In their view, Black Power advocates, New Left activists, countercultural dropouts, and the Johnson administration had all strayed from core American ideals. Often Jewish or Catholic, these commentators lamented that Johnson and his party had followed the "adversary culture" in the universities by trading universalism, individualism, and meritocracy for a group-based egalitarianism that was neither attainable nor fair. Like the fusionists, libertarians, and traditionalists to their right (now dubbed paleoconservatives), neoconservatives argued that mainstream American values were universally valid and sufficient to solve the problems of the world, if only the recalcitrant nations abroad and minorities at home could be brought into the cultural fold. Many of these figures warmed to free-market policies in the 1970s and supported Reagan in the 1980s.[20]

Many of the early neoconservatives—Daniel Patrick Moynihan, Seymour Martin Lipset, James S. Coleman, Edward C. Banfield, James Q. Wilson— were themselves trained social scientists who believed their disciplinary colleagues had led the nation astray. Above all, they targeted theories that stressed the economic roots of human behavior and the capacity of the state to change such behavior through regulatory initiatives or infusions of cash. Whether these neoconservatives were discussing global communism or urban poverty and racial discrimination, they argued that the needed solutions were primarily moral and cultural, not economic. In their view, the accelerating turn toward social-structural explanations for individual achievement and malfeasance—crime, delinquency, and so forth—missed the central roles of character and culture in human action. Such theories ran counter to both American values and human nature itself. Still, these neoconservatives were immersed in the social sciences and hopeful about the political promise of research in those fields, if they could be detached from unrealistic goals and faulty conceptions of individual motivation and social relations. They sought to mobilize the cultural authority of the social sciences on behalf of what they considered a commonsense interpretation of the human world.

Looking back, in fact, Irving Kristol traced neoconservatism's powerful impact on public debates and the national policy climate to the fact that "most of us were social scientists," and "the best use of social science is to refute false social science." In truth, Kristol argued, exploring history or casually observing the social scene would have produced the same conceptions of human behavior. But "in an age when 'experts' are overvalued," politicians and media outlets had latched onto these dissidents in the social sciences. Viewed from this angle, the rise of neoconservatism represents one strand in the proliferation of what could be called "sciences of inaction" in the 1970s. Whereas earlier forms of social science had typically justified governmental action to ameliorate social problems, the new sciences of inaction authorized restraint on the part of government, and perhaps even society at large. Moynihan expressed hope that social scientists would increasingly "assert the *absence* of knowledge" on pressing public questions, leaving them "less popular but more useful." Social science, in this view, should counsel caution in the face of reformers' irresponsible demands for immediate and total change. The problem with an activist state was not that it violated personal liberties, as many conservatives argued, but rather that it followed many social scientists in misreading reality itself, which dictated that governmental activism simply could not work in many areas. The growing desire for social manipulation had far outstripped the available knowledge base—and perhaps any imaginable knowledge base.[21]

Thus, neoclassical economics, the best known of the sciences of inaction, called for economic deregulation and the restoration of market-based approaches. For their part, neoconservative analysts of poverty, crime, and educational inequality argued that the solutions lay beyond the reach of government policy, in areas such as family structure. These problems, they insisted, had powerful cultural and moral dimensions that could not be addressed through a combination of apolitical expertise and utilitarian statecraft. More controversially, the handful of social scientists who argued for fixed racial differences in intelligence and the larger group of theorists who traced human behavior—often including a gendered division of labor—to evolutionary imperatives challenged the underlying rationale for educational reforms, affirmative action policies, and other egalitarian projects. Only in certain areas of economics and education did the sciences of inaction come to dominate the academic conversation. But these modes of argumentation proved highly influential outside the universities, where sympathetic pundits and politicians embraced them while ignoring or denigrating the scholarly alternatives. The sciences of inaction also flourished in conservative think tanks, which operated independently of conventional academic practices.[22]

Unlike many other practitioners of the sciences of inaction, however, neoconservatives sometimes traced the origins of technocratic liberalism back to the Enlightenment era. The Catholic Moynihan, for example, argued that Rousseau and other eighteenth-century theorists had "loosed upon the world a moral fury that has wrought as much evil, in contrast to the mere brutality of the past, as mankind has ever known." That evil, he added, "may yet destroy us." Most of the time, Moynihan focused on his main contention: that urban problems stemmed from an entrenched "culture of poverty" whose members rejected the goals and values of the prosperous middle class. Indeed, he believed that most social scientists grasped this truth; the problem lay with self-seeking reformers and bureaucrats who had cherry-picked from the scholarly literature an alternative theory that rooted poverty in an absence of opportunities for advancement and blamed the cultural pathologies of the poor on material deprivation rather than the other way around. Moynihan, echoing the anti-ideological tenor of postwar consensus liberalism, credited that theory of poverty to a self-defeating attempt to realize spiritual goals through political means. He identified the prevailing form of liberalism as a secularized "quest for divinity," generated by "a religious crisis of large numbers of intensely moral, even godly, people who no longer hope for God." Government, Moynihan insisted, could not answer the call; it offered no solution for "the crisis of values that is sweeping the Western world." The belief that it could had produced the totalitarian states of the twentieth century.[23]

Kristol, who focused more heavily on philosophical issues, traced the utopian quest for perfection through politics back to the early modern rise of "millenarianism, rationalism, and what Professor Hayek calls 'scientism.'" First, he explained, the Reformation made millenarianism the keynote of Christianity. Then, beginning with Francis Bacon, "the messianic impulse was secularized," and "science and reason and technology took over the promise of redemptive power—of transforming this dismal world into the place it 'ought' to be." Still, Kristol contended, such utopian ambitions had long been disciplined by the cultural force of bourgeois individualism, embedded in common wisdom and "immensely strengthened by the widespread belief in an afterlife." Only in the twentieth century had the crisis revealed its shape, when both religion and the "bourgeois ethos" weakened. But it would not fully abate until social scientists, politicians, and citizens threw off "the political metaphysics of modernity" and began "the long trek back to pre-modern political philosophy—Plato, Aristotle, Thomas Aquinas, Hooker, Calvin, etc."[24]

Even as they deplored modern society's cultural tenor, however, these neoconservatives also employed a populist language that exempted the masses and traced the influence of "radical-utopian rationalism" to a cadre of political and cultural elites. (Kristol famously urged fellow neoconservatives to "explain to the American people why they are right, and to the intellectuals why they are wrong.") By the 1970s, they increasingly described technocratic liberalism as the ideology of a distinct class—intellectuals, professionals, managers, experts—that had taken the reins of power in the twentieth century, replacing the commercial bourgeoisie as the dominant economic and cultural force in industrialized societies. This "new class" analysis actually spanned the political spectrum. It was first developed by disaffected communists and then discussed among Students for a Democratic Society leaders in the 1960s. The underlying idea could also be found in statements by law and order conservatives such as Nixon, Agnew, and George Wallace. But in the 1970s, American sociologists and political commentators began to use the term more precisely and systematically as an analytic category to name the technocratic, managerial professionals whose influence so many on the left and right decried. The new class concept embodied a critique of the assumption that a struggle between the working class and the bourgeoisie defined modern societies: Marx had not foreseen the emergence of what other 1970s scholars called the "professional-managerial class." A flurry of works in 1979 shone a spotlight on the new class concept, which had come to play a key role in neoconservative analyses. The idea allowed neoconservatives to argue that meddling, self-seeking reformers had fueled the egalitarian impulses of the 1960s as part of a self-interested campaign to wean Americans from their bourgeois ideals and make them dependent on the state. Kristol thus charged that new class bureaucrats had "a hidden agenda: to propel the nation from that modified version of capitalism we call 'the welfare state' toward an economic system so stringently regulated in detail as to fulfill many of the traditional anticapitalist aspirations of the Left."[25]

The new class idea created important points of convergence with other critics of scientism and rationalism, including theorists with communitarian tendencies. The Lutheran émigré sociologist Peter L. Berger, who had authored pioneering works on secularization in the 1960s, argued in 1979 that the "modern scientific worldview," or "secular humanism," represented the ideology of the new class, as well as a "secular theodicy" that helped moderns cope with the loss of traditional verities. For Berger, the new class reflected the impact of childrearing experts who had pushed beyond even moral relativism to cognitive relativism: letting each child create "his own symbolic universe," one that was inevitably "rather pleasant and 'humanistic'" given modern tendencies. Backed by the latest psychology and highly attractive to those engaged in "the production and distribution of knowledge," secular humanism altered reality itself to fit the mind-set of the spoiled grown children who increasingly ran the universities and other American institutions. Behind all this, Berger saw a structure of domination. "The class interests of the New Class," he wrote, "are masked by appeals to compassion and by the claim that they contribute to the welfare of the downtrodden."[26]

Many other theorists of the 1970s likewise connected theological and political questions as they thought about recent cultural changes. The Catholic neoconservative Michael Novak, who excoriated the new class's "lust for power" in 1972, argued in the same year that the recent resurgence of "ethnic consciousness" among young Americans of Southern and Eastern

European descent reflected a far broader "cultural revolution" against a society rooted in the quixotic dream of total societal and personal rationalization. "Science, reason, progress, economics, planning, strong central government, social engineering, behavior control, productivity, quantification, expertise": these, Novak argued, had been the core values of American progressives for decades, pushing the country ever closer to its Stalinist counterpart. But the new generation, he believed, had experienced the full force of a rationalized society ("No other group of young people in history was ever brought up under a more intensive dose of value-free discourse, quantification, analytic rationality, meritocratic competition, universal standards," and the like) and had developed "a profound starvation for a denser family life, a richer life of the senses, the instincts, the memory." Declaring oneself Italian, or Polish, or Slovak (in Novak's case) entailed nothing less than breaking through the arid, abstract rationalism that dominated contemporary culture to the depths of authentic human experience below.[27]

But many analysts agreed with Berger that the rising generations shared the values of the new class, which set the tone for the culture at large. For Christopher Lasch, those values constituted what Philip Rieff had earlier called a "therapeutic ethos." In his widely read books *Haven in a Heartless World* (1977) and *The Culture of Narcissism* (1979), Lasch dissected marriage guides, childrearing books, and other offshoots of the social sciences. These works, he argued, used a patina of scientific neutrality to obscure the fact that authority was steadily flowing away from parents toward bureaucrats. Beneath a seemingly permissive culture, according to Lasch, lay an unprecedented new form of societal organization in which parents and other individuals voluntarily disciplined themselves in accordance with the dictates of experts, feeling subjectively free all the while. He identified "industrial sociology, personnel management, child psychology" and related fields as "the organized apparatus of social control" in a permissive-repressive society resting on therapeutic relations. Modern society, Lasch concluded, no longer invoked morality as a source of authority: "It demands only conformity to the conventions of everyday intercourse, sanctioned by psychiatric definitions of normal behavior." In such a degraded culture, individuals used psychological insights to feather their own emotional nests while allowing shared norms and the common good to wither. Though still a self-identified radical, Lasch influenced an array of neoconservative and communitarian theorists in the 1970s and 1980s. His writings, like those of many other thinkers on both the left and the right, reflected a growing tendency among critics of scientism to find its main source in a new class of professionals, managers, experts, and bureaucrats.[28]

SCIENCE AND SOCIAL JUSTICE

The proliferating social movements of the 1970s and the growing bodies of theoretical analysis associated with those movements often featured sharp challenges to modern science as well. Whereas some critics sought merely to open scientific careers to women and other minority groups, others targeted the applications, the theoretical content, and the underlying philosophical and methodological assumptions of the scientific disciplines. As feminists, Black Power advocates, and other activists gained footholds in the universities, an array of theorists both within and beyond the academy worked to understand the dynamics of repression in modern societies. Grappling with Marxism, radical psychology, and other explanatory resources, they often identified the cultural influence of the sciences as a significant part of the problem. Few, as yet, wrote off the entire scientific enterprise as merely an instantiation of Western values. But the charge that science encoded and advanced ideologies of domination such as patriarchy and racism circulated widely in the radical movements of the era.

As activists and theorists surveyed the landscape of the 1970s, the social sciences once again came in for sustained criticism. Biology, which was supplanting physics as the cutting edge of the scientific enterprise, also drew considerable attention, not only for its potential to facilitate gene-level manipulation but also for its unstated assumptions about race, gender, and other forms of difference. Indeed, a small but vocal minority of social scientists now jettisoned the postwar era's cultural explanations of human behavior and sought to explain the social world in evolutionary terms. Such initiatives heightened the sense among many radicals that biology had joined the sciences of inaction, thwarting the cause of human liberation by ascribing unequal traits and capacities to different social groups. In response, radical critics charged that modern science was infused with a powerful social ideology holding that the social hierarchies of 1950s America exemplified human freedom, while their critics were psychologically imbalanced and out of step with nature itself. These radicals argued that the scientists' language of detachment and objectivity obscured the profoundly ideological character of their theories.

But those who saw science as a tool of social domination often singled out social scientists. In her 1970 book *Sexual Politics,* the radical feminist writer Kate Millett argued that medical researchers had abandoned "the threadbare tactic of justifying social and temperamental differences by biological ones," and religion alone could not have fueled the counterrevolution against women's emancipation since the 1920s. "The new formulation of old attitudes," she reasoned, "had to come from science and particularly from the emerging social sciences of psychology, sociology, and anthro-pology—the most useful and authoritative branches of social control and manipulation." These disciplines gave modern society its characteristic means of propping up invidious social distinctions, allowing it to drape restrictive gender norms "in the fashionable language of science."[29]

The "psy-sciences" were especially frequent targets for both movement activists and their academic counterparts in the 1970s. In *The Myth of the Hyperactive Child* (1975), the journalists Peter Schrag and Diane Divoky contended that a new system of social control had emerged in recent years. More sinister and effective than its predecessors because it hid the identity of the oppressors, the new system referred all social judgments back to "an overarching, cosmic and impersonal" source of authority. By such means, wrote Schrag and Divoky, "the power to manipulate is immeasurably en-hanced": "This is science talking, it is the natural order of things; what we are doing to you has nothing to do with the arbitrary decisions of school administrators or cops or the social bias of the community." In fact, they contended, research on the human mind featured "more scientism than sci-ence" and mainly served to cloak "an ideology which sees almost all non-conformity as sickness." Science, in their view, sustained social control by authorizing a regime of a diagnosis and treatment—increasingly, through psychotropic drugs—based on "an almost totally arbitrary set of defini-tions" that often amounted to "the bias of teachers or cops translated into scientific jargon." In short, modern science enabled the medicalization of social deviance.[30]

Whereas Schrag and Divoky identified B. F. Skinner as the avatar of postwar science, feminists often concerned themselves primarily with Freudianism. Decades earlier, figures such as Simone de Beauvoir had already identified Freud's work as a key vector for patriarchal norms. Betty Friedan's iconic *The Feminine Mystique* (1963) argued that the commonplace assertion of women's essential difference from men "derived its power from Freudian thought," which had generated a "scientific religion" and even a "new su-perego that paralyzes educated modern American women." Friedan's radical successors generally agreed. To be sure, many feminists of the late 1960s and 1970s attacked both mainstream psychoanalysis and behaviorism, even as they drew on the humanistic "third way" of Carl Rogers and his allies. Within the field of psychology itself, Phyllis Chesler brought together all of these critiques in *Women and Madness* (1972). Other theorists, how-ever, focused on Freud in particular. "In America," wrote Millett, "the influ-ence of Freud is almost incalculable."[31]

Black activists aced a different obstacle in the late 1960s and 1970s: the "damage thesis" that pervaded the postwar behavioral sciences. The 1954

Brown v. Board of Education decision, mandating the integration of American public school systems, reflected the increasingly widespread argument that slavery and segregation had stunted Black Americans' psychological growth, leaving them with distorted personalities and thereby perpetuating their economic and social marginality. In the 1960s, that argument meshed neatly with the culture of poverty concept and Moynihan's emphasis on the devastating effects of female-headed households in Black communities. To Black activists and scholars, these theories represented the ideological use of scientific authority to justify oppression. The psychologist Joseph White, in his pioneering article "Toward a Black Psychology" (1970), explained that mainstream theories were "developed by white psychologists to explain white people" and to promote "achievement within an Anglo middle class frame of reference." Little wonder, said White, that the authors of such studies saw departures from the preferred cultural norms as instances of deprivation. Yet this tendency was as harmful as it was predictable, since it erased Black families from the prevailing portraits of society and consigned them to the annals of pathological deviance. "The knowledge that becomes institutionalized within a society," declared the psychologist James Banks, "is often designed to support the *status quo* and to legitimize the position of those in power."[32]

Similar critiques emerged across the social sciences in the early 1970s, as Joyce A. Ladner announced "the death of white sociology" and professional associations of Black scholars took shape, typically as radical caucuses within disciplinary groups and then as stand-alone organizations. (Thus, the Caucus of Black Anthropologists appeared in 1968 and birthed the Association of Black Anthropologists in 1975, while the Caucus of Black Sociologists formed in 1970 and became the Association of Black Sociologists in 1976.) Here, too, the damage thesis symbolized the wider failure of the social sciences to express either the interests or the experiences of Black Americans. In his 1971 article on "The Myth of the Impotent Black Male," which lambasted the damage thesis and lauded the virility of Black men in a manner that many feminists found retrograde, the sociologist Robert Staples identified American social science as "a form of ideology, a propaganda apparatus which serves to justify racist institutions and practices." Such research functioned as "a means of social control exercised by white America to retain its privileges in a society partially sustained by this ideology."[33]

Meanwhile, feminists targeted the gendered assumptions that anchored the behavioral sciences. Despite their differences, Friedan and Millett each pivoted from Freud to the functionalist sociology of Talcott Parsons and his allies. In general, they argued, functionalist approaches—those asserting that norms and institutions persisted because they served important social functions—offered powerful rationalizations of the status quo. By definition, such theories could not sustain social criticism because they portrayed whatever existed as necessary and good. Feminists singled out gender norms as a key example of this quietistic tendency. According to Friedan, even the anthropological champions of cultural relativism—indeed, even Margaret Mead—reinforced the assertions of innate sexual difference underpinning the work of Parsons and other postwar behavioral scientists. "Less a scientific movement than a scientific word-game," she wrote, "functionalism put American women into a kind of deep freeze" by giving the imprimatur of empirical truth to a timeless—and profoundly passive—conception of "women's role." Millett, too, identified functionalism as a kind of all-purpose device for justifying oppression. "Functionalist description inevitably becomes prescriptive," she wrote. "The discovery that a mode is functional tends to grant it prescriptive authority." Though less overtly restrictive than Freudianism, functionalism still "points an accusatory finger of maladjustment at any woman who fails to conform." By declaring themselves thoroughly disinterested, Millett explained, behavioral scientists buttressed "the vast gray stockades of the sexual reaction."[34]

Even as these critics deplored the reduction of social inequalities to innate biological differences, the late 1960s and early 1970s witnessed a resurgence of biological explanations of human behavior in the social sciences. This tendency moved in several directions politically. Some of its advocates, like the psychologist Jensen, the physicist William Shockley, and others who discerned fixed racial differences in intelligence, took direct aim at Great Society initiatives. Meanwhile, popular 1960s best sellers by the likes of Robert Ardrey and Desmond Morris challenged the era's radical movements, tracing male aggression and the gendered division of labor back to humanity's emergence on the African savannah. By contrast, the advocates of sociobiology and evolutionary psychology, much more scholarly movements that took shape in the mid-1970s, identified themselves as thoroughgoing egalitarians. The Harvard biologist E. O. Wilson, who launched the field of sociobiology in a 1975 book of that name, and the British biologist Richard Dawkins, who popularized the new theories of kin selection and reciprocal altruism in *The Selfish Gene* (1976), both leaned to the left. The anthropologist Tiger was a critic of capitalism who believed that the postwar emphasis on the cultural roots of human behavior had allowed American elites to become as manipulative as their Soviet counterparts. And Robert Trivers, who laid much of the theoretical groundwork for sociobiology in the early 1970s, joined the Black Panther Party in 1979, having befriended Huey Newton after leaving Harvard for Santa

Cruz. But other movement activists and radical scholars charged that sociobiology would inevitably reinforce Jensen's racism and Ardrey's sexism. Tracing social patterns to biological roots, it would become a powerful new science of inaction. In this view, all biological explanations of social behavior tended to reinforce structures of domination by placing them beyond human control. Whatever its advocates' intentions, sociobiology would serve the interests of the powerful, taking the country back to the days of eugenics and biological racism.[35]

The sociobiology controversy drew in critics of the natural sciences as well as the social sciences. As it took shape, Marxists, feminists, and Black activists, some from within the field of biology itself, mobilized to challenge the new mode of social explanation. The Cambridge-based Science for the People spun off a Sociobiology Study Group—including two of Wilson's Harvard colleagues, the Marxist geneticist Richard C. Lewontin and the New Left–inspired evolutionary theorist and popularizer Stephen Jay Gould—that aimed to refute his work and drew sustained public attention to the controversy. The Boston area also featured the Genes and Gender Collective, which brought a feminist lens to bear on the critique of sociobiology. Although some feminist biologists, such as Sarah Blaffer Hrdy, turned the new mode of evolutionary explanation to their own purposes, most worked to sustain the foundational distinction between sex as a biological phenomenon and gender as a cultural pattern. Dorothy Burnham, a Black Marxist biologist, merged the various streams of criticism by working with both the Genes and Gender Collective and Freedomways, a group of Black radicals.[36]

Radicals in the natural sciences grappled with issues beyond sociobiology as well. One was immediate and practical: the lack of career opportunities for women, Black researchers, and other minority groups in the scientific disciplines. Related questions about the direction and the application of scientific research also demanded attention, both in themselves and as they interacted with theoretical tendencies. "Does culture make biology?" wondered the feminist biologist Riti Arditti in 1973. "Does the scientist furnish the culture with new ideas about women, stretching the range of her potential, or does the culture 'dictate' to the scientist what he is going to find and how he is going to present it?" Biomedicine drew considerable attention, given its intimate ties to women's bodily experiences and its capacity to connect feminist research to the women's health movement.[37]

Larger questions sometimes came to the fore as well, inspiring broad analyses of science's trajectory over the centuries. For example, feminists such as the mathematical biologist Evelyn Fox Keller noted the age-old equation of science with masculinity. How did it matter that rationality and objectivity were gendered male? Of course, Keller was engaged in scientific research herself. Like most radical theorists of the day, she did not want to leave science to the dominant groups. The critics of the 1970s typically sought to find a place in, and thereby redirect, science and other powerful institutions. Still, some kinds of scholars and activists did include science in sweeping indictments of the modern world. As the counterculture flowered, for example, some of its denizens felt that they stood entirely outside Western civilization and its prevailing modes of rationality. And groups of Black militants argued that their communities represented internal colonies with greater affinities to Africa than to Washington, New York, or even Berkeley.[38]

By the late 1970s, an emerging group of ecofeminists had set themselves against the modern scientific enterprise as a whole. Theorists such as the theologian Rosemary Radford Ruether, the writer Susan Griffin, and the historian Carolyn Merchant connected the domination of nature to the domination of women, finding these paired dynamics at the heart of Western religion, science, and capitalism alike. Men, Griffin explained, excluded women from science by definition when they claimed that it required an "objective, detached and bodiless" stance. This was one of innumerable ways that Western thought located men and women on opposite sides of sharp, binary "separations," such as "mind from emotion" and "body from soul," and valorized the masculine qualities. For her part, Merchant emphasized the mechanistic quality of scientific thought. "As the unifying model for science and society," she wrote, "the machine has permeated and reconstructed human consciousness so totally that today we scarcely question its validity." The result, she argued, was "the death of nature": its reduction to "a system of dead, inert particles moved by external, rather than inherent forces," which accompanied a capitalist "framework of values based on power." For ecofeminists, the masculine bias of science lay in its methodological and philosophical principles—the call for objective detachment, the mechanical metaphors—as well as its specific contents.[39]

Some Native American activists likewise grounded their struggle for liberation in a comprehensive refutation of European and American patterns, including science. They, too, connected their own oppression to the systematic destruction of the natural environment. The Standing Rock Sioux thinker Vine Deloria Jr. raised this challenge to Western rationality especially sharply and consistently. Originally trained in a Lutheran seminary, Deloria came to blame Christianity for the destruction of both native environments and indigenous peoples. He blasted the entire edifice of modern science as well. At times, Deloria wrote in a scholarly idiom as a critical ethnographer, challenging the bureaucratized, results-oriented character of the mainstream social sciences. Typically, he decried the colonial character of anthropological research instead. Overall, Deloria argued that scientists' relentless abstraction could never accommodate the orientation toward concrete places that anchored native cultures. In time, Deloria would even reject Darwinism. From the beginning, he upheld tribal origin stories against prevailing scientific accounts, especially the "land bridge" theory holding that native peoples had come to North America from Asia. In his later years he would position himself as an Indian creationist, arguing that only ideological groupthink could explain absurdities such as the land bridge story and the theory of natural selection. Deloria's campaign against Western science began to close the circle between radical, liberationist activism and the worldview of the burgeoning Christian Right.[40]

BY THE 1970S, critics of science's cultural influence were far more likely to stand on the left or the right, and to have in mind detailed programs of political change, than were their interwar and postwar predecessors. Amid the pitched battles of the day, these critics found science complicit with specific norms, institutions, and policies that they opposed. It was much less common than in the 1950s to hear sweeping generalizations about the deleterious influence of science and rationality across the centuries. For the most part, conservatives worried about particular developments over the previous few decades, such as the sexual revolution, the expansion of the welfare state, or the power of the new class. Left critics targeted forms of domination that dated back centuries or even millennia, but they, too, focused mainly on specific twentieth-century formations when thinking about science's role in the story. Above all, it was the particular style of technocratic liberalism endemic to the postwar decades that inspired much of the criticism in the 1970s.

Although the era's challenges to scientific authority were narrower in scope than earlier iterations, they often had significant political effects. Today, advocates of science remember the 1970s not just for the efflorescence of astrology and creationism, but also for the confluence of those cultural shifts with a significant decline in the scientific community's political power and funding levels. At the local level, citizens worked to block applications of science that they saw as threats to their health and wellbeing. When the city council in Cambridge asserted its right to oversee certain forms of genetic research at Harvard and MIT, even after the geneticists had decided to voluntarily regulate themselves at the 1975 Asilomar Conference, something had clearly changed. And it was not just local groups that asserted their right to shape the conditions of scientific research in the

1970s. Federal authorities increasingly held scientists to account as well, tightening research budgets—especially for the previously dominant physical sciences—while requiring additional paperwork from grantees and demanding more immediate, practical, and clearly demonstrable outcomes. Beginning in the late 1960s, the growing resistance of science administrators to the war in Vietnam destroyed the assumption of harmony between their interests and those of military leaders, causing Nixon to eliminate the President's Science Advisory Committee in 1973. Later in the decade, the end of the Vietnam conflict and the détente with the Soviet Union led Congress to reorient its funding priorities toward global economic competitiveness, while the maverick Democratic senator William Proxmire used his monthly Golden Fleece Awards to shine a media spotlight on federally funded research that seemed to either replicate or defy common sense.[41]

And there was more to come. After 1980, Christian conservatives would find prominent roles in the Reagan administration and work diligently to subordinate expertise to faith, while the president stoked the embers of the Cold War and embraced a controversial missile defense project that most scientists deemed unworkable. In the universities, meanwhile, critical scholars would declare that knowledge was socially constructed, objectivity was impossible, and science was simply the local folklore of Europeans and their descendants. Indeed, an array of theorists across the humanities and social sciences would join religious critics in issuing broadsides against "Western modernity" and the "Enlightenment project." As academic critics began to deconstruct scientific theories alongside films and novels, many researchers and their champions would conclude that rationality, and with it civilization, stood on the brink of destruction.

11

SCIENCE AS CULTURE

IN THE 1980S AND 1990S, a style of criticism that had been partially sidelined since the early 1960s again took center stage. During the World War II and postwar years, myriad commentators had discerned a systematic crisis of modernity in the twentieth century. They had identified a scientific worldview as the philosophical backbone of the modern age, dating back to the Enlightenment, the Scientific Revolution, or even the Renaissance and underpinning all manner of national and global problems. Versions of that sweeping, holistic argument carried through the upheavals of the 1960s and 1970s, especially among traditionalists, Christian conservatives, and theorists of the counterculture. Overall, however, claims about science's influence became more politically specific and temporally bounded in those years. Most critics targeted the practices and institutions of the postwar liberal establishment, even when they traced its technocratic sensibility to earlier roots. But in the 1980s, as liberalism ceded ground to Reaganite conservatism, the critique of scientific modernity returned in full force. Commentators of varied political and religious persuasions, working in universities, think tanks, and religious organizations, once again discerned the hegemony of scientific rationality and secular liberalism behind the beliefs, norms, and structures of the modern age.

New themes also emerged within this broad frame. Some of these had begun to take hold in the 1960s and 1970s, including the argument that putatively value-neutral science was not the most dangerous ideology of all because it countenanced relativism but because it claimed to be universal: science's proponents hid its status as a product of human judgments and presented it as unmediated knowledge of an objective reality. Post-1970s critics also joined their radical predecessors in singling out capitalism as a core feature of modern society. At the same time, these theorists added a distinctive, pluralistic sensibility of their own. Rather than aiming to replace the hegemony of science with a universal framework of morality, they often sought instead to place multiple, competing belief systems on equal ground. Late twentieth-century arguments about science, liberalism, and modernity featured a widespread retreat—sometimes principled, sometimes merely tactical—from the conventional assertion of universal human values.

Such arguments reflected the intersection of several broad tendencies. One was the "cultural turn" that swept through the universities in the wake of the 1960s. Arguing that structural analyses, whether liberal or Marxist, missed the constitutive role of meaning-making practices, wave upon wave of scholars shifted their attention to culture in the 1970s and 1980s. Some simply thought that the prevailing social-scientific interpretations focused too heavily on institutions and ignored the power of beliefs and values. Others, however, turned to culture in order to understand Marxism's theoretical and predictive limitations. As the revolutionary fervor of the late 1960s faded, the New Left's sensitivity to the inner, subjective dimensions of oppression came to the fore among theorists seeking to understand the sources and intensity of resistance to social change.

Although many of these interpreters examined popular culture as a field of social conflict, others traced cultural products to deeper philosophical commitments. Theorists in numerous disciplines contended that everyday practices and institutions reflected underlying assumptions about human beings and their world. To be sure, these figures often decried "foundationalism": the view that one could discover a single, foundational form of knowledge on which to base all judgments. At the same time, many of them traced human thoughts and expressions to a different kind of foundation: established traditions of reasoning, each resting on competing, nonempirical presuppositions. Like interwar and postwar purveyors of this philosophical reduction, commentators in the 1980s and 1990s often identified scientific naturalism and one or more varieties of theism as the main contenders.

Whereas their predecessors typically declared that one philosophy, theology, or political framework was correct, however, late twentieth-century critics increasingly championed pluralism instead. No one, they argued, should pay a social cost for the presuppositions they adopted, because each set was equally unprovable—a product of faith, not empiricism. This position flowed naturally from the views of radical theorists, for whom the incommensurable value sets reflected the perspectives of social groups such as races, genders, or classes. But many religious critics now reasoned in this vein as well. They insisted that room should be made for all of the competing frameworks, provided their adherents agreed not to force their commitments on others. Modern science, in this view, threatened society because its claim to universality undermined pluralism. Such critics often charged that liberalism reflected this scientific assumption as well, embodying an implicitly totalitarian refusal to recognize the existence and validity of multiple frameworks of reasoning, each with its own set of first principles.

These overlapping tendencies had an outcome: As the New Deal order and the mainline Protestant establishment eroded in the United States and an age of political and theological conservatism dawned, a range of commentators insisted that liberalism still ruled the roost—indeed, that its sway was stronger than ever. Across the disciplines, and often in the think tanks and churches as well, critics argued that secular, liberal modernity exerted a grip on the world that no other ideology had ever matched. Of course, this argument reflected a shift from colloquial, everyday definitions of liberalism—support for a welfare state in the United States, advocacy of a limited state in Europe—to that of political theorists. In this understanding, liberalism combined individualism with secularism, identifying human persons as atomistic individuals whose politics concerned only their worldly interests.

Those who defined liberalism in this broad, philosophical manner typically identified the sharp rightward turn of the Reagan and Thatcher years as a minor modulation within the liberal frame. Both welfare-state and free-market regimes, in this view, rested on liberal foundations: an individualistic view of human beings and a secular conception of politics. Thus, critical scholars deplored liberalism's assumption of "possessive individualism"; a growing number also singled out its inability to pursue substantive equality between social groups. Meanwhile, religious commentators increasingly argued that even the Christian Right's ascendance had failed to change liberalism's defining feature—its refusal to acknowledge the existence, let alone the constitutive character, of prepolitical, collective identities and obligations. Various critics thus charged that liberal regimes, claiming a spurious neutrality, failed to account for human groups and their moral dimensions. As the religion scholar Jeffrey Stout has noted, these observers looked at the rich, messy, contested terrain of democratic politics and saw only the thin, procedural liberalism of a theorist such as John Rawls. And behind that form of liberalism, for many critics, lay the icy hand of modern science.[1]

SCIENCE AND HEGEMONY

Although many of science's defenders in the late twentieth century felt besieged by the academic left, the radicals of the 1960s and 1970s had actually been slow to abandon the idea that science could transcend cultural particularities. As social critics found niches in the universities, relatively few believed that individuals and groups, once liberated from their social shackles, would find themselves with clashing, incommensurable views of the world. Most assumed that science could be shorn of ideology, producing universally valid knowledge—either truly neutral knowledge that comported with all of the possible value systems or truly moral knowledge that aligned with a universal framework of human values. Though these critics sought to align the disciplines with liberation movements, they typically viewed science, like the university and other institutions, as a form of power that could be wrested away from the establishment. Although "science, religion, language and psychoanalysis" had all been used against women, the feminist psychologist Phyllis Chesler warned, they were not "hopelessly tainted." Radical scholars often lamented the problems that researchers chose and the social uses of their findings while exempting the research process itself from charges of ideological corruption. Thus, when *The Black Scholar*'s editors declared in 1974 that the "myth of 'pure' science" had been exploded and all research reflected "the prevailing social ideologies," they questioned science's directions and applications, not its findings or methods. "No struggle can be undertaken without knowledge," they told their readers. Even scientists' long complicity with racism did not justify an "anti-science, anti-intellectual attitude." Although some radical theorists targeted Western rationality as a whole, then, most saw science as central to the movement, given its capacity to generate instrumentally useful knowledge. This assumption paralleled the search for a liberatory technology, as seen in Shulamith Firestone's proposal to replace childbearing with artificially assisted reproduction. Science and technology, in this view, were indispensable forms of power that could be taken over, taken back; they were not intrinsically corrupting.[2]

By the 1980s, however, another response to ideological tendencies in the sciences had emerged. Calls for a feminist biology or a Black social science could also authorize deeper challenges to science's quest for neutrality. As the academic left took shape, its emphasis on social conflict and its valorization of oppositional cultures raised the possibility that science, too, was just one of many partial, interested perspectives. Since the 1960s, several kinds of critics had rejected the contention of both liberals and mainstream Marxists that human beings, freed from arbitrary restraints, would converge on a single culture or framework of values. Those who emphasized the centrality of group identities and tied them to divergent cultural expressions included Black nationalists, Third World activists, and Afrocentrists, who often essentialized racial groupings; "difference feminists," who identified distinct male and female cultural traits; and those postcolonial critics who asserted a fundamental disjuncture between Western and non-Western mentalities. These groups increasingly argued that they had their own, distinctive ways of knowing as well as valuing, eating, dressing, speaking, and making art. A possible corollary was that no amount of experimental rigor, observational acuity, or peer review could enable human beings to see the world from a truly detached, God's-eye view. And if science could not shake itself free from cultural contexts, then critical scholars and movement leaders should either create their own, alternative sciences or look elsewhere for interpretive resources.[3]

By the 1980s, moreover, the liberation movements and affirmative action initiatives of the 1970s had spawned the broad sensibility known as multiculturalism. Although the multiculturalist rendering of minority communities as distinct, equally valuable, and perhaps incommensurable cultures said little about science per se, it opened the door to identifying science as the product of one of the local cultures, not a form of knowledge that transcended such particularities. Even when theorists said nothing about science, multiculturalism's strong assertions of human difference challenged the universalism that underlay prevailing conceptions of science. But foregrounding the beliefs, actions, and experiences of groups from different backgrounds than the white Europeans who paraded through history textbooks and literary anthologies also raised the possibility of questioning the other forms of knowledge those white Europeans had developed. If the writings of Toni Morrison and Chinua Achebe were no less masterworks than those of Melville, Proust, or Joyce, might alternative knowledge systems hold equivalent value as well?

Multiculturalism represented one expression of a wide-ranging shift in the universities of the 1970s and 1980s. The post-1960s emphasis on difference as a constitutive feature of human experience intersected with the cultural turn, which emphasized the power of cultural meanings to shape that experience. Critical scholars working to turn the disciplines into agents of social change increasingly came to see the cultural domain as a key battleground. With hopes of rapid transformation fading and the conservative backlash accelerating, scholars exploring race, class, gender, sexuality, and colonialism identified culture as a site of domination and resistance, power and agency. Indeed, some viewed culture as the main locus of conflict between economic or social groups. Recovering the voices and cultural expressions of the downtrodden took on a new urgency, as these came to be seen as forms of resistance in their own right. Meanwhile, the cultural focus also helped to explain why the movements of the 1960s had encountered so much opposition: because the entire cultural matrix encoded domination.[4]

This cultural turn spun off numerous projects of intellectual self-purification in the 1970s and 1980s. Critical scholars sought not only to change the culture at large but also to root out concepts and methods in their own disciplines that reinforced patterns of domination. In doing so, they explored a host of alternative theoretical frames. Many looked to Britain, where the postwar theorists E. P. Thompson and Raymond Williams had developed heterodox versions of Marxism. Thompson emphasized the role of cultural expressions in forging working-class consciousness during the early industrialization period, while Williams developed a materialist approach to the study of cultural products. Their work fed into the burgeoning field of cultural studies, as scholars such as the Jamaican-born Stuart Hall analyzed youth subcultures, television shows, and other cultural expressions. Critics attuned to the cultural domain also looked to the writings of Frantz Fanon and his postcolonial successors, who identified the mind itself as a primary site of colonization and resistance. Likewise, feminists had long found patriarchal tendencies throughout Western cultures. On all sides, culture increasingly appeared as a site of struggle, not an expression of universal values.[5]

Despite many points of disagreement—how easily dominated subjects could throw off their mental chains, how these related to material oppression—the new emphasis on the cultural dimensions of power led innumerable commentators to adopt the "cultural hegemony" concept that the Italian Marxist Antonio Gramsci had developed after World War I. Seeking to explain the predictive failures of Marxist theory, Gramsci had argued that ruling classes retained power by instilling their values in the dominated classes, thus mentally disarming them. For critical scholars in the 1970s and 1980s, Gramsci's model transposed agency and resistance onto the field of culture, emphasizing contests of interpretation rather than armed struggle or sabotage. It raised the possibility of counterhegemonic blocs, led by "organic intellectuals" who challenged official subterfuge and developed alternative frameworks, potentially sparking system-wide change. Like cultural studies approaches, the cultural hegemony concept offered both a new interpretive vocabulary and a new imperative to examine the social effects of prevailing belief systems—and to see critical potential in marginalized alternatives.[6]

These versions of the cultural turn had the potential to dramatically reshape scholarly interpretations by portraying science as an instrument of hegemony, enlisting those who would otherwise resist domination. For many decades, and with growing frequency since the mid-1960s, some critics had argued that a scientific culture enforced the rule of a particular social group, whether it was New Dealers, the bourgeoisie, new class technocrats, white men, or Europeans and their descendants. By giving new credence to the charge that prevailing cultural values encoded the interests of the powerful, the cultural turn raised pressing questions about science. Did scientific knowledge cut through cultural conflicts and provide resources for navigating them? Or was science itself a cultural product, caught up in power relations and legitimizing domination? If culture was a domain of struggle, then a great deal hinged on whether science reflected or transcended the cultural contexts in which it arose. And by the 1980s, critical scholars increasingly tied science to those contexts.

Even beyond the precincts of the academic left, many formulations of the cultural turn defined science as a product of particular cultures rather than a form of knowledge that rose above them. One key development was the rise of interpretivism in the social sciences. Taking shape in the late 1960s and early 1970s, especially with the ethnographer Clifford Geertz's work, interpretivism portrayed human beings as interpreters of texts and signs, weaving the webs of meaning that structured their lives. Many interpretivists distinguished sharply between the natural sciences, where a reductive, quantitative approach matched the contours of the phenomena under study, and the social sciences, where the meaning-making dimension of human affairs required a methodology rooted in acts of interpretation, like the practices of humanities scholars. (The subfields reshaped by interpretivism and related approaches came to be called the "humanistic social sciences.") Properly understood, social inquiry was merely an organized, informed, and contextually sensitive version of everyday interpretive processes. Interpretivists thus embedded the social sciences in their cultural settings. The natural sciences might produce universal, decontextualized, noninterpretive knowledge, but the social sciences traded only in particular, embedded cultural meanings.[7]

Although interpretivism's architects typically favored liberalism or communitarianism, radical scholars likewise deplored "positivistic," value-neutral approaches in the social sciences. In itself, interpretivism said nothing about the real-world effects of alternative theories of knowledge. But in radical circles, methodological dissidents traced potent social consequences to the quantitative, reductive tendencies of the modern disciplines. In the 1970s and 1980s, an array of critical histories charged that twentieth-century social scientists had reduced themselves to blinkered empiricists, or even "service intellectuals" who did the bidding of the powerful. Since the early twentieth century, in this view, a value-neutral approach to knowledge had sustained social inequalities by disabling the possibility of morally grounded social criticism. Such critics echoed elements of the Gramscian hegemony framework and often followed the interpretivists in suggesting that positivistic approaches suited the natural sciences but not social inquiry.[8]

By contrast, the nascent field of science and technology studies (STS) considered the natural sciences just as culture-bound as the social sciences. Nowhere could one find neutral, prepolitical truths, stripped of human subjectivity and reflecting only external realities. Going beyond many activists, STS scholars argued that social factors shaped science's findings as well as its overall direction and application. Early on, Marxist theories figured prominently. In Britain, for example, the historian Robert M. Young traced parallels between Malthusian economics and Darwinism, while the "strong programme" of sociologists such as Barry Barnes and David Bloor held that social interests shaped even mathematics. By the 1980s, as Marxist overtones weakened, scholars explored how certain claims and theories came to be called "science" while others did not. These theorists bracketed conventional understandings and applied the same kinds of social explanations on each side. Thus, Trevor Pinch argued that only the assent of professional scientists distinguished science from belief in the paranormal, a prominent feature of 1970s skepticism. This culturalist, relativist approach came to be associated with the phrase "social construction": science bore the imprint of its creation by specific people, through complex social processes in settings such as laboratories, field sites, and conferences.[9]

Social-constructionist approaches also spread among critical scholars elsewhere. In fields such as history and sociology, some argued that modern science and technology reflected the logic of capitalism, providing both instrumental aid and ideological cover. David F. Noble's *America by Design* (1977), originally a dissertation under Christopher Lasch, explained how engineers and corporate reformers, backed by the ideological power of the social sciences, ensured that "corporate industry has taken on a scientific aura and capitalism has assumed the appearance of reason itself." A decade later, the critical sociologist Stanley Aronowitz emphasized science's ideological role. A "mechanical and reductionist worldview," he argued, with its empiricist and quantitative bent, view of the world as raw material for manipulation, and claim to strict neutrality, had infected even Marx and Engels and made science as intellectually "coercive" as religious systems. Rejecting a narrow focus on the applications of science alone, Aronowitz targeted the assumption of both "Marxist scientism" and liberalism that science and technology followed "logical necessity" alone, as "neutral instruments that can be separated from the context in which they are developed." Indeed, Aronowitz argued that growing public skepticism since the 1960s reflected scientists' failure to acknowledge their social allegiances, even though few of the critics got to the root of the problem in science's "metaphysical foundations."[10]

Scholars combined the social constructionist insight with many other political programs as well. Feminists such as Evelyn Fox Keller and Anne Fausto-Sterling explored the gendered assumptions and implications of the sciences, as the "science question in feminism" fueled a growing controversy by the early 1980s. Could a form of knowledge-making infused with patterns of domination nevertheless serve the cause of liberation? Even those who believed that it could often assumed that science reflected underlying social and cultural conditions and sought a distinctively feminist model of research or theory of knowledge. For example, the philosopher of science Sandra Harding developed a conception of "feminist standpoint theory." Adapted from Marxist precedents, it held that women, as an oppressed group, stood in a privileged epistemological position: a science shaped by the realities of women's lives would provide a more truly objective understanding of the world than one in thrall to male domination. Other feminists argued that scientific reasoning itself reflected patriarchal prerogatives. Indeed, one observer noted that every school of feminist theory in the 1980s except the mainstream, liberal version challenged "the fundamental assumptions underlying the scientific method, its corollaries of objectivity and value neutrality, or its implications."[11]

Another style of critical analysis explored the history of science's identification as a politically and religiously neutral form of knowledge. A controversial book by the STS scholars Steven Shapin and Simon Schaffer put the Scientific Revolution in dialogue with early modern political disputes. They set the chemist Robert Boyle's assertion that experimental methods offered a path to socially purified knowledge against the political philosopher Thomas Hobbes's claim that experimental results inevitably bore the marks of their creators. "Hobbes was right," Shapin and Schaffer concluded in the book's final sentence: science is "conventional and artifactual," so that "it is ourselves and not reality that is responsible for what we know." The influential French theorist Bruno Latour likewise linked the construction of knowledge to the construction of the social order while rejecting some of the deepest conventions of modern political thought. The interpretivists were wrong to separate social inquiry from natural science, Latour insisted; each featured socially embedded, fundamentally interpretive practices. "All sciences are the offspring of biblical exegesis," he wrote, and their practitioners diverged "only in the source of their texts, not in the hermeneutic skill they deploy." Invoking Shapin and Schaffer's book,

Latour equated modernity with the postulation of a sharp divide between science and politics. Early modern scientists, he argued, had muted the voices of natural entities and arrogated the authority to speak for them, in a move comparable to denying "voting rights to proletarians, or to women." Like Shapin and Schaffer, Latour targeted deep-seated views of science and identified them as the source of unquestioned social and political understandings. These writers announced the dissolution of a "settlement" or "constitution" that had defined the Western world for centuries.[12]

Latour also drew on Michel Foucault, whose writings became widely influential in the 1980s. An array of American theorists embraced the work of Foucault and his French contemporaries Jacques Derrida, Jean-François Lyotard, Jean Baudrillard, and Gilles Deleuze. For all the differences between these French theorists, American interpreters saw in their writings a shared framework of "poststructuralism" that emphasized the primacy of language in human affairs and promised a critical alternative to the prevailing forms of Marxism. Like Reinhold Niebuhr before them, poststructuralists sought to ground a new kind of radical critique in tenets more typical of the traditionalist right, challenging universalism, rationalism, and the possibility of a neutral perspective in the sciences or anywhere else. By the 1970s, local traditions had increasingly figured as irreplaceable sources of wholeness and meaning in radical thought. Now, poststructuralists combined that communal emphasis with a frank skepticism toward claims of genuine knowledge and disinterested judgment. Indeed, they looked to a cadre of German antimodernists that even Niebuhr had eschewed: figures such as Friedrich Nietzsche, Martin Heidegger, and Carl Schmitt, who had inspired right-wing nationalists in Europe. From this new vantage point, poststructuralists saw common threads between Enlightenment rationalists and their humanistic critics. Even Marxism and its offshoots, they emphasized, shared with liberalism—and, of course, science—a commitment to neutrality, rationality, and universalism that the lessons of human experience simply could not sustain.[13]

Poststructuralism appealed to numerous activist scholars at a time when previously marginalized populations were gaining footholds in the academy. Its earliest adherents included many feminists and postcolonial critics. Both groups sought to understand forms of oppression that long predated, and still permeated, both welfare-state liberalism and Marxism. Both, too, had often been denied the capacity for scientific rationality and defined as collective entities, not autonomous individuals like white men. From this angle, even versions of humanistic Marxism—for example, the Frankfurt School's call for a reflexive social science that cut through the ideological claims of the official sciences—could seem overly rationalistic. Meanwhile, interpretivism had little to say about the social character and functions of the natural sciences. Poststructuralism, which also emphasized the interpretive character of all human experience and highlighted texts and narratives, provided a ready alternative. Critical scholars found new resources in French theory for thinking through patriarchy, homophobia, racial subjugation, colonialism, and other forms of coercion.[14]

Poststructuralists routinely implicated science as they worked to dismantle the essentialism and universalism they saw at the heart of Western thought. In the case of essentialism, they argued that linguistic categories—especially collective nouns delineating groups of people—did not correspond to actual entities in the world; they were representations, not realities. Indeed, such categories were expressions of power, serving to marginalize and exclude. This approach extended earlier conceptions of human behavior as a product of environmental forces into the domain of identity and belonging, while adding an emphasis on group conflict. (Human history, in Foucault's famous dictum, "has the form of a war rather than that of a language: relations of power, not relations of meaning.") Binary distinctions also came in for particular scrutiny, especially antinomies long associated with Western science and often used by the privileged to distinguish themselves from despised groups: reason/emotion, culture/nature, science/religion, health/sickness. Indeed, the anti-essentialist tenor of poststructuralist thought challenged the entire scientific enterprise, given that science purported to grasp external, nonsubjective realities. Poststructuralists suggested that distinctions, categories, and labels considered natural were actually deeply responsive to power relations, reflecting contingent human choices rather than inevitable facts or external realities.[15]

The turn against universalism likewise reflected a long history of Western abuses in the name of scientific rationality. The universal principles touted by liberals and Marxists in the 1950s looked suspiciously like the values of white, straight, male, middle-class Westerners. Moreover, many of those patterns persisted through the revolts of the 1960s and 1970s. By the 1980s, poststructuralists argued that the very concept of deep, structural universals froze power relations in place by delegitimizing dissent. Social change would require dismantling not only the prevailing worldview but also the underlying sense that one worldview would work for everyone. There were no final answers, only people in conflict, waging their struggles on domains that ranged from the highest reaches of philosophical abstraction to the lowest, most mundane forms of daily activity. Among poststructural theorists, attacks on the universal values proposed by the postwar generation broadened into a critique of the whole idea of universals, holding that claims to universality represented exertions of power. Like anti-essentialism, the campaign against universalism identified science as a kind of meta-power, a uniquely potent weapon for disarming those who would resist other operations of power.[16]

Even more than STS research, poststructuralism defined how critical scholars thought about science and its social meanings in the late twentieth century. The intimate ties to power suggested by Foucault's term "power/knowledge," the rejection of essentialism and universalism, and the movement's assertions about the centrality of conflict all challenged conventional understandings of science. So, too, did poststructuralists' emphasis on the critical power of subjugated forms of knowledge. These commitments shaped the development of the academic left as its influence grew in the 1980s and 1990s, infusing the social-constructionist model with poststructuralist understandings of language, power, and selfhood. New styles of criticism joined with old, as poststructuralists denied that anyone could attain what the philosopher Thomas Nagel famously termed the "view from nowhere" and the feminist scholar Donna Haraway called the "god trick." Even the forms of bottom-up objectivity espoused by many Marxists and by feminists in Sandra Harding's vein ran afoul of this critique. Haraway sought to ground the capacity for genuine insight in self-consciously partial viewpoints. With no single, liberatory framework available, she argued, an array of "situated knowledges" offered the only alternative to the false objectivity of mainstream science. As the presumption of universalism—and thus a common moral framework—faded, a new emphasis on difference prevailed across much of the academic left. "The postmodern world-view entails the dissipation of objectivity," wrote the Polish-British theorist Zygmunt Bauman: "the slow erosion of the dominance once enjoyed by science over the whole field of (legitimate) knowledge," leading to the emergence of multiple, competing systems of truth.[17]

As the oft-invoked term "postmodern" indicates, many poststructuralists set themselves against the modern world, often implicating science in its worst features. Whereas Marxists and other groups of critics sought to mobilize Western institutions and ideals—mass production, the labor movement, individual subjectivity and creativity—against patterns of domination, many poststructuralists found such formations unredeemable and aimed to hasten their demise. The postmodern label signaled a sense of an epochal shift in the late twentieth century, a passage out of modernity. To be sure, poststructuralists renounced the humanism of modernity's postwar critics, which they equated with liberal individualism: the modernist fiction of an autonomous subject. Yet they, too, issued sweeping denunciations of modernity, identifying the "Enlightenment project"—a rage to impose rational order on all forms of existence—as its defining feature and often dating that phenomenon back to Descartes.[18]

In these instances, the postmodern sensibility echoed a long-standing tendency among modernity's critics to reduce social existence to philosophical tenets. Like their 1950s predecessors, though in a more anticapitalist and antiliberal tone, many poststructural theorists identified scientific rationality as the keynote of the modern. In this view, capitalism, liberalism, colonialism, and much else rested on the philosophical base created by Enlightenment rationalism and individualism. Even the political phase of the Enlightenment in eighteenth-century Europe faded into the background, in such analyses appearing mainly as a site for the intensified application of a conceptual framework. Similarly, the domination of the European commercial classes over workers, racialized others, and colonial subjects also seemed to reflect philosophical principles: above all, the presumptions of a neutral standpoint and a universal form of knowledge, which simultaneously embodied and masked the bourgeoisie's drive to power by making its parochial conceits appear inevitable.

As in the postwar years, then, critiques of modernity often took the form of sweeping, stylized philosophical histories, stocked with ideal types and designed to explain how the conceptual seeds of modern existence had taken root, grown, and flowered. These accounts again implicated scientific rationality in a remarkable array of social ills, up to and including the Holocaust. Bauman, for example, equated modernity with the political project of "social engineering," one of his most frequently used terms. Aiming at "an artificial social order," he explained, this social engineering impulse generated racism when it encountered populations that did not fit the plan. The Holocaust was simply another "exercise in the rational management of society" through "applied science," informed by the racism that such programs constantly spun off. Meanwhile, scientists themselves, elevated to the status of deities and expected to radically transform both nature and society, had become fully complicitous in modernity's violence by ruling out considerations of value. German scientists, Bauman wrote, had "cleared the way to genocide through sapping the authority, and questioning the binding force, of all normative thinking, particularly that of religion and ethics." After all, there was research to be done, in both the Nazified universities and the death camps themselves. "What scientists want," Bauman declared, "is merely to be allowed to go where their thirst for knowledge prompts them," regardless of the human costs.[19]

Bauman emphasized that the cult of science had paved the way for genocide, not simply failed to prevent it. The "extermination of a whole people" could not even be imagined, he reasoned, without "the engineering approach to society, the belief in the artificiality of social order," and the social-scientific institutions and practices that made those commitments manifest. "Inhumanity is a matter of social relations," Bauman summarized. "As the latter are rationalized and technically perfected, so is the capacity and the efficiency of the social production of inhumanity." Many poststructural thinkers agreed with Bauman that modern societies were, almost by definition, far more destructive, repressive, and violent than their premodern counterparts. Thus, meaningful social change awaited a comprehensive, collective project of mental demodernization. Science could be of little use, given its alignment with modernity's violent campaigns to discipline a fractious world.[20]

Nor could the modern state, many poststructuralists argued. Even those theorists who did not link modernity to genocide often identified the state as the embodiment of modern errors—and ignored many other kinds of dynamics shaping modern societies, including economic relations. A common argument held that national states encoded a depersonalized, fictional "modern subject": the liberal, autonomous, choosing self that many critics also saw behind science's illusory quest for a thoroughly detached standpoint. Interpreters in this vein sought to transcend the abstracted, impersonal character of modern policy and law, sometimes by resuscitating bygone forms of premodern politics. As usual, this political critique often implicated science, which had long served to legitimate depersonalized rule. Of course, science had instrumentally aided modern regimes as well, through its contributions to public health, economic management, and other forms of policy that poststructuralists tended to find deeply manipulative.[21]

Poststructural insights occasionally appealed to more mainstream progressives as well as radical critics. One was the philosopher Richard Rorty, who inherited a social-democratic sensibility from his father, the radical journalist James Rorty, and his maternal grandfather, the Social Gospel leader Walter Rauschenbusch. Politically, Rorty advocated "postmodernist bourgeois liberalism" rather than root-and-branch criticism of modern practices. He believed that a regime of individualism, constitutional rights, and tolerance of difference would appeal to all peoples if offered for consideration on its practical merits rather than imposed from on high or presented as universally valid according to some spurious method of proof. Rorty's metaphor for social relations was conversation, not war; the goal was to avoid shutting down dialogue by invoking capital-T "Truth" or other "conversation-stoppers." Philosophically, however, Rorty echoed numerous critiques of scientific modernity and added some of his own. Turning against analytic philosophy in the 1960s, Rorty helped to spearhead a "neopragmatist" revival of the philosophies of William James and

John Dewey. Unlike James and especially Dewey, however, late twentieth-century neopragmatists distrusted the social sciences, and in Rorty's case the natural sciences as well. Equating science with a narrow, reductive positivism and aligning themselves with the humanities, they positioned themselves as critics of science's cultural overreach, not as champions of mental modernization.

Rorty's writings created important bridges between neopragmatism, interpretivism, and poststructuralism, while also doing much to give the term "postmodern" its familiar meaning. The key point of overlap was his thoroughgoing rejection of foundationalism—the belief that there was some neutral, nonsubjective plane of reality or form of experience to which interpreters could refer in seeking to establish knowledge claims. Despite their differences, Rorty shared with more radical poststructuralists a sense that all knowledge is a matter of partial interpretation from particular locations. Indeed, he sought to eliminate "the notion of *discovering the truth* which is common to theology and science." Rorty's writings helped to launch the term "postmodern" on its American career, to attach that term to the work of French theorists, and to advance the associated claim that an epochal turn was under way in the late twentieth century, from a mode of thinking defined by the faulty, modernist presumption of scientific objectivity toward a dramatically altered, postmodern understanding of knowledge and politics.[22]

SCIENCE WARS AND CULTURE WARS

Despite such sweeping claims, few of the critics denied that scientific procedures could teach us something about the world. Poststructuralists often focused on broad, philosophical frameworks associated with science rather than particular disciplines and theories. And STS scholars looking at specific cases often used the language of social construction to make a point about human knowledge, not the external world: namely, that scientists' social locations and cultural commitments shaped their work, even at the level of perception and observation. The social construction metaphor usually served to compare science with, say, a house. Scientists did not dream up their theories; they built with the natural materials around them, though for purposes and in forms that reflected their social contexts. As the events of the late 1990s and early 2000s would show, STS scholars had little patience for climate denial or—with a few exceptions—the campaign against Darwinism.[23]

Still, some STS researchers shared a flamboyant style with poststructuralists in other disciplines. They issued sweeping claims about science's inability to grasp reality in itself and the disastrous effects of believing that it could. Meanwhile, science's defenders often joined critical scholars in linking science closely to liberalism and modernity, causing them to fear that deflationary accounts of truth heralded the imminent destruction of society. They read even the most detailed and sophisticated examples of social constructionism as assaults on the very foundation of the scientific enterprise. Of course, more provocative and sweeping formulations drew the most attention anyway. When Shapin and Schaffer elevated Hobbes's authoritarian skepticism over Boyle's cool rationalism, or Latour and his collaborators lamented the denial of agency to scallops and even rocks, or poststructuralists argued that the antimodernist Nietzsche epitomized genuine democracy, many of science's defenders concluded that the universities had gone off the deep end into nihilism.[24]

In the mid-1990s, this controversy blew up into a series of polemical "science wars." Science's champions published high-profile books deploring the "higher superstition" and "fashionable nonsense" in the humanities and social sciences. One of the most vocal combatants, the physicist Alan Sokal, sought to prove his point by publishing a fabricated article on the social construction of quantum gravity in a cultural studies journal edited by the likes of Stanley Aronowitz. The lesson was clear, for Sokal and his allies: postmodernists had replaced the quest for truth—and meaningful social change—with self-serving faddishness. All of the old enemies ("emotion, feeling, introspection, intuition, autonomy, creativity, imagination, fantasy, and contemplation," one observer enumerated) had been let back through the gates.[25]

In response, critical scholars sometimes dubbed Sokal and his allies "conservatives in science" who would subvert democracy in the interests of capital. Many of them, including Sokal and Norman Levitt, were actually committed radicals who embraced the universalistic sensibility of the socialist tradition. But each side in the science wars insisted that the other buttressed Reagan-era conservatism—and each agreed that science anchored modern society, underpinning its central practices and institutions. They advocated different bases for a progressive movement, disagreeing especially on the roles of science and the state. For Sokal, Levitt, and their allies, a viable left needed a firm grounding in reality, not the arcane, fuzzy, and dispiriting rhetoric of the poststructuralists. A humane social order required an understanding of truth as something out there, accessible by all rational minds.[26]

Although this internecine split on the left also figured in the broader "culture wars" of the late 1980s and 1990s, those battles mostly broke along different lines. Yet they also implicated science, sometimes quite directly.

One conflict centered on education, where the academic left and its curricular initiatives sparked resistance from conservatives and neoconservatives as well as socialists who found identity-based movements divisive. Critics charged that radical professors were destroying the foundations of the social order by challenging the primacy of Western cultural ideals. Many took inspiration from the University of Chicago philosopher Allan Bloom, who defended the Western tradition in his surprise 1987 best seller *The Closing of the American Mind*. Like other followers of Leo Strauss, Bloom excoriated value-neutral approaches in the social sciences. He portrayed a steady descent from cultural relativism to poststructural nihilism, "the last, predictable stage in the suppression of reason and the denial of the possibility of truth." Although scientific thinkers often saw poststructuralism as a mortal threat, other commentators believed that it merely sharpened science's obliteration of human values.[27]

In the wider cultural arena, disagreements over religion's public role anchored the conflicts of the late 1980s and 1990s. Christian conservatives took aim at secular and liberal tendencies in the media and politics as well as academia. Here, even more than in curricular disputes, the critics traced poststructuralism to the relativistic implications of science—and not just the twentieth-century social sciences. Darwinism also came in for renewed attack from those who wanted creationism taught in schools. When the Supreme Court ruled against the swelling creationist movement in 1987, it birthed a potent offshoot. The new theory of intelligent design jettisoned biblical references and targeted a single scientific principle: methodological naturalism, which prohibited supernatural explanations. Darwinism, it contended, failed to account for a number of crucial organic structures, which featured the complex interplay of multiple parts that otherwise had no survival value and could only have been designed by an intelligent being. In this stripped-down form, the campaign against Darwinism threatened to finally dislodge natural selection from its legally secure but culturally precarious place in education. It became a central cause for Christian Right leaders, who blamed divorce, promiscuity, and an array of other outrages on secular worldviews.[28]

Scientists alarmed by the intelligent design movement could find plenty of additional reasons to fear that the American public had turned decisively against them in the 1980s and 1990s. The culture wars followed their own dynamics, but they also served as proxies for profound economic and technological changes in the last two decades of the twentieth century. These shifts, and the responses to them, convinced many of science's defenders that they were besieged on all sides by Christian conservatives, postmodernists, and the mushrooming New Age movement. Older concerns about scientists' complicity with the military-industrial complex and penchant for manipulative projects of social engineering did not disappear in the late twentieth century, but a series of potent new controversies that took shape in those years led many scientists to conclude that public support for their work had utterly collapsed.

These controversies involved immediate, mundane matters as well as broad cultural dynamics. Some hit scientists close to home by reshaping the landscape of research grants. In the 1970s and 1980s, international economic competition largely replaced the Cold War as the primary rationale for federal spending on research, although the push for high-tech weaponry continued. The new orientation favored projects with immediate, practical results, not basic, blue-sky research. When combined with the skepticism of William Proxmire and other congressional critics, it also added the insult of intensified bureaucratic oversight (and the resulting paperwork) to the injury of tighter budgets. Leading scientists also saw their voices ignored in key funding decisions, especially Reagan's choice to focus on a missile defense system that few scientists believed could work and the 1993 cancellation of the massive Superconducting Super Collider project in Texas.[29]

New areas of research also generated intense public controversy by the 1990s. These disputes drew in broader segments of the public than the antinuclear movement or even the creationist campaign. For example, the emergence of the tech industry and cyberculture, whose Silicon Valley titans combined the iconoclasm of the counterculture with the economic libertarianism spreading rapidly in Washington, produced both lofty flights of techno-utopian fancy and a strong current of cultural unease that broke through in "cyberpunk" science fiction and elsewhere. Critics of varied persuasions charged that a computerized world would be culturally deadening, implicitly totalitarian, or both. Meanwhile, leaps in robotics and artificial intelligence—IBM's Deep Blue computer defeated world chess champion Garry Kasparov in 1997—challenged foundational assumptions about humanity's relationship to machines, which increasingly performed cognitive tasks as well as physical ones. As before, commentators often associated these emerging technologies with scientists' abandonment of values and purposes. Thus, the cultural critic Neil Postman discussed the errors of scientism at length in *Technopoly,* his slashing 1993 attack on a machine-obsessed, expert-dominated society.[30]

The ascendant biosciences generated even more fireworks, especially by the 1990s. Concerns about genetic engineering had simmered among both religious thinkers and critical humanists since the 1970s. Now, a series of dramatic announcements—the birth of a cloned sheep, the image of a laboratory mouse with a synthetic human ear on its back—brought that field into the spotlight. The quest to map the human genome, launched in 1990, heightened fears about the dislocating potential of genetic engineering. Philosophers and cultural critics sparred publicly over the dangers of biotechnology, while films such as *Jurassic Park* (1993) and *Gattaca* (1997) illustrated dystopian possibilities. Meanwhile, new ways of understanding and manipulating the human brain also provoked resistance. Evolutionary psychologists, who traced behavioral traits back to survival imperatives in the deep human past, could now draw on scanning techniques that correlated activity in various parts of the brain with specific actions and thoughts. Yet progressives' charge that biological accounts of behavior buttressed inequality gained credence when the neoconservatives Richard J. Herrnstein and Charles Murray published *The Bell Curve* (1994), a massive, influential study claiming to definitively prove the existence of innate racial differences in intelligence. Critics on the left also decried other applications of the biosciences, including "academic capitalism"—scientists' monetization of their own research through start-ups and patents—as well as the plunder of the global South for pharmaceutical patents and the impact of genetically modified crops, which soon drew suspicion from the wider public as well. In psychiatry, the ongoing shift from therapy to drugs as the first line of defense, even for mild personality disorders, brought concerns about profit-driven overdiagnosis and inspired a flood of popular works exploring Prozac's impact on the human condition.[31]

While the biosciences fueled much of the cultural ferment and creationism drew headlines in the legal sphere, political challenges to science increasingly centered on environmental policy. The post-1970 regime of regulatory science, much of it sponsored by the Environmental Protection Agency, had always rankled conservatives. Within the environmental movement, meanwhile, the 1980s brought more radical tactics—liberating lab animals, confronting whalers, sabotaging construction equipment—and sharper attacks on industrial society as a whole. Anticapitalist social ecologists and nature-centered deep ecologists charged that the scientific establishment marginalized ecological concerns and abetted industrial capitalism. Some conservatives seized the opportunity to push back against hated regulations by portraying environmentalism as a dangerous brew of terrorism, totalitarianism, and misanthropy that would take human beings back to the Stone Age while stripping away their jobs and their freedoms. Increasingly, they also questioned the findings of environmental research, even in its mainstream forms. When global warming came into focus in the 1990s, conservative think tanks and pundits mobilized to portray climate scientists as political hacks driven by liberal groupthink. The ascent of Newt

Gingrich's conservative faction within the Republican Party, and then in Congress after 1994, locked in climate change as a partisan issue. Like Darwinism, but with the immense added weight of economic consequences, climate research fueled conflicts over science's findings, theories, and methods as well as its political and cultural implications.[32]

SCIENCE AS SECULARISM

The conservative ascendancy also reinforced a growing tendency in American public culture to conflate the term "religion" with the theology and politics of the Christian Right. The culture wars frame employed by media commentators and some groups of scholars identified two parties to the struggle: religious conservatives and secular liberals, each working diligently to impose their vision of the nation on the other. That analysis omitted huge segments of the population, including religious Americans of all political stripes who supported a strict reading of church-state separation, secular conservatives, politically liberal theists, and secular liberals who did not want to see religion stamped out. The culture wars narrative also implied that theological liberals were religiously inauthentic, failing to grasp the true meanings of their faith traditions. Of course, theological conservatives had long contended that they held a monopoly on genuine religion, whereas their counterparts kowtowed to science and other secular forces. The culture wars frame reinforced this narrow, contentious definition of religious authenticity by identifying members of the Christian Right as the truly religious Americans and rendering all other theological positions anomalous, unstable, or simply invisible.[33]

A related conception of religious authenticity operated among scholars of religion in the late twentieth century, inflecting numerous disciplines and reshaping conceptions of science's cultural influence along the way. Through the preceding decades, mainline Protestants had dominated the study of religion in the leading American universities. Whether theological liberals or Christian realists in the vein of Reinhold Niebuhr, they had tended to equate genuine religion with their own approaches, while placing even moderate evangelicals such as the postwar revivalist Billy Graham beyond the pale. But in the 1980s, a new cadre of scholars who combined more traditional theologies—evangelical, Calvinist, Catholic, Mormon—with relatively cosmopolitan cultural and political sensibilities claimed the mantle of religious authenticity for themselves. Like many academic leftists, they represented social groups that entered the elite universities in the wake of the 1960s and worked to bring their perspectives into scholarly and public debates. As they did, they challenged the tenor of modern thought in terms that often paralleled multiculturalism and sometimes even poststructuralism. Their version of the critique of universalism identified liberal theology as a conduit for Enlightenment rationalism, which had monopolized American public culture and obliterated genuine pluralism. In this view, science was the controlling force behind secular politics and culture, liberal Protestantism, and even Niebuhr's critique of liberal theology.[34]

Of course, these new critics diverged from radical scholars in important ways. Members of the academic left understood human relations as a matter of conflict between social groups: racialized communities, for example, or colonizers and colonized. The new religious critics, by contrast, saw religious and secular belief communities jostling for position. Yet they joined the champions of social diversity in decrying the hegemony of the dominant group and arguing that it needed to give the others a seat at the table. As they did, they mapped religious identities onto other forms of social difference, equating their religious traditions with the perspectives, subcultures, and collective experiences of, say, women, or the Black community. In this case, however, the hegemonic power was not white men of European descent—which these critics generally were themselves—but rather liberal Protestants or secularists, groups that they viewed as largely interchangeable. Conversely, they identified their own religious traditions as marginalized, subjugated knowledges that the centuries-long embrace of scientific naturalism had systematically excluded from the public sphere, through a form of conceptual violence that had serious real-world effects for those on the outside looking in.

These scholars crafted sophisticated new formulations of the argument, offered in the mid-twentieth century by Catholic commentators and Protestant champions of the theological renaissance, that every belief system rested on nonempirical articles of faith—and that modernity featured near-total domination by a naturalistic perspective that denied it had such presuppositions. Unlike past critics, however, they followed the academic left in rejecting the presumption that the various communities of belief would ever converge on a common framework. Abandoning the universalism that had so often characterized Western Christianity, they called for an open-ended system of pluralism. The new religious critics steered a course between theological liberalism and Protestant fundamentalism, portraying these as equally dangerous, coercive forms of universalism and identifying their own pluralistic model as the only equitable—and authentically religious—path. They aligned themselves with the "post-Enlightenment" orientation of late twentieth-century academia, arguing that "there is no universal scientific or moral vision that will unite the race."[35]

Like poststructuralists and STS researchers, in short, these theorists of religion described science as a product of particular cultures and challenged its claim to universalism in the name of social justice. Many charged that both academic discourse and public culture in the United States were profoundly intolerant, despite the constant paeans to diversity. Neither the secular climate of academia nor the regime of strict church-state separation could ever be religiously neutral, they contended. Quite the opposite; secularism meant forced conformity to the dominant framework through the official establishment of scientific naturalism and the marginalization of all other worldviews. Modern science, like the secular institutions it authorized, was intrinsically oppressive because it refused to acknowledge the legitimacy of alternative perspectives.[36]

The new religious critics thus adopted a stance we might call "multifoundationalism." They held that modern societies featured multiple, discrete communities of interpretation, each following its own foundational presuppositions, and their adherents should have equal standing in the schools, the universities, and public discourse. In this view, the ground rules of scholarship and politics systematically discriminated against everyone except scientific naturalists and theological liberals, who were essentially naturalists in disguise. Science reigned supreme over modern thought, culture, and institutions. Yet it was neither universally valid nor religiously neutral. Indeed, it encoded a form of religious domination that even the champions of diversity typically ignored. Only by rejecting the assumption that science offered a privileged road to truth could societies extend equal treatment to all of their members and achieve genuine justice.[37]

Multifoundationalists argued that scientific naturalists began with materialistic premises, not empirical data, and that it was no less rational to work from theistic presuppositions. In 1988 the philosopher Alasdair MacIntyre, a recent convert to Catholicism with a Marxist background, influentially formulated this argument in *Whose Justice? Which Rationality?* MacIntyre had long decried liberal individualism, most recently in *After Virtue* (1981), a celebrated account of moral reasoning that rejected social-scientific determinism and drew on Philip Rieff and other critics of therapeutic culture. Now, he identified rationality as a "tradition-constituted" phenomenon, always operating within one of a number of discrete traditions of interpretation. Although MacIntyre allowed that experience sometimes falsified particular theories, the main lesson was that no tradition held a monopoly on rationality—or rather, that "there are rationalities rather than rationality." None of the leading traditions of moral reasoning could be dismissed as fundamentally irrational.[38]

Protestant theorists, too, questioned the privileged status of scientific reasoning. At Calvin College, the philosophers Nicholas Wolterstorff and Alvin Plantinga developed a system of "Reformed epistemology" that focused on the question of warrant. On what grounds could one legitimately hold a belief? It was entirely warranted, wrote Plantinga in an influential 1983 article, to ground "basic beliefs" in immediate, personal insights and experiences rather than sensory evidence, so long as one could adequately meet all of the currently known objections to those beliefs. Emotions such as gratitude and awe, for example, warranted belief in God if their holder addressed the possible rebuttals. Like MacIntyre, Wolterstorff and Plantinga assumed that experience constrained the possible foundational commitments but upheld the sincerely held, rationally defended presuppositions of many religious traditions as well as empiricism. Again, there were multiple, equally rational communities of belief.[39]

Such arguments appealed to a growing number of Christian scholars in the 1980s and 1990s. These philosophical claims also intersected with church-state battles outside the universities. Christian Right leaders occupied key positions in the Reagan administration, while the less devout Bush faced a major primary challenge on the right from Pat Buchanan in 1992. Yet Christian conservatives still feared that they were losing cultural ground. Although they managed to reshape many areas of policy and practice, especially at the state and local levels, much of the public accepted divorce, women's equality, and homosexuality, in line with academic and media trends. Meanwhile, basic legal patterns—particularly the federal right to abortion and a relatively strict interpretation of church-state separation—persisted. Popular works by the Lutheran-turned-Catholic priest Richard John Neuhaus (*The Naked Public Square*, 1984) and the Yale law professor and novelist Stephen L. Carter (*The Culture of Disbelief*, 1993) captured in powerful phrases the belief of many Christian conservatives that American institutions systematically repressed their views. Viewing strict separationism as the forced imposition of a secular philosophy rather than a practical, negotiated response to religious diversity, such critics charged that a monolithic liberal regime prevented religious believers from even expressing their views in public. American culture had been stripped of all religious and moral content, leaving only a thin, utilitarian language for discussing collective decisions.[40]

As religious traditionalists came to see themselves as persecuted outsiders, their accounts of social difference and science's political roles partially converged with those of the academic left. Since the 1970s, Christian Right leaders and their moderate counterparts had employed both a majoritarian argument—that the American people embraced Christian or Judeo-Christian faith and their public institutions should follow suit—and a minoritarian alternative that identified them as despised outcasts, suffering under the yoke of science and secularism. Liberals actively excluded traditional religious communities, the latter argument ran, even as they worked to include every other imaginable social group. Now, the minoritarian strategy gained ground, especially in legal and academic circles. Yet the terms of the argument had changed. In the 1970s, figures such as Jerry Falwell, Francis Schaeffer, and C. Everett Koop had urged their readers to rise up against secularism and restore the country's Christian foundations. By the 1990s, religious traditionalists increasingly advocated a pluralistic solution. Rather than insisting they possessed universal truths, they used multi-foundational understandings of knowledge and belief to argue that simple justice demanded their inclusion at the cultural table, as an oppressed minority alongside many others. They often joined the academic left in portraying science as a form of knowledge that suppressed alternative viewpoints by erasing its origins in particular cultural contexts and claiming universal validity.[41]

The minoritarian defense of traditional religion spread across numerous domains and intersected with a host of intellectual changes. By the early twenty-first century, it would drive a resurgence of accommodationism in jurisprudence: the view that the First Amendment allowed the federal government to aid all religions equally, whereas strict separationism established an official, secular faith. In the 1980s and 1990s, however, much of the ferment centered on political theory. There, John Rawls's iconic book *A Theory of Justice* had set the terms of debate since its appearance in 1971. Critics of varied persuasions took aim at Rawls's work, often in ways that implicated science. A common target, for example, was the concept of "public reason," which Rawls began to flesh out in his essays after 1980 and the German social theorist Jürgen Habermas later articulated in a different idiom. Rawls's version held that public decisions could be justified only through arguments that all reasonable persons could accept because they did not derive from contested sources such as religion. To many critics—including Neuhaus himself—this conception produced the naked public square: a domain where both law and convention forced religious believers to police their language and suppress their views.[42]

Michael Sandel, Charles Taylor, and others who came to be called "communitarians" in the 1980s targeted a related aspect of Rawls's work: the thin model of selfhood that they detected behind his "original position" thought experiment. Rawls had proposed that citizens should strive toward the rules and institutions they would design if they stood behind a "veil of ignorance," knowing nothing of the particular personal and social attributes they would bear in the imagined society and thus seeking to maximize justice for persons of all types. Religious critics welcomed Sandel's assault on the concept of an "unencumbered self," bereft of all forms of social belonging. He argued that religious commitments amounted to prepolitical, collectively binding obligations or duties, not mere subjective preferences. Taylor, in a monumental interpretation of the "sources of the self," fleshed out such challenges to the idea of an autonomous, rationally choosing individual by tracing the career of this modern or liberal self as it had emerged from, and eventually replaced, more communal understandings of personhood. Such studies often linked value-neutrality in the sciences to modernity's arid individualism. Taylor, for example, had dissected B. F. Skinner's behaviorism in his dissertation and helped to define interpretivism in a 1971 article before he turned to conceptions of selfhood. He would go on to write an equally massive and influential account of secularization.[43]

The minoritarian understanding of traditional religion also inspired new historical interpretations of religion's relationship to science in the United States. Much of this work concerned religion's shifting role in higher education. Critics described the marginal place of religious traditionalists in the universities as both a microcosm and a cause of their exclusion from American public life in general. Thus, the historian George M. Marsden argued that all "ideals for the social good" were equally "sectarian," and so too were "the sciences that serve those ideals." Operating on the basis of naturalistic presuppositions, he contended, mainstream academic institutions often discriminated against those who were "seriously Christian in academically unpopular ways." At the same time, they allowed nontheists to express their views openly and accommodated liberal Protestantism in its secular guise as "political progressivism." In short, modern universities disingenuously used the language of tolerance to suppress dissenters.[44]

Citing Reformed epistemology and MacIntyre's work, Marsden advocated "a Madisonian approach that thwarts the tyranny of the majority" in both the universities and American public life more broadly. He envisaged the universities, for example, as "federations for competing intellectual communities of faith and commitment," each with its own "dogmatic framework" and its corresponding principles of freedom and rationality. Marsden and his allies backed this proposal with detailed historical explications of the secularization of American academia. Through the 1990s and into the new millennium, they published a series of works arguing that the liberal Protestants of the nineteenth and early twentieth centuries had undermined the classic Christian enterprise and eventually oriented the universities toward exclusively naturalistic principles. Marsden characteristically described modern secularism as "a liberal Protestantism with the explicit Christianity removed," enforcing the rule of experts and featuring John Dewey as its "high priest." Such portraits of secular scholarship as an oppressive and hegemonic force, establishing science as an official faith, shaped much subsequent work on American religious life as well as the history of the research universities.[45]

By the 1990s, "postsecular" perspectives were gaining ground on the left as well. When critical theorists identified popular culture as a site of resistance, or STS scholars explored paths not taken in science and politics, or Foucaultians valorized subjugated knowledges, or postcolonial critics rejected Eurocentric categories, religion cropped up again and again. Widespread cultural conventions also held that science and religion stood in a zero-sum relationship and that one or the other anchored any possible worldview. Meanwhile, the communitarian argument about the prepolitical character of religious belonging paralleled many radical critics' claim that liberal regimes failed to account for social groups, structural oppression, and collective justice. As a result, it became more common on the academic left to move from the charge that science and liberalism encoded domination to the claim that secularism represented a form of hegemony, or at least that religion offered potent and perhaps irreplaceable resources for challenging social hierarchies. In short, critical scholars increasingly associated both science and secularism with elite rule.[46]

Indeed, some radical theorists had argued since the 1960s that aligning science with human values was a religious project. Some of the sharpest critics of science and modernity, including the historians Theodore Roszak and Morris Berman, sought mystical alternatives to the dead-end rationalism of modern science. But others, especially women and Black theorists, saw radical possibilities in the established religious traditions. Figures such as Mary Daly, James H. Cone, Emilie Townes, and Cornel West gave Christianity a prominent place in the struggle. In fact, some commentators saw the Christian Right as a powerful counterforce to hegemonic scientism and secularism, even as they deplored its social and political values. The activist rabbi Michael Lerner and his associates at *Tikkun,* who arrayed their "spiritual progressivism" against scientific naturalism, offered a prominent example. Although Christian creationists were theologically "absurd" and politically "evil," *Tikkun* associate editor Peter Gabel wrote in 1987, "there is something correct and admirable in their refusal to accept the hegemony of science as a privileged source of truth."[47]

Concern with religion also flourished among postcolonial theorists, at a time when scholars with backgrounds outside the Judeo-Christian faiths increasingly made their mark on the Western academy. Although the post-secular discourse would not fully flower until after 2000, interpreters of colonialism began to join religious studies scholars in developing the insight that "religion" itself was a Western category, co-constituted with "science" and "the secular"—and with the domination of all whose beliefs, values, practices, and institutions did not fit the bill. In its global, postcolonial iterations, postsecularism emphasized how secular regimes reflected Europe's self-conception as Christendom, as well as their tendency to repress local traditions in colonized areas, including the age-old foe of Islam. A 1998 anthology entitled *Secularism and Its Critics* featured statements by Taylor and Sandel alongside pieces on colonial and postcolonial settings, especially India. In those areas, secularism and religion had mapped especially neatly onto patterns of social domination, including both colonization itself and the history of forced secularization under nationalist regimes in Egypt, Turkey, and elsewhere. In the burgeoning postcolonial discourses of the 1990s, the economic, social, and political concerns of the academic left converged with the minoritarian claims of religious traditionalists, despite the different sensibilities that inspired those arguments. As the century drew to a close, many kinds of critics identified secular modernity, sprawling across the global North and South, as the shared enemy.[48]

AS IN THE POSTWAR YEARS, critical portraits of modernity in the 1980s and 1990s fueled convergences across political and religious divides. Once again, perceptions of science's character and cultural influence provided a key point of overlap. A host of late twentieth-century critics framed modernity as scientific and secular, as well as liberal and capitalist. They identified science as a central cause or vector of modern forms of domination, lacking any critical potential and reflecting the will to power of elites. Rather than charging that science fostered relativism and enabled manipulative programs of social engineering, as had postwar critics and many of their 1960s successors, they tended to indict it on the opposite grounds: that its claim to neutrality undermined intellectual and cultural diversity by allowing proponents to silently side with one of the competing perspectives. Science provided an unparalleled ideological resource because it portrayed the dominant group's views as neutral, universal, inevitable, and unquestionable.

These arguments often rested on a sharp distinction between science and all other cultural systems, including religion, philosophy, the humanities, and political thought. These appeared as the conservators of genuine personhood, tilting against all forms of rationalism—and certainly not sharing in it themselves. Such forms of reasoning, many critics implied, presented themselves as situated human judgments; they did not claim to offer objective, universal truths. Indeed, the assertion that science, alone among human practices, erased its own conceptual foundations seemed to imply that every other form of thought in human history had eschewed strong truth claims and openly stated its presuppositions—and would thus authorize a robust regime of cultural diversity if it replaced science in the driver's seat. It was easy to conclude that simply choosing the humanities or a religious tradition over science represented a significant blow for inclusion and justice.

The new modes of critique intersected with a wider societal embrace of cultural, moral, and religious difference. By the start of the twenty-first century, arguments about belief, identity, and inclusion in the United States increasingly proceeded in a pluralistic idiom. On all sides, "We're right and you're wrong" was giving way to "Our voices should be heard too," especially for public purposes. The pluralization of both academic and political understandings would continue to shape perceptions—and even the content—of science, politics, religion, and much else as the early twenty-first century unfolded. For centuries, conflicts between competing cultural programs in the West had largely been defined by the assumption of universalism. Americans had often been especially fervent defenders of universalism, insisting that a single body of truth or knowledge would prove valid for all of humanity. By the end of the twentieth century, however, that assumption had begun to recede.

Still, old habits die hard. Despite the new emphasis on human difference, few commentators saw merely a congeries of clashing groups and perspectives around them. Instead, most discerned a broad, big-tent alliance that included the majority of humanity and stood under threat from a recalcitrant minority that was almost constitutionally incapable of joining the fold. Multiple versions of this quasi-universalism flourished by the 1990s. Liberals and progressives often identified cultural pluralism itself as the shared human ground, while treating differences in language, dress, foodways, and religious observance as relatively superficial and assuming a high degree of commonality beneath them. Here, the enemies of tolerance—especially Christian and Muslim "fundamentalists"—represented the threat. Meanwhile, critical scholars and activists often pitted the entire array of social and cultural minorities against a hegemonic few. They assumed that marginalized groups of all kinds shared a common vision of a future without coercive power, in which structures of cultural and political authority had been broken down. A third quasi-universalist stance portrayed most human beings as "seriously religious," defining every aspect of their lives in theological terms and seeking to reshape social and political institutions accordingly by throwing off the secular shackles imposed by liberal elites.

Each of these quasi-universalist perspectives normalized particular values and practices while portraying the alternatives as irrational, pathological, and socially dangerous—indeed, as the product of a desire to dominate the right-thinking masses. Each also tied science to liberalism and placed both squarely on one side of the divide: with the tolerant majority in the first case and with the repressive elite in the other two views. Of course, many conservative evangelicals still saw a single, unified belief community on the horizon, despite a host of challengers, both secular and religious. But many other late twentieth-century commentators combined the era's intense emphasis on cultural difference with one of the quasi-universalist frameworks, each of which directly implicated science in the posited clash between a minority seeking to exert cultural dominance and a majority pursuing its own forms of flourishing and self-expression. These normalizing visions of culture and power would deeply inflect understandings of science in the new millennium.

CONCLUSION

Humanizing Science

AS I BEGAN RESEARCHING THIS BOOK, a fracas over scientism broke out. At a Brandeis University graduation ceremony, *New Republic* literary editor Leon Wieseltier told humanities majors that they represented "the resistance" in a society dominated by "instrumental reason": "the twin imperialisms of science and technology." Wieseltier sounded all the familiar themes—the enslavement of human beings to machines, the tyranny of numbers, the depredations of scientism and "technologism," the unchallenged dominance of "utility, speed, efficiency, and convenience" in modern culture—and identified the humanities as the antidote. The evolutionary psychologist Steven Pinker fired back, charging that petulant humanists welcomed science when it cured disease but not when it impinged on their professional fiefdom. The march of science and enlightenment had vastly improved the human condition, Pinker declared. Moreover, it had invalidated all traditional religions along the way. Only science, Pinker insisted, could address "the deepest questions about who we are, where we came from, and how we define the meaning and purpose of our lives." Humanities scholars would remain irrelevant, he contended, until they embraced the scientifically informed humanitarianism that constituted the "de facto morality" of the modern world, despite waves of irrationalism. The ensuing controversy stretched through the summer and fall.[1]

As I finished the book, a global pandemic gripped the world. Societies faced immediate, practical, life-or-death questions about how to incorpo-

rate science and expertise into their collective decisions. Yet the old refrains could still be heard. Writing in *Commentary,* the conservative commentator Sohrab Ahmari responded to the outbreak by arguing that "the
ideology of scientism" had plunged the world into "a half-millennial funk."
In the face of a deadly virus, Ahmari wrote, moderns lacked any sense of
why "life is worth living and passing on"; they could not even assert that
"being is preferable to nonbeing." Meanwhile, a *Los Angeles Times* reader
traced the spread of rumors and misinformation about COVID-19 to the
fact that Americans, unlike Europeans, were neither educated nor secular:
being "singularly gullible," they accepted any proffered lie "on faith." Pinker
chimed in as well, arguing that political decisions favoring economic well-
being over bodily health reflected the "malignant delusion" of evangelicals'
"belief in the afterlife," which "devalues actual lives."[2]

This tired pattern of sweeping charges and countercharges has recurred
throughout the early twenty-first century, as in prior decades. Bitter public
controversies have swirled around climate change, intelligent design, genetically modified foods, vaccines, psychotropic drugs, race and medicine,
reproductive technologies, cloning, the possibility of genetically enhanced
embryos, artificial intelligence, nanotechnology, social media, workplace
automation, surveillance cameras, data mining, and many dozens of related
or subsidiary issues. In response, cultural critics have reiterated their familiar positions—usually the lament that a soulless science dominates
modern life or the fear that a rising tide of unreason will return humanity
to the dark ages. Massive abstractions still abound, as commentators invoke the usual broad categories: science, scientism, rationalism, the Enlightenment, the humanities, humanism, religion, faith, irrationality, the West,
modernity.[3]

Yet such large-scale abstractions have proven remarkably unhelpful in
the controversies of our day. Each of the issues we face has its own distinctive contours, involving the complex interrelations of scientific and technical developments with social norms, practices, and institutions. Despite
the fiery statements of combatants and worried onlookers, the scientific enterprise as a whole is not at stake in debates over vaccination, genetic engineering, or climate change. Rather, these controversies involve particular
scientific findings, theories, techniques, devices, and practices, as they relate to the deeply held (and often directly conflicting) values of many different groups. We will struggle to address the vexing questions of the
twenty-first century if we continue to use the blunt interpretive tools of the
nineteenth and twentieth centuries. Those tools were forged in polemics between clashing cultural elites over science's extension into new domains—
first the history of life on earth and then human relations—and its place in

high schools and colleges. Such blanket injunctions to place our trust in science, or religion, or the humanities, or any other broad framework, offer remarkably little guidance on how to respond to the social possibilities raised by particular scientific or technical innovations.

The challenges to scientific authority that have circulated in the United States since the 1920s are not wrong in every detail. Science is a messy, thoroughly human enterprise that does not, and cannot, address many of the issues we face. Indeed, most scientists share that assessment themselves. But the kinds of critics described in this book have either ignored such instances of modesty and self-reflection or treated them as rare exceptions to the rule. Generalizing broadly, they have asserted that science as a whole claims the ability to answer all questions and solve all problems. They have also contended that science reigns supreme in modern societies, determining the basic contours of our thought. Finally, these critics have traced a host of specific social problems to science's cultural influence.[4]

The second of these assumptions—that science sets the tone of modern culture—anchors the rest of the argument and deserves special scrutiny. If science is not culturally dominant, then it does not particularly matter what science's advocates claim it can accomplish. Similarly, if science is not culturally dominant, then it cannot have caused the litany of ills often laid at its door. Much depends on how we think scientific authority functions in modern societies. So it is particularly important to consider this question carefully, rather than simply tracing the familiar grooves of stock interpretive habits. Thanks in part to the arguments described in the preceding chapters, we all "know" that modern societies differ fundamentally from their predecessors, and that much of the difference concerns the rise of science. But is science truly so influential? Or do we instinctively blame it for problems that mainly reflect basic human qualities such as pride and greed? What, exactly, is scientific about our world?

Following the common, though problematic, practice of using the United States to exemplify modernity, we can certainly say that science serves important public functions. Biologists and physicists wield forms of authority in courtrooms that religious leaders and literary critics do not possess. The Federal Reserve draws on economic experts, not the Bible or Melville. The Environmental Protection Agency takes cues from the natural sciences, the Department of Education from the social sciences. Public high schools can teach Darwinism but not creationism or intelligent design. Looking at these instances, we might conclude that science enjoys a unique position of privilege in American public culture.

Yet other experts also share in such privileges. Although critics often charge that science has monopolized the very term "knowledge," we constantly rely on the knowledge of historians, journalists, jurists, and eyewitnesses, among others, even though we do not consider their work scientific. Science's elevated position turns out to be partly a matter of selective exclusion: In keeping with the First Amendment, public institutions in the United States refrain—rightly or wrongly, consistently or inconsistently—from treating theological tenets as established truths. Meanwhile, even the most fervent champions of literature and the arts have rarely claimed that these offer forms of knowledge that should be used in courtrooms or policy decisions.

But the locus of expertise is not what usually drives the argument that modern cultures are scientific. Rather, it is a matter of values. Looking at the world, who does not find it wanting in innumerable respects? This is as close to a universal human experience as one can imagine. Encountering numerous people with values different from our own, and an array of institutions and practices that also clash with our values—sometimes in ways that harm us—what can we conclude? Over the past few centuries, a growing number of Westerners have identified causes other than a deity's wrath, human sinfulness, or the innate inferiority of other social groups. Instead, they have habitually looked to culture: the domain of collective beliefs, norms, and practices. There, they have found the sources of the world's perversity, calling these sources heresy, ignorance, superstition, or simply error. The arguments have gone back and forth: We need this kind of religion, or that one. We need this philosophical gloss on science, or that one. We need a transcendent moral framework, or a humanistic sensibility.[5]

Ironically, the detachment of public institutions from religious agencies has made it far easier for religious and humanistic critics to press their cases against science, although it has sometimes hindered their ability to find a hearing as well. Particularly as religious communities became more tolerant of one another, the shared enemy—formerly heathenism, now materialism, naturalism, or secularism—presented an obvious target. Surely secular institutions had resulted from secular philosophies, and surely science had produced those philosophies. As scientists ratcheted up their claims of value-neutrality, more and more critics drew a causal connection. Science, they argued, had brought forth a world dominated by shallow, materialistic values, or by a purely instrumental mind-set that obscures the very existence of values—or perhaps by the values of a hegemonic social group, painted with the brush of neutrality.

Is it really the case, however, that secular institutions and practices reflect the cultural dominance of science? Some countries have certainly witnessed the active imposition of "scientific" worldviews by militantly secular regimes. Even in such cases, however, the practices of science did not necessarily align with the philosophies marching under its banner. In other contexts, including that of the United States, the relations between science, philosophy, and secularization have been even more complex and indirect—not least because science is an institutionalized set of knowledge practices, not a philosophical system. It is a gross simplification to claim that science "has authored our world," or that contemporary social life is largely "the concrete and dramatic re-enactment of eighteenth-century philosophy."[6]

Many features of the modern world are secular but not scientific, even if one grants for a moment the conflation of scientific practices with science-inspired philosophies. Law, bureaucracy, capitalism, consumerism, journalism, education, sports: these modes of practice, like many others, reflect in part the waning control of religious institutions. But they do not share a single, common philosophical foundation with science. Like all social formations, each takes much of its shape from age-old human traits and from conflicts between particular groups. And each, in turn, generates a distinctive array of cultural assumptions, values, and behaviors. Such practices and institutions may rely instrumentally on the results of scientific research, but they often stand in significant tension with science as well as with one another.[7]

This is not to say that foundational presuppositions are historically irrelevant. Beliefs about the existence and nature of a deity, the sources of human action, and the composition of the world can make a huge difference, although these are inevitably compounded with many other forces. Most practices and institutions in the modern West—and, in fact, many of their premodern and non-Western counterparts as well—seem pointless or even harmful to those who assume that a deity determines our worldly fortunes, that our earthly actions matter primarily in relation to our otherworldly fate, or both. The typical patterns of exertion in modern societies fit much better with the view that one's earthly well-being largely reflects one's earthly actions, and that these actions matter primarily for that reason.

Although the spread of that emphasis on the here and now has been highly consequential, it is neither secular nor scientific in itself. Of course, it is widespread among nontheists, although some have deemed human action essentially meaningless. But it also comports with a broad range of religious understandings, even as it clashes with others. Indeed, one of the points of contention in many debates around science and modernity is the legitimacy of these comparatively worldly forms of religion. It is important to remember that the protean category of "the secular" is always tied to specific understandings of religion. For each group, the secular is what lies beyond the boundaries of those religions it recognizes as genuine or authentic. The claim that modernity is secular or scientific often accompanies the contention (or the unstated assumption) that theologically liberal forms of theism are not, in fact, legitimately religious; they simply drape science or secularism in religious garb. Narratives about science's disastrous cultural effects allow critics of liberal theologies to boost their own religious frameworks by blaming widely deplored social phenomena on the influence of alternative views, as many champions of the humanities have also done.[8]

Again, there are important questions at stake here, with real consequences. Is there a God that intervenes in our affairs? Should our educational system emphasize biology, literature, or the Bible? The answers matter a great deal. But how we frame our arguments also matters. It is unjust and socially harmful to push all the results of human frailty onto our opponents' ledgers. That approach breeds resentment and distrust, including skepticism toward our own cultural programs when it becomes clear that we have vastly overstated the real-world effects of the competing perspectives—and that ours is no cure for human foibles either. Meanwhile, tracing social problems back to philosophical disagreements leaves us unable to address those problems themselves, both by misrepresenting their main causes and by convincing us that we must litigate our intellectual conflicts before we can take meaningful action. Without overstating the degree of shared ground, we should work to build coalitions where possible, even if we believe our own views will shine through in the end. Many of the ideas that profoundly shape social behavior—ideas about racial equality, for example, or the need for economic regulation—cut across religious and secular perspectives. Our competing patterns of apologetics should not forestall collective action in these areas.

Science's champions, like its critics, have often gone to absurd lengths to discredit worldviews they considered harmful. But here—as with religion and the humanities, for that matter—it is important to distinguish between science as a set of practices and institutions and the philosophies that have gone under its name. The critics described in this book often took the salutary step of differentiating science from these philosophies: materialism, naturalism, positivism, and so on. But they did so in a manner that placed highly valued scientific practices on their own side of the philosophical line and blamed the ills of the world on science-oriented outlooks. We would be wise to build in a different way on that distinction, which working researchers have often drawn as well. There are three separate stories to be told, although they intersect at many points: one story about the social roles of scientific practices and findings, a second about the trajectories and entanglements of science-inspired philosophies, and a third about the development of secular patterns and institutions. To insist on differentiating these levels of analysis is not to claim that science is intrinsically pure, operating in glorious isolation from the human world. Quite the opposite; science is profoundly shaped by social and cultural dynamics. Yet distinguishing between science, philosophy, and the secular, rather than conflating them, would allow us to understand their historical entanglements more clearly—and to grapple more effectively with the implications of empirical research in our own day.[9]

Over time, in fact, a more charitable and nuanced assessment of science might help us liberate researchers from the extravagant assertions of disinterestedness that envelop their work. It is not their claims alone, but also the arguments and actions of many other groups, that have trapped scientists in the cage of absolute value-neutrality. Critics often declare that science eschews considerations of value, in order to blame it for doing so. Some go farther, contending that genuine science provides absolutely certain knowledge—not models, not probabilities, not calculations of risk—and must do so before we can act on it. And most of us, when confronted with either findings or applications that we dislike, demand that the researchers in question demonstrate their complete disinterestedness before we will take them seriously. The integration of scientific disputes into party politics over the past few decades has entrenched this societal demand for neutrality even more deeply.[10]

This cycle must be broken if we are to recognize science for what it truly is: a thoroughly human practice like any other, yet one that produces remarkable outcomes. Rather than arguing that science's validity depends on the personal neutrality of individual researchers—a view that piggybacks on Christian understandings of benevolence, sacrifice, service, and humility—we could instead learn to value scientific findings for their reliability. Indeed, we could improve scientific procedures by adding new features, such as forms of citizen participation, to help ensure the reliability of the results. But this may prove impossible if critics continue to view science as a monstrous cultural presence and blame it for humanity's faults, rather than simply assessing its strengths and weaknesses. It is time to get past the polemics, acknowledge that science is a central feature of our world, and decide what we will make of it.

NOTES

ACKNOWLEDGMENTS

INDEX

NOTES

INTRODUCTION

1. Cary Funk, Meg Hefferon, Brian Kennedy, and Courtney Johnson, "Trust and Mistrust in Americans' Views of Scientific Experts," Pew Research Center, August 2, 2019, https://www.pewresearch.org/science/2019/08/02/trust-and -mistrust-in-americans-views-of-scientific-experts; Stephanie Fine Sasse and Lucky Tran, eds., *Science Not Silence: Voices from the March for Science* (Cambridge, MA: MIT Press, 2018).

2. Of course, Richard Hofstadter famously argued in 1963 that American culture was intrinsically hostile to intellectual activity. But "anti-intellectualism" is too broad a category to capture the particularities of arguments about science's cultural authority. Moreover, recent interpreters have tended to reverse Hofstadter's concerns and emphasize the unchecked power of experts, especially in the mid-twentieth century.

3. Nancy R. Pearcey, "Darwin Meets the Berenstain Bears: Evolution as a Total Worldview," in *Uncommon Dissent: Intellectuals Who Find Darwinism Unconvincing,* ed. William A. Dembski (Wilmington, DE: ISI Books, 2004), 53. Since I began this project, sociologists have started to address the centrality of moral visions in debates over science. See especially John H. Evans, *Morals Not Knowledge: Recasting the Contemporary U.S. Conflict between Religion and Science* (Berkeley: University of California Press, 2018).

4. Jacques Barzun, *Science: The Glorious Entertainment* (New York: Harper and Row, 1964). Barzun himself argued that social scientists had been "secreting . . . something akin to a poison" into the world by extending the scientific assumption

of "purposelessness" to human behavior, so that the physicist's mechanistic outlook had become the "image of our inner life." Jacques Barzun, in "Discussion," *Proceedings of the Academy of Political Science* 28, no. 2 (April 1966): 16; Barzun, "Science as a Social Institution," in ibid., 11, 13.

5. In my previous book, I explored a variety of ways in which advocates of science linked it to Protestantism and to democracy: Andrew Jewett, *Science, Democracy, and the American University: From the Civil War to the Cold War* (New York: Cambridge University Press, 2012). On the science-technology link, see Ronald Kline, "Construing 'Technology' as 'Applied Science': Public Rhetoric of Scientists and Engineers in the United States, 1880–1945," *Isis* 86, no. 2 (June 1995): 194–221.

6. Barzun, *Science;* Victor H. Noll, "The Habit of Scientific Thinking," *Teachers College Record* 35, no. 1 (January 1933): 1–9. Variations in these usages will partly reflect linguistic differences: the denotative and connotative distinctions between, for example, the English "science," the French *sciences,* and the German *Wissenchaft,* to say nothing of their rough analogs in other areas of the world.

7. Of course, these meanings intersect with the constructions and resonances of categories closely linked to "science," such as "religion" and "politics." Understanding such phenomena is not just a matter of looking for "high" ideas in "low" places. Rather, vernacular contexts tend to produce distinctive formulations that often require additional, contextually specific knowledge to interpret.

8. For helpful overviews of popular stereotypes, especially in the crucial postwar period, see Marcel C. LaFollette, *Making Science Our Own: Public Images of Science, 1910–1955* (Chicago: University of Chicago Press, 1990); and Aaron Lecklider, *Inventing the Egghead: The Battle over Brainpower in American Culture* (Philadelphia: University of Pennsylvania Press, 2013). Paul S. Boyer discussed both cultural images and intellectual arguments in *By the Bomb's Early Light: American Thought and Culture at the Dawn of the Atomic Age* (New York: Pantheon, 1985). An important study of postwar debates around science and religion is James A. Gilbert, *Redeeming Culture: American Religion in an Age of Science* (Chicago: University of Chicago Press, 1997).

9. This fact helps explain an otherwise confounding phenomenon. Despite the proliferating challenges to science, other measures suggest that it enjoys widespread public confidence. Asked whether they trust various institutions, Americans consistently rank science and medicine near the top, behind only the military—and far ahead of schools, corporations, government agencies, the media, and even religious organizations. Indeed, a poll taken shortly before the COVID-19 pandemic showed science pulling ahead of the military; it had risen steadily while the military remained constant. Significant majorities, across all religious traditions, have also favored generous federal funding for scientific research; this contingent surely includes many who dislike government spending in general. Funk et al., "Trust and Mistrust in Americans' Views of Scientific Experts." On regulatory questions, meanwhile, Americans have been far more likely than Europeans to leave decisions about potentially dangerous technologies or substances to scientists and private companies

rather than government. Sheila Jasanoff, *Designs on Nature: Science and Democracy in Europe and the United States* (Princeton, NJ: Princeton University Press, 2005).

10. These kinds of perceptions play a significant role in this book, which contains innumerable iterations of terms such as "seemed" and "viewed." The God's-eye view of the past that historians often adopt does not match how human beings encounter their worlds in their own times. We can learn a great deal by writing from multiple, embedded perspectives instead, seeking to reconstruct what particular kinds of people saw and what they took it to mean. On this practice, see K. Healan Gaston, "Reinscribing Religious Authenticity: Religion, Secularism, and the Perspectival Character of Intellectual History," in *American Labyrinth: Intellectual History for Complicated Times,* ed. Andrew Hartman and Raymond Haberski Jr. (Ithaca, NY: Cornell University Press, 2018), 223–238.

11. Everett Ross Clinchy, ed., *The World We Want to Live In* (Garden City, NY: Doubleday, 1942), 80.

12. Of course, science faced additional challenges in the mid-twentieth century as well. Conservative Protestants targeted Darwinism, as they had since the 1920s. Other groups emphasized the destructive potential of nuclear energy. But historians know far less about the broader, and arguably more influential, mode of criticism detailed here.

13. "Feature 'X,'" *America* 92, no. 5 (October 30, 1954): 125–127; Byron E. Eshelman, "Tyranny Can Be Subtle," *Christianity and Crisis* 12, no. 4 (March 17, 1952): 27–28; Andrew Hacker, "The Specter of Predictable Man," *Antioch Review* 14, no. 2 (July 1954): 195–207; Waldo Frank, "The Central Problem of Modern Man," *Commentary* 1, no. 8 (June 1946): 51; *America* quoted in Paul S. Boyer, *By the Bomb's Early Light: American Thought and Culture at the Dawn of the Atomic Age* (New York: Pantheon, 1985), 209; Dwight Macdonald quoted in Boyer, *By the Bomb's Early Light,* 235. Timothy Melley has written insightfully on the "agency panic" that gripped many postwar Americans as they sought to come to terms with new patterns of social causation (*Empire of Conspiracy: The Culture of Paranoia in Postwar America* [Ithaca, NY: Cornell University Press, 2000]). Mark Greif sees a comprehensive "crisis of man" in this period (*The Age of the Crisis of Man: Thought and Fiction in America, 1933–1973* [Princeton, NJ: Princeton University Press, 2015]). For additional material on post–World War II challenges, see Boyer, *By the Bomb's Early Light;* David Paul Haney, *The Americanization of Social Science: Intellectuals and Public Responsibility in the Postwar United States* (Philadelphia: Temple University Press, 2008); Mark Solovey, *Shaky Foundations: The Politics–Patronage–Social Science Nexus in Cold War America* (New Brunswick, NJ: Rutgers University Press, 2013); David A. Horowitz, *America's Political Class under Fire: The Twentieth Century's Great Culture War* (New York: Routledge, 2003); and Stephen Schryer, *Fantasies of the New Class: Ideologies of Professionalism in Post–World War II American Fiction* (New York: Columbia University Press, 2011).

14. At the time, many scholars viewed postwar critiques with great alarm. The files of the sociologist Robert K. Merton groaned with hundreds of examples he

collected over the years. Other commentators also worried about challenges to science and expertise: e.g., Bernard Mausner and Judith Mausner, "A Study of the Anti-Scientific Attitude," *Scientific American* 192, no. 2 (February 1955): 35–39; Margaret Mead and Rhoda Metraux, "Image of the Scientist among High-School Students," *Science* 126 (August 30, 1957): 384–390.

15. Walter Lippmann, *An Inquiry Into the Principles of the Good Society* (Boston: Little, Brown, 1937), 387. On the centrality of philosophical anthropology, see also Greif, *The Age of the Crisis of Man.*

16. On these earlier views, see Jewett, *Science, Democracy, and the American University.*

17. "The Crisis of the Individual: A Series," *Commentary* 1, no. 2 (December 1945): 1–2.

18. James B. Conant, *Science and Common Sense* (New Haven, CT: Yale University Press, 1951), 259.

19. Walter Lippmann, *Essays in the Public Philosophy* (Boston: Little, Brown, 1955), 125; Peter Viereck, *Conservatism Revisited* (1949; repr., New York: Free Press, 1962), 33–35; Frank S. Meyer, "Retreat from Relativism," *National Review* 1, no. 4 (December 14, 1955): 24–25. Stephen Schryer explores this understanding of the humanities as an antidote to science's corruption of modern culture in *Fantasies of the New Class.*

20. Reinhold Niebuhr, *The Nature and Destiny of Man: A Christian Interpretation,* 2 vols. (New York: Charles Scribner's Sons, 1941–1943); Hans Morgenthau, *Scientific Man vs. Power Politics* (Chicago: University of Chicago Press, 1946), esp. 5–6. Such postwar arguments reflected a broader push toward idealist accounts of history among American thinkers, at a time when historical materialism smacked of communism and any acknowledgement of persistent social conflicts seemed a threat to national unity. A common version of the argument, then as now, described all human thought and behavior as outgrowths of specifically theological commitments, between which no neutrality was possible. "Man thinks and acts in accordance with what he worships," explained one group. Clinchy, *The World We Want to Live In,* 2. See also Andrew Jewett, "Science and Religion in Postwar America," in *The Worlds of American Intellectual History,* ed. Joel Isaac, James T. Kloppenberg, Michael O'Brien, and Jennifer Ratner-Rosenhagen (New York: Oxford University Press, 2017), 237–256.

21. When Barzun published his book-length indictment of scientism in 1964, the philosopher A. R. Louch castigated him for viewing "human problems generally through a screen of academic problems." Though no friend to scientism, Louch saw only a "punning likeness" between mechanistic science and "mechanized lives." Americans, he insisted, were in thrall to pecuniary values, not empiricism or positivism—in short, to big business rather than science. Louch, "Idols of the Lab," *Commentary* 38, no. 2 (August 1964): 73–74. Barzun himself had noted the inward-looking character of postwar debates back in 1959. "The intellectuals' chief cause of anguish are one another's works," he observed in a rare self-reflective moment. "The intellectuals do not attack bars and

bowling alleys, or Tammany Hall and the chemical trust; they attack the philosophy of John Dewey and the State Department's policy on modern art. The humanists inveigh against science, and the scientists decry 'vague inspirational subjects.'" Barzun, *The House of Intellect* (New York: Harper & Brothers, 1959), 2.

22. Jeffrey P. Moran, *American Genesis: The Antievolution Controversies from Scopes to Creation Science* (New York: Oxford University Press, 2012); Jonathan Scott Holloway, *Confronting the Veil: Abram Harris Jr., E. Franklin Frazier, and Ralph Bunche, 1919–1941* (Chapel Hill: University of North Carolina Press, 2002); Molly Ladd-Taylor, *Mother-Work: Women, Child Welfare, and the State, 1890–1930* (Urbana: University of Illinois Press, 1994); Julia Grant, *Raising Baby by the Book: The Education of American Mothers* (New Haven, CT: Yale University Press, 1998); Miller quoted in Holloway, *Confronting the Veil*, 82; Lecklider, *Inventing the Egghead*, esp. 69–115.

23. Meanwhile, critics have often used the writings of methodological dissidents in the sciences—those who rejected value-neutrality, materialism, reductionism, or related ideals—as proof that science indeed adhered to such principles, rather than seeking to broaden the prevailing images of science itself.

24. Jewett, *Science, Democracy, and the American University*. A recent study of the subsequent career of this value-neutral ideal is Audra J. Wolfe, *Freedom's Laboratory: The Cold War Struggle for the Soul of Science* (Baltimore: Johns Hopkins University Press, 2018). For boundary work, see Thomas F. Gieryn, *Cultural Boundaries of Science: Credibility on the Line* (Chicago: University of Chicago Press, 1999).

25. Will Herberg, *Judaism and Modern Man: An Interpretation of Jewish Religion* (Philadelphia: Jewish Publication Society of America, 1951), 21.

26. More broadly, the style of criticism described in this book has entrenched a particular understanding of religion that centers on its perceived capacity to provide the robust, nonempirical, and relatively fixed moral principles that scientific thinking is said to threaten. Religion, in this view, primarily offers cognitive claims about moral obligation and only incidentally trades in mystery, contemplation, consolation, community, or various other elements. Likewise, this critical tradition casts literature and the arts as founts of moral tenets, not sources of aesthetic pleasure, self-understanding, or collective belonging.

27. Gordon Keith Chalmers, *The Republic and the Person: A Discussion of Necessities in Modern American Education* (Chicago: Regnery, 1952), 43. This was not the only mode of criticism that the New Deal state faced, even outside conservative circles. Many other Americans, then as now, thought the federal government was militantly secular and wanted to stamp out religion, whatever its economic policies. In both cases, historians should incorporate understandings of science alongside those of economics, race, and crime when thinking about how broad cultural tendencies have intersected with political sensibilities.

28. This new emphasis could be seen already in the 1920s and has recurred frequently since, among groups with varied cultural ideals and political commitments. In 1952, for example, the former socialist Reinhold Niebuhr deemed

businessmen far superior to scientific experts because they knew when to set aside ideologies and seek compromises, whereas experts insisted on trying to remake the world in their own image. Niebuhr, *The Irony of American History* (New York: Charles Scribner's Sons, 1952).

1. MENTAL MODERNIZATION

1. This section draws on Andrew Jewett, *Science, Democracy, and the American University: From the Civil War to the Cold War* (New York: Cambridge University Press, 2012). On nineteenth-century contrasts to Europe, see esp. Daniel T. Rodgers, *Atlantic Crossings: Social Politics in a Progressive Age* (Cambridge, MA: Belknap Press of Harvard University Press, 1998).

2. On the rise of methodological naturalism and its impact on views of morality and religion, see esp. George M. Marsden, *The Soul of the American University: From Protestant Establishment to Established Nonbelief* (New York: Oxford University Press, 1994); Julie A. Reuben, *The Making of the Modern University: Intellectual Transformation and the Marginalization of Morality* (Chicago: University of Chicago Press, 1996); Jon H. Roberts and James Turner, *The Sacred and the Secular University* (Princeton, NJ: Princeton University Press, 2000); and Marsden, *The Twilight of the American Enlightenment: The 1950s and the Crisis of Liberal Belief* (New York: Basic Books, 2014).

3. Sinclair Lewis, *Babbitt* (New York: Harcourt, Brace, 1922). As Christopher G. White has shown, many religious liberals appropriated resources from the new psychology, even as other religious leaders recoiled from it. White, *Unsettled Minds: Psychology and the American Search for Spiritual Assurance, 1830–1940* (Berkeley: University of California Press, 2009).

4. Few of these thinkers were as radical as their statements sounded, especially on moral questions. The consumerism, sexual experimentation, and political quietism of the 1920s presented them with a series of conundrums. Many welcomed the new generation's frank recognition of human needs and the social costs that accrued from failing to address those needs. Some of the mental modernizers nevertheless thought that young Americans had crossed the line into mindless hedonism, or even nihilism. Most favored more wholesome, socially minded avenues for the realization of personal needs than those associated with bathtub gin and petting parties. Still, the celebrants of the modern mind embraced new understandings of the status and justification of moral principles. One who went further, questioning the content of Christian ideals, was the Johns Hopkins biologist Raymond Pearl, a confidant of H. L. Mencken and a champion of pleasurable connoisseurship in all realms (Raymond Pearl, *To Begin With: Being Prophylaxis against Pedantry* [New York: Knopf, 1927]; Sharon Kingsland, "Raymond Pearl: On the Frontier in the 1920's," *Human Biology* 56, no. 1 [February 1984]: 1–18).

5. Quoted in John C. Burnham, "The Encounter of Christian Theology with Deterministic Psychology and Psychoanalysis," *Bulletin of the Menninger Clinic* 49, no. 4 (1985): 330.

6. Bertrand Russell, "A Free Man's Worship," in *The Collected Papers of Bertrand Russell,* vol. 12, ed. Andrew Brink, Margaret Moran, and Richard A. Rempel (Boston: George Allen and Unwin, 1985); Joseph Wood Krutch, *The Modern Temper* (New York: Harcourt, Brace, 1929), 249. On Russell's views, see esp. Peter H. Denton, *The ABC of Armageddon: Bertrand Russell on Science, Religion, and the Next War, 1919–1938* (Albany: State University of New York Press, 2001); on Krutch's views, see Peter Gregg Slater, "The Negative Secularism of *The Modern Temper:* Joseph Wood Krutch," *American Quarterly* 33, no. 2 (July 1981): 185–205.

7. John B Watson, *Psychology from the Standpoint of a Behaviorist* (Philadelphia: Lippincott, 1919); Kerry W. Buckley, *Mechanical Man: John Broadus Watson and the Beginnings of Behaviorism* (New York: Guilford, 1989); Louis Berman, quoted in Burnham, "The Encounter of Christian Theology with Deterministic Psychology and Psychoanalysis," 344. See also R. Laurence Moore, "Secularization: Religion and the Social Sciences," in *Between the Times: The Travail of the Protestant Establishment in America, 1900–1960,* ed. William R. Hutchison (New York: Cambridge University Press, 1989), 236.

8. Jon H. Roberts has explored religious responses to Freud in "Psychoanalysis and American Christianity," in *When Science and Christianity Meet,* ed. David C. Lindberg and Ronald L. Numbers (Chicago: University of Chicago Press, 2003), 225–244; see also Roberts, "Psychology in America," in *The History of Science and Religion in the Western Tradition: An Encyclopedia,* ed. Gary B. Ferngren (New York: Garland, 2000), 575–581; and Burnham, "The Encounter of Christian Theology with Deterministic Psychology and Psychoanalysis."

9. Dewey's major works in the 1920s included *Reconstruction in Philosophy* (New York: Holt, 1920); *Human Nature and Conduct: An Introduction to Social Psychology* (New York: Holt, 1922); *Experience and Nature* (Chicago: Open Court, 1925); *The Public and Its Problems* (New York: Holt, 1927); and *The Quest for Certainty: A Study of the Relation of Knowledge and Action* (New York: Minton, Balch, 1929).

10. Baker Brownell, *The New Universe: An Outline of the Worlds in Which We Live* (New York: Van Nostrand, 1926), 347–348, 359. For a representative argument by a biologist, see C. Judson Herrick, *Fatalism or Freedom: A Biologist's Answer* (New York: Norton, 1926). On Dewey's influence and the arguments described in the following paragraphs, see esp. Jewett, *Science, Democracy, and the American University.*

11. Vernon L. Kellogg, *Headquarters Nights: A Record of Conversations and Experiences at the Headquarters of the German Army in France and Belgium* (Boston: Atlantic Monthly Press, 1917); Philip J. Pauly, *Biologists and the Promise of American Life, from Meriwether Lewis to Alfred Kinsey* (Princeton, NJ: Princeton University Press, 2000); Edward J. Larson, *Summer for the Gods: The Scopes Trial and America's Continuing Debate over Science and Religion* (New York: Basic Books, 1997).

12. On concerns about mass consumption in general, see Daniel Horowitz, *The Morality of Spending: Attitudes toward the Consumer Society in America, 1875–1940* (Baltimore: Johns Hopkins University Press, 1985).

13. David O. Levine, *The American College and the Culture of Aspiration, 1915–1940* (Ithaca, NY: Cornell University Press, 1986); John Black Johnston, *The Liberal College in Changing Society* (New York: Century, 1930).

14. Lynn Dumenil, *The Modern Temper: American Culture and Society in the 1920s* (New York: Hill and Wang, 1995); Paul V. Murphy, *The New Era: American Thought and Culture in the 1920s* (Lanham, MD: Rowman and Littlefield, 2012).

15. Accounts focused mainly on technological and economic changes included William F. Ogburn, *Social Change, with Respect to Culture and Original Nature* (New York: B. W. Huebsch, 1922); Charles Beard, ed., *Whither Mankind: A Panorama of Modern Civilization* (New York: Longman, 1928); Beard, ed., *Toward Civilization* (New York: Longman, 1930); and Charles and William Beard, *The American Leviathan: The Republic in the Machine Age* (New York: Macmillan, 1930).

16. Joan Shelley Rubin, *The Making of Middlebrow Culture* (Chapel Hill: University of North Carolina Press, 1992), 209–265; Jewett, *Science, Democracy, and the American University*.

17. James Harvey Robinson, *The Mind in the Making: The Relation of Intelligence to Social Reform* (New York: Harper, 1921); Robinson, *The Humanizing of Knowledge* (New York: George H. Doran, 1923); John Herman Randall Jr., *The Making of the Modern Mind: A Survey of the Intellectual Background of the Present Age* (Boston: Houghton Mifflin, 1926); Randall, *Our Changing Civilization: How Science and the Machine Are Reconstructing Modern Life* (New York: Frederick A. Stokes, 1929). Provocative statements on the social sciences included George A. Dorsey, *Why We Behave like Human Beings* (New York: Harper & Brothers, 1925); H. A. Overstreet: *About Ourselves: Psychology for Normal People* (New York: Norton, 1927); and Raymond B. Fosdick, *The Old Savage in the New Civilization* (Garden City, NY: Doubleday, Doran, 1928). On the idea of "the West," see esp. Gilbert Allardyce, "The Rise and Fall of the Western Civilization Course," *American Historical Review* 87 (June 1982): 695–725.

18. John B. Watson and William McDougall, *The Battle of Behaviorism: An Exposition and an Exposure* (London: K. Paul, Trench, Trubner & Co., 1928).

19. Buckley, *Mechanical Man*. "Watsonism has become gospel and catechism in the nurseries and drawing rooms of America," lamented the philosopher Mortimer Adler. Quoted in Ann Hulbert, *Raising America: Experts, Parents, and a Century of Advice About Children* (New York: Knopf, 2003), 125.

20. Daniel Patrick Thurs, *Science Talk: Changing Notions of Science in American Popular Culture* (New Brunswick, NJ: Rutgers University Press, 2007).

21. Daniel J. Kevles, *In the Name of Eugenics: Genetics and the Uses of Human Heredity* (New York: Knopf, 1985); Alexandra Minna Stern, *Eugenic Nation: Faults and Frontiers of Better Breeding in Modern America*, 2nd ed. (Oakland: University of California Press, 2016). On the varying responses to eugenics, see Christine Rosen, *Preaching Eugenics: Religious Leaders and the American Eugenics Movement* (New York: Oxford University Press, 2005). To many critics, Freud's emphasis on innate drives echoed the claims of biological de-

terminists, while Watson's theory, despite its disavowal of innate drives, still traced behavior to material causes.

22. Dorothy Ross, *The Origins of American Social Science* (New York: Cambridge University Press, 1991); Charles E. Merriam, *New Aspects of Politics* (Chicago: University of Chicago Press, 1925); Harold D. Lasswell, *Psychopathology and Politics* (Chicago: University of Chicago Press, 1930); Walter Lippmann, *Public Opinion* (New York: Harcourt, Brace, 1922); Lippmann, *The Phantom Public* (New York: Harcourt, Brace, 1925). On the potential for friction between neo-positivist social science and religion, see esp. Moore, "Secularization: Religion and the Social Sciences."

23. Morton J. Horwitz, *The Transformation of American Law, 1870–1960: The Crisis of Legal Orthodoxy* (New York: Oxford University Press, 1992); John H. Schlegel, *American Legal Realism and Empirical Social Science* (Chapel Hill: University of North Carolina Press, 1995); Richard Handler, "Boasian Anthropology and the Critique of American Culture," *American Quarterly* 42, no. 2 (1990): 252–273; Elazar Barkan, *The Retreat of Scientific Racism: Changing Concepts of Race in Britain and the United States between the World Wars* (New York: Cambridge University Press, 1992); Vernon J. Williams, *Rethinking Race: Franz Boas and His Contemporaries* (Lexington: University Press of Kentucky, 1996).

24. Frederick Lewis Allen, *Only Yesterday: An Informal History of the Nineteen-Twenties* (New York: Blue Ribbon, 1931), 240; John A. Farrell, *Clarence Darrow: Attorney for the Damned* (New York: Doubleday, 2011); Andrew Edmund Kersten, *Clarence Darrow: American Iconoclast* (New York: Hill and Wang, 2011).

25. Jeffrey P. Moran, "'Modernism Gone Mad': Sex Education Comes to Chicago, 1913," *Journal of American History* 83 (1996): 481–513; Pauly, *Biologists and the Promise of American Life*; Dumenil, *The Modern Temper*; K. Healan Gaston, *Imagining Judeo-Christian America: Religion, Secularism, and the Cold War Redefinition of Democracy* (Chicago: University of Chicago Press, 2019).

26. James Colgrove, "'Science in a Democracy': The Contested Status of Vaccination in the Progressive Era and the 1920s," *Isis* 96, no. 2 (June 2005): 167–191; Buck v. Bell, 274 US 200 (1927); Paul A. Lombardo, *Three Generations, No Imbeciles: Eugenics, the Supreme Court, and Buck v. Bell* (Baltimore: Johns Hopkins University Press, 2008).

27. Larson, *Summer for the Gods.*

28. Farrell, *Clarence Darrow: Attorney for the Damned*; Kersten, *Clarence Darrow: American Iconoclast.*

29. Conversely, the Deweyans' view of science seemed so obvious to them, and was so common in the American universities, that they sometimes failed to notice the capacity for misunderstanding. The philosopher Max Carl Otto repeatedly warned Dewey that when he called for "the application of scientific method to the conduct of life," most audiences thought he meant viewing "human conduct as the scientist looks upon the dance of atoms." Otto knew that Dewey simply had in mind "apply[ing] the same loyalty to truth, the same devotion

to experimental method, the same criticism to beliefs and institutions, as the scientist applies in his field." Science, Dewey explained, was simply "a certain attitude and quality of mind," not the specific, mechanistic worldview of classical physics. Despite Otto's repeated prodding, however, Dewey routinely ignored the possibility "that I would be thought to set up physics as a model in any respect" beyond the "operational character of its concepts" and "its emphasis on relations instead of inner essences." Otto to John Dewey, February 2, 1926; Dewey to Otto, March 9, 1923; and Dewey to Otto, October 1, 1930, in *The Correspondence of John Dewey, 1871–2007, Electronic Edition,* http://www.nlx.com/collections/132.

2. RESISTING THE MODERN

1. John Dewey, *Reconstruction in Philosophy, Enlarged Edition* (Boston: Beacon Press, 1948), xxxv.
2. For examples, see Christopher G. White, *Unsettled Minds: Psychology and the American Search for Spiritual Assurance, 1830–1940* (Berkeley: University of California Press, 2009); Andrew Jewett, *Science, Democracy, and the American University: From the Civil War to the Cold War* (New York: Cambridge University Press, 2012).
3. Walter Lippmann, *A Preface to Morals* (New York: Macmillan, 1929), 52.
4. James Turner, *The Liberal Education of Charles Eliot Norton* (Baltimore: Johns Hopkins University Press, 1999); Ralph Adams Cram, *The Nemesis of Mediocrity* (Boston: Marshall Jones, 1921). Cram and several other figures discussed in this chapter appear in Robert M. Crunden, *The Superfluous Men: Conservative Critics of American Culture, 1900–1945,* 2nd ed. (Wilmington, DE: ISI Books, 1999).
5. J. David Hoeveler Jr., *The New Humanism: A Critique of Modern America, 1900–1940* (Charlottesville: University Press of Virginia, 1977); Norman Foerster, *Humanism and America: Essays on the Outlook of Modern Civilisation* (New York: Farrar and Rinehart, 1930); C. Hartley Grattan, *The Critique of Humanism: A Symposium* (New York: Brewer and Warren, 1930). The New Humanists viewed modern literature in its various manifestations as either an outgrowth of modern materialism or an equally mindless, Rousseauian reaction against that materialism. Rather than "wander[ing] irresponsibly in a region quite outside of normal human experience," Babbitt argued, artists should seek out and live by "what is normal and central in human experience," especially the "higher reality" that provided moral guidance (Babbitt, *Rousseau and Romanticism* [Boston: Houghton Mifflin, 1919], xxii, 354–355).
6. Paul Elmer More, "The Demon of the Absolute," in *The Essential Paul Elmer More,* ed. Byron C. Lambert (New Rochelle, NY: Arlington House, 1972), 196, 205; Irving Babbitt, "Humanism: An Essay at Definition," in Foerster, *Humanism and America,* 44, 32–33; Babbitt, *Rousseau and Romanticism,* 372.
7. Irving Babbitt, *Democracy and Leadership* (Boston: Houghton Mifflin, 1924), 312; Babbitt, *Rousseau and Romanticism,* 388.

8. Walter Lippmann, "Humanism as Dogma," in Grattan, *The Critique of Humanism*, 818–819; Louis Trenchard More, "The Pretensions of Science," in Foerster, *Humanism and America*, 7, 16–17.

9. Paul Elmer More, "The Demon of the Absolute," 204–205; Irving Babbitt, "Humanism," 40. Louis Trenchard More sought to protect the study of social phenomena against naturalistic incursions by espousing a rigidly antimetaphysical view of science that he traced back to Isaac Newton and also found in the recent writings of the physicist P. W. Bridgman ("The Pretensions of Science," 14–15, 23). See also Louis Trenchard More, *The Dogma of Evolution* (Princeton, NJ: Princeton University Press, 1925).

10. Louis Trenchard More, "The Pretensions of Science," 8, 17, 4; Babbitt, "Humanism," 39.

11. Babbitt, "Humanism," 48, 51, 40.

12. Paul Elmer More, "The Demon of the Absolute," 194; Babbitt, *Rousseau and Romanticism*, 346; Babbitt, "Humanism," 43.

13. A classic study is Paul K. Conkin, *The Southern Agrarians* (Knoxville: University of Tennessee Press, 1988).

14. Twelve Southerners, "Introduction: A Statement of Principles," in *I'll Take My Stand* (1930; repr., Baton Rouge: Louisiana State University Press, 1977), xxxix–xl, xliii–xliv; Lyle H. Lanier, "A Critique of the Philosophy of Progress," in *I'll Take My Stand*, 136–153. Although the Southern Agrarians differed on religious questions, many of them equated science with industrial capitalism and Christianity with the submerged agrarian alternative.

15. Lewis Mumford, "Towards an Organic Humanism," in Grattan, *The Critique of Humanism*, 354–356. For Frank's views, see Waldo Frank, *The Re-Discovery of America: An Introduction to a Philosophy of American Life* (New York: Charles Scribner's Sons, 1929); and Casey Nelson Blake, *Beloved Community: The Cultural Criticism of Randolph Bourne, Van Wyck Brooks, Waldo Frank, and Lewis Mumford* (Chapel Hill: University of North Carolina Press, 1990).

16. Mumford, "Towards an Organic Humanism"; Thomas P. Hughes and Agatha C. Hughes, eds., *Lewis Mumford: Public Intellectual* (New York: Oxford University Press, 1990), 303–305.

17. This term comes from Douglas Sloan, who explores the development of the new understanding in *Faith and Knowledge: Mainline Protestantism and American Higher Education* (Louisville, KY: Westminster John Knox Press, 1994).

18. Helpful studies include Ronald H. Stone, *Politics and Faith: Reinhold Niebuhr and Paul Tillich at Union Seminary in New York* (Macon, GA: Mercer University Press, 2012); Daniel F. Rice, *Reinhold Niebuhr and His Circle of Influence* (New York: Cambridge University Press, 2013).

19. Quoted in Daniel F. Rice, *Reinhold Niebuhr and John Dewey: An American Odyssey* (Albany: State University of New York Press, 1993), 97; Reinhold Niebuhr, *Does Civilization Need Religion? A Study in the Social Resources and Limitations of Religion in Modern Life* (New York: Macmillan, 1927), 11, 16, 10, 13.

20. Niebuhr, *Does Civilization Need Religion?*, 21, 59, 238, 224, 172, 192, 229.

21. Reinhold Niebuhr, *Moral Man and Immoral Society: A Study in Ethics and Politics* (New York: Scribner's, 1932), xx, xiii, 44.

22. Shailer Mathews, *The Faith of Modernism* (New York: Macmillan, 1924), 23, 114–115, 8, 5–6.

23. W. P. King, ed., *Behaviorism: A Battle Line* (Nashville, TN: Cokesbury Press, 1930).

24. John Wright Buckham, *Personality and Psychology: An Analysis for Practical Use* (New York: Doran, 1924), 15, vii, 23–24. See also Buckham, "Beyond Science," *Journal of Religion* 9, no. 4 (October 1929): 505–522.

25. Quoted in Michael Kazin, *A Godly Hero: The Life of William Jennings Bryan* (New York: Knopf, 2006), 289; James H. Leuba, *The Belief in God and Immortality* (Boston: Sherman, French, 1916); Jon H. Roberts, "Conservative Evangelicals and Science Education in American Colleges and Universities, 1890–1940," *Journal of the Historical Society* 5, no. 3 (Fall 2005): 297–329; Edward J. Larson, *Summer for the Gods: The Scopes Trial and America's Continuing Debate over Science and Religion* (New York: Basic Books, 1997); Adam Laats, *Fundamentalism and Education in the Scopes Era: God, Darwin, and the Roots of America's Culture Wars* (New York: Palgrave Macmillan, 2010); Michael Lienesch, "Abandoning Evolution: The Forgotten History of Antievolution Activism and the Transformation of American Social Science," *Isis* 103 (December 2012): 692–693.

26. Quoted in Willard H. Smith, *The Social and Religious Thought of William Jennings Bryan* (Lawrence, KS: Coronado, 1975), 188. Of course, conservative evangelicals directed much of their immediate fire at theological liberals. But liberals, in their view, erred above all in accepting the authority of modern science.

27. Carlton J. H. Hayes, *Essays on Nationalism* (New York: Macmillan, 1926); K. Healan Gaston, *Imagining Judeo-Christian America: Religion, Secularism, and the Cold War Redefinition of Democracy* (Chicago: University of Chicago Press, 2019).

28. On Catholic views of education, see esp. Thomas E. Woods, *The Church Confronts Modernity: Catholic Intellectuals and the Progressive Era* (New York: Columbia University Press, 2004); Gaston, *Imagining Judeo-Christian America*.

29. William M. Halsey, *The Survival of American Innocence: Catholicism in an Era of Disillusionment* (Notre Dame, IN: University of Notre Dame Press, 1980), 144 (quoted), 145, 150. Not all American Catholic thinkers adopted this neo-Thomistic critique. James Joseph Walsh, head of Fordham's medical school, celebrated modern progress and took a hopeful view of the future, while identifying science as one of innumerable cultural fruits that the medieval Church had bequeathed to the West: e.g., Walsh, *A Catholic Looks at Life* (Boston: Stratford, 1928).

30. Fulton J. Sheen, *God and Intelligence in Modern Philosophy* (New York: Longmans, Green, 1925). See also Sheen, *Philosophy of Science* (Milwaukee, WI: Bruce, 1934). For the full sweep of Sheen's career, see Thomas C. Reeves, *America's Bishop: The Life and Times of Fulton J. Sheen* (San Francisco: Encounter Books, 2001); Kathleen L. Riley, *Fulton J. Sheen: An American Catholic Re-*

sponse to the Twentieth Century (Staten Island, NY: St. Paul's/Alba House, 2004).

31. George N. Shuster, *The Catholic Spirit in America* (New York: Dial Press, 1927), 75, 16–17, 23, 69, 14; see also Thomas E. Blantz, *George N. Shuster: On the Side of Truth* (Notre Dame, IN: University of Notre Dame Press, 1993). Williams spelled out his views in *Catholicism and the Modern Mind* (New York: Dial Press, 1928).

32. Halsey, *The Survival of American Innocence,* 117; T. S. Eliot, "The Humanism of Irving Babbitt," in *Selected Essays, 1917–1932* (New York: Harcourt, Brace, 1932), 384–385, 391; Babbitt, "Humanism," 38–40.

33. Babbitt, *Rousseau and Romanticism,* xxiii. Gorham Munson, a younger member of the group, argued that the New Humanism promised liberation from superstitious enthrallment to scientists; its widespread adoption would complete the push beyond the medieval mentality by leading the average citizen "to regard science in the same way he regards art and philosophy—with respect, but not sentimentally or superstitiously." Each field, Munson stated, was "man-made and therefore fallible"; it did not require a divine foundation (Munson, *The Dilemma of the Liberated: An Interpretation of Twentieth Century Humanism* [New York: Coward-McCann, 1930], 167–168).

34. T. S. Eliot, "Literature, Science, and Dogma," *Dial* 82, no. 3 (March 1927): 239–243; Eliot, *After Strange Gods: A Primer of Modern Heresy* (New York: Harcourt, Brace, 1934); Eliot, *The Idea of a Christian Society* (New York: Harcourt, Brace, 1940); Bernard Iddings Bell, *Right and Wrong after the War* (Boston: Houghton Mifflin, 1918); Bell, *Postmodernism, and Other Essays* (Milwaukee, WI: Morehouse, 1926); Bell, *Common Sense in Education* (New York: W. Morrow, 1928); Harvey Wickham, *The Misbehaviorists: Pseudo-Science and the Modern Temper* (New York: L. MacVeagh, The Dial Press, 1928); Wickham, *The Impuritans: A Glimpse into that New World Whose Pilgrim Fathers Are Otto Weininger, Havelock Ellis, James Branch Cabell, Marcel Proust, James Joyce, H. L. Mencken, D. H. Lawrence, Sherwood Anderson, et id genus omne* (New York: L. MacVeagh, The Dial Press, 1929); Wickham, *The Unrealists: James, Bergson, Santayana, Einstein, Bertrand Russell, John Dewey, Alexander and Whitehead* (New York: L. MacVeagh, The Dial Press, 1930).

3. SCIENCE AND THE STATE

1. Albert Jay Nock, *Our Enemy, the State* (New York: William Morrow, 1935); Walter Lippmann, *An Inquiry into the Principles of the Good Society* (Boston: Little, Brown, 1937). Perhaps surprisingly, one of the earliest and best-known analyses of bureaucratic influence—*The Managerial Revolution: What Is Happening in the World* (New York: John Day, 1941), by the disaffected Marxist James Burnham—said almost nothing about science, despite Burnham's earlier insistence that a "Religion of Science" dominated the popular mind: Burnham and Philip Wheelwright, *Introduction to Philosophical Analysis* (New York: Holt, 1932).

2. I take the phrase "rational reformers" from John M. Jordan, *Machine-Age Ideology: Social Engineering and American Liberalism, 1911–1939* (Chapel Hill: University of North Carolina Press, 1994).

3. Franklin D. Roosevelt, "The Responsibility of Engineering," *Science*, n.s., 84 (October 30, 1936): 393; Henry A. Wallace, "The Social Advantages and Disadvantages of the Engineering-Scientific Approach to Civilization," *Science*, n.s., 79 (January 5, 1934): 3–4.

4. William F. Ogburn, *Social Change, with Respect to Culture and Original Nature* (New York: B. W. Huebsch, 1922); Jordan, *Machine-Age Ideology*. For Ogburn's sociology, see Robert C. Bannister, *Sociology and Scientism: The American Quest for Objectivity, 1880–1940* (Chapel Hill: University of North Carolina Press, 1987).

5. William F. Ogburn, "The Folk-Ways of a Scientific Sociology," *Scientific Monthly* 30, no. 4 (April 1930): 300–306; quoted in Barbara Laslett, "Unfeeling Knowledge: Emotion and Objectivity in the History of Sociology," *Sociological Forum* 5, no. 3 (September 1990): 422.

6. *Recent Social Trends in the United States; Report of the President's Research Committee on Social Trends* (New York: McGraw-Hill, 1933); Natural Resources Committee, Subcommittee on Technology, *Technology Trends and National Policy* (Washington, DC: Government Printing Office, 1937), 10; Jordan, *Machine-Age Ideology*; Patrick D. Reagan, *Designing a New America: The Origins of New Deal Planning, 1890–1943* (Amherst: University of Massachusetts Press, 1999).

7. Lippmann, *An Inquiry into the Principles of the Good Society*, 387.

8. Walter Lippmann, *A Preface to Morals* (New York: Macmillan, 1929).

9. Walter Lippmann, *The Method of Freedom* (New York: Macmillan, 1934); Lippmann, *An Inquiry into the Principles of the Good Society*, 112; Barry D. Riccio, *Walter Lippmann—Odyssey of a Liberal* (New Brunswick, NJ: Transaction Books, 1994), 124–125.

10. Lippmann, *An Inquiry into the Principles of the Good Society*, 203–210, 373, 348, 22–23, 387. Lippmann rejected the usual claim that technological innovation had produced the industrial economy. Rather, he traced the new economic world to the creation of the modern corporation in the nineteenth century (Lippmann, *An Inquiry into the Principles of the Good Society*, 13).

11. Lippmann, *An Inquiry into the Principles of the Good Society*, 346, 351, 379.

12. Lippmann, *An Inquiry into the Principles of the Good Society*, 379–383. For Lippmann's identification of the humanities as the proper basis for modern education, see his "Education vs. Western Civilization," *American Scholar* 10, no. 2 (Spring 1941): 184–193, and "The State of Education in This Troubled Age," *Vital Speeches of the Day* 7, no. 7 (January 15, 1941): 200, as well as Craufurd D. Goodwin, *Walter Lippmann: Public Economist* (Cambridge, MA: Harvard University Press, 2014), esp. 7–8, 253–254.

13. John Dewey, *Freedom and Culture* (New York: G. P. Putnam's Sons, 1939), 118.

14. The classic study is Edward A. Purcell Jr., *The Crisis of Democratic Theory: Scientific Naturalism and the Problem of Value* (Lexington: University Press

of Kentucky, 1973). More recent accounts include Benjamin L. Alpers, *Dictators, Democracy, and American Public Culture: Envisioning the Totalitarian Enemy, 1920s–1950s* (Chapel Hill: University of North Carolina Press, 2003); Michaela Hönicke Moore, *Know Your Enemy: The American Debate on Nazism, 1933–1945* (New York: Cambridge University Press, 2010); Mark Greif, *The Age of the Crisis of Man: Thought and Fiction in America, 1933–1973* (Princeton, NJ: Princeton University Press, 2015); and K. Healan Gaston, *Imagining Judeo-Christian America: Religion, Secularism, and the Cold War Redefinition of Democracy* (Chicago: University of Chicago Press, 2019).

15. George S. Counts, *Dare the School Build a New Social Order?* (New York: John Day, 1932); Harry Elmer Barnes, *Society in Transition* (New York: Prentice-Hall, 1939), 56, 661–663, 653–654; Lindeman, quoted in Peter J. Kuznick, *Beyond the Laboratory: Scientists as Political Activists in 1930s America* (Chicago: University of Chicago Press, 1987), 81.

16. John Dewey, "Liberalism and Social Action (1935)," in *Later Works of John Dewey, 1925–1953,* vol. 11, ed. Jo Ann Boydston (Carbondale: Southern Illinois University Press, 1987), 54; Dewey, *Freedom and Culture,* 154, 151; Joseph Ratner, "Introduction to John Dewey's Philosophy," in *Intelligence in the Modern World: John Dewey's Philosophy,* ed. Ratner (New York: Modern Library, 1939), 200.

17. Everett Ross Clinchy, ed., *The World We Want to Live In* (Garden City, NY: Doubleday, Doran, 1942), 81; Carlton J. H. Hayes, *A Generation of Materialism, 1871–1900* (New York: Harper, 1941); John A. Ryan, "Religion, the Indispensable Basis of Democracy," *Vital Speeches of the Day 5,* no. 21 (August 15, 1939): 667–670; Thomas F. Woodlock, *The Catholic Pattern* (New York: Simon and Schuster, 1942).

18. Reinhold Niebuhr, *Moral Man and Immoral Society: A Study in Ethics and Politics* (New York: Scribner's, 1932); Niebuhr, *Reflections on the End of an Era* (New York: Scribner's, 1934); Niebuhr, *An Interpretation of Christian Ethics* (New York: Harper, 1935); Niebuhr, *Beyond Tragedy: Essays on the Christian Interpretation of History* (New York: Scribner's, 1937); Niebuhr, *Christianity and Power Politics* (New York: Scribner's, 1940); Niebuhr, *The Nature and Destiny of Man: A Christian Interpretation,* 2 vols. (New York: Charles Scribner's Sons, 1941/1943).

19. Thomas Woodlock, John Courtney Murray, Wilfrid Parsons, Louis J. A. Mercier, and many other Catholic thinkers spelled out their critiques in *Man and Modern Secularism: Essays on the Conflict of the Two Cultures* (New York: National Catholic Alumni Federation, 1940). A later statement is John LaFarge, *Secularism's Attack on World Order* (Washington, DC: Catholic Association for International Peace, 1944). On the discourse of secularism, see esp. Gaston, *Imagining Judeo-Christian America.*

20. Gaston, *Imagining Judeo-Christian America.* Sheen and his producers received letters and transcript requests by the tens of thousands, and one historian has suggested that 30 percent came from non-Catholics. Mark S. Massa, *Catholics and American Culture: Fulton Sheen, Dorothy Day, and the Notre Dame Football Team* (New York: Crossroad, 1999), 90. On this phenomenon, see

also Thomas C. Reeves, *America's Bishop: The Life and Times of Fulton J. Sheen* (San Francisco: Encounter, 2001), 110; and Sheen, *Treasure in Clay: The Autobiography of Fulton J. Sheen* (Garden City, NY: Doubleday, 1980), 73. On the Protestant side, conservative evangelicals found the New Deal even more alarming than their liberal and moderate counterparts, but science did not anchor their critique: Matthew Avery Sutton, "Was FDR the Antichrist? The Birth of Fundamentalist Antiliberalism in a Global Age," *Journal of American History* (March 2012): 1052–1074.

21. Andrew Jewett, *Science, Democracy, and the American University: From the Civil War to the Cold War* (New York: Cambridge University Press, 2012); Wolfgang Köhler, "The Naturalistic Interpretation of Man (The Trojan Horse)," in *The Selected Papers of Wolfgang Köhler,* ed. Mary Henle (New York: Liveright, 1971), 355.

22. Thom Weidlich, *Appointment Denied: The Inquisition of Bertrand Russell* (Amherst, NY: Prometheus Books, 2000); Mortimer Adler, "God and the Professors," in *Science, Philosophy, and Religion: A Symposium* (New York: Conference on Science, Philosophy, and Religion, 1941), 128; quoted in James Gilbert, *Redeeming Culture: American Religion in an Age of Science* (Chicago: University of Chicago Press, 1997), 75. On the CSPR, see also Fred W. Beuttler, "Organizing an American Conscience: The Conference on Science, Philosophy and Religion, 1940–1968" (PhD diss., University of Chicago, 1995), and Jan C. C. Rupp, "The Cultural Foundations of Democracy: The Struggle between a Religious and a Secular Intellectual Reform Movement in the American Age of Conformity," in *Religious and Secular Reform in America,* ed. David K. Adams and Cornelis A. van Minnen (Edinburgh: Edinburgh University Press, 1999), 231–247.

23. Gaston, *Imagining Judeo-Christian America;* F. Ernest Johnson, *The Church and Society* (New York: Abingdon Press, 1935); Johnson, *The Social Gospel Re-Examined* (New York: Harper & Brothers, 1940); J. Douglas Brown et al., "The Spiritual Basis of Democracy," in *Science, Philosophy, and Religion: Second Symposium* (New York: Conference on Science, Philosophy, and Religion, 1942), 255; Hocking, quoted in Fulton J. Sheen, *Philosophies at War* (New York: Scribner's, 1943), 53. Not all of these figures identified scientific thinking as the leading cause of modern changes. In 1945, Theodore M. Greene, a member of the Princeton group, portrayed moderns as utterly enslaved to their machines but held that naturalistic philosophies found little support outside academic circles; the main problem was technology, not science per se (Greene, "Christianity: Its Secular Alternatives," in Paul Tillich et al., *The Christian Answer* [New York: Scribner's, 1945], 45–90).

24. John H. Hallowell, *The Decline of Liberalism as an Ideology with Particular Reference to German Politico-Legal Thought* (Berkeley: University of California Press, 1943), vii, 100, 112, 71–72, 113, 110, 23. Hallowell further developed his views in *The Moral Foundations of Democracy* (Chicago: University of Chicago Press, 1954).

25. Julia E. Johnsen, *Plans for a Post-War World* (New York: H. W. Wilson, 1942); Max Carl Otto, *The Human Enterprise: An Attempt to Relate Philosophy to*

Daily Life (New York: Crofts, 1940), 137 (emphasis removed); Charles E. Merriam, *What Is Democracy?* (Chicago: University of Chicago Press, 1941), 12–14. For universal ethics, see E. G. Conklin, "Science and Ethics," in *Science and Man,* ed. Ruth Nanda Anshen (New York: Harcourt, Brace, 1940), 436–452; cf. Carl J. Friedrich, *The New Belief in the Common Man* (Boston: Little, Brown, 1942), 313–314. On 1940s debates over the nation's global role and aims, see esp. Frank A. Warren, *Noble Abstractions: American Liberal Intellectuals and World War II* (Columbus: Ohio State University Press, 1999) and John Fousek, *To Lead the Free World: American Nationalism and the Cultural Roots of the Cold War* (Chapel Hill: University of North Carolina Press, 2000).

26. Gardner Murphy, ed., *Human Nature and Enduring Peace* (Boston: Houghton Mifflin, 1945), 271, 255, 455–457 (emphasis removed); Lawrence K. Frank, "Society as the Patient," *American Journal of Sociology* 42 (November 1936): 335–344; Margaret Mead, *From the South Seas: Studies of Adolescence and Sex in Primitive Societies* (New York: W. Morrow, 1939), 272. A recent study of Mead in the wartime context is Peter Mandler, *Return from the Natives: How Margaret Mead Won the Second World War and Lost the Cold War* (New Haven, CT: Yale University Press, 2013).

27. Alfred Korzybski, *Manhood of Humanity: The Science and Art of Human Engineering* (New York: E. P. Dutton, 1921); Ross Evans Paulson, *Language, Science, and Action: Korzybski's General Semantics: A Study in Comparative Intellectual History* (Westport, CT: Greenwood Press, 1983); Allen Walker Read, "Changing Attitudes toward Korzybski's General Semantics," *General Semantics Bulletin* 51 (1984): 15; S. I. Hayakawa, *Language in Action* (New York: Harcourt, Brace, 1941), Irving J. Lee, *Language Habits in Human Affairs* (New York: Harper & Brothers, 1941); Korzybski, *General Semantics Seminar 1937: Transcription of Notes from Lectures in General Semantics Given at Olivet College* (Lakeville, CT: Institute of General Semantics, 1964), 84–86. Another important popular treatment was Stuart Chase, *The Tyranny of Words* (New York: Harcourt, Brace and World, 1938).

28. Thurman Arnold, *The Folklore of Capitalism* (New Haven, CT: Yale University Press, 1937); I. A. Richards, "Basic English," *Fortune* 23 (1941): 89–91, 111–112; Rudolf Carnap, *The Unity of Science* (London: K. Paul, Trench, Trubner, 1934); W. V. O. Quine, "Relations and Reason," *Technology Review* 41 (1939): 300; Edmund C. Berkeley, "Conditions Affecting the Application of Symbolic Logic," *Journal of Symbolic Logic* 7 (December 1942): 162; Rudolf Carnap, "Logic," in *Factors Determining Human Behavior* (Cambridge, MA: Harvard University Press, 1937): 107–118. See also Susanne K. Langer, *An Introduction to Symbolic Logic* (Boston: Houghton Mifflin, 1937). For representative challenges to general semantics and other scientific theories of language, see John Crowe Ransom, "Editorial Notes: The Arts and the Philosophers," *Kenyon Review* 1, no. 2 (Spring 1939): 194–199; Eliseo Vivas et al., "The New Encyclopedists: A Symposium," *Kenyon Review* 1, no. 2 (Spring 1939): 159–182; "The Tyranny of Semantics," *Christian Century* 55, no. 10 (March 9, 1938): 296–297; and James M. Gillis, "Craze: Semantics," *Catholic World* 146, no. 729 (March 1938): 646–649.

29. Quoted in Fulton J. Sheen, *Philosophies at War* (New York: Scribner's, 1943), 39–40; quoted in "Mgr. Sheen Assails Worldly Philosophy," *New York Times,* November 16, 1942, 22.
30. Sheen, *Philosophies at War,* 76, 53, 51, 66, 69–70.
31. Dewey, *Freedom and Culture,* 6; Reinhold Niebuhr, *Does Civilization Need Religion? A Study in the Social Resources and Limitations of Religion in Modern Life* (New York: Macmillan, 1927), 203.

4. SOCIAL ENGINEERING

1. Ralph W. Tyler, "Implications for Science Teaching," in *Papers of the Fourth Conference on Scientific Manpower* (Washington, DC: National Science Foundation, 1955), 16; quoted in "Dangerous Scientists," *Time* 66, no. 13 (September 26, 1955). Tyler and other commentators drew heavily on Dael Wolfle's study *America's Resources of Specialized Talent: A Current Appraisal and a Look Ahead* (New York: Harper, 1954). See also Spencer Weart, *Nuclear Fear: A History of Images* (Cambridge, MA: Harvard University Press, 1988); Marcel C. LaFollette, *Making Science Our Own: Public Images of Science, 1910–1955* (Chicago: University of Chicago Press, 1990); David Kaiser, "The Atomic Secret in Red Hands? American Suspicions of Theoretical Physicists during the Early Cold War," *Representations* 90, no. 1 (Spring 2005): 28–60; Aaron Lecklider, *Inventing the Egghead: The Battle over Brainpower in American Culture* (Philadelphia: University of Pennsylvania Press, 2013), esp. 160–220; and Paul Rubinson, *Redefining Science: Scientists, the National Security State, and Nuclear Weapons in Cold War America* (Amherst: University of Massachusetts Press, 2016). A popular, satirical treatment from the time is Anthony Standen's *Science Is a Sacred Cow* (New York: Dutton, 1950); see also George Guion Williams's *Some of My Best Friends Are Professors: A Critical Commentary on Higher Education* (New York: Abelard-Schuman, 1958).
2. In addition to B. F. Skinner's works, critics of social engineering often singled out a Carnegie-sponsored survey of the social sciences by a popular writer associated with the New Deal's technocratic side: Stuart Chase, *The Proper Study of Mankind: An Inquiry into the Science of Human Relations* (New York: Harper, 1948). As for scientism, the naturalistic philosopher Max Carl Otto complained about the ubiquity of that epithet as early as 1943 (Otto, "Scientific Humanism," *Antioch Review* 3, no. 4 [December 1943]: 534). The term had begun to find a purchase decades earlier, as in works by Charles Gray Shaw, a Christian philosopher at New York University: e.g., Shaw, *The Ego and Its Place in the World* (New York: Macmillan, 1913), 34, 39; Shaw, *The Ground and Goal of Human Life* (New York: New York University Press, 1919), esp. 62–83. But its real takeoff occurred among specialists in the 1940s and then wider audiences in the 1950s. Catholic theorists and conservative émigrés used the term especially frequently in the 1940s: Fulton J. Sheen, *Philosophies at War* (New York: Scribner's, 1943), 51; John James Wellmuth, *The Nature and Origins of Scientism* (Milwaukee: Marquette University Press, 1944);

Edward B. Jordan, "A Recrudescence of Scientism," *Catholic World* 163, no. 4 (August 1946): 414–419; Alan Devoe, "Man Versus Scientism," *Commonweal* 51, no. 24 (March 24, 1950): 632–635; F. A. Hayek, "Scientism and the Study of Society. Part I," *Economica* 9, no. 35 (August 1942): 267–291; Hayek, "Scientism and the Study of Society. Part III," *Economica* 11, no. 49 (February 1944): 27–39; Eric Voegelin, "The Origins of Scientism," *Social Research* 15, no. 4 (December 1948): 462–494.

3. Everson v. Board of Education of the Township of Ewing, 330 US 1 (1947); Illinois ex rel. McCollum v. Board of Ed. of School Dist. No. 71, Champaign County, 333 US 203 (1948); Zorach v. Clauson, 343 US 306 (1952); K. Healan Gaston, *Imagining Judeo-Christian America: Religion, Secularism, and the Cold War Redefinition of Democracy* (Chicago: University of Chicago Press, 2019), esp. 142–146.

4. Gaston, *Imagining Judeo-Christian America*; Whittaker Chambers, *Witness* (New York: Random House, 1952), 9–10, 16–17.

5. On the events of the period, see Audra J. Wolfe, *Competing with the Soviets: Science, Technology, and the State in Cold War America* (Baltimore: Johns Hopkins University Press, 2012). Some critics worried that American children might join their Soviet counterparts in worshiping science as a new religion (Julia Mickenberg, *Learning from the Left: Children's Literature, the Cold War, and Radical Politics in the United States* [New York: Oxford University Press, 2005], 179). But at the federal level, a desire to defeat the enemy on the scientific battleground prevailed. Helpful recent excursions into the world of postwar behavioral science and the military-industrial complex include Joel Isaac, *Working Knowledge: Making the Human Sciences from Parsons to Kuhn* (Cambridge, MA: Harvard University Press, 2012); Joy Rohde, *Armed with Expertise: The Militarization of American Social Research during the Cold War* (Ithaca, NY: Cornell University Press, 2013); Paul Erickson et al., *How Reason Almost Lost Its Mind: The Strange Career of Cold War Rationality* (Chicago: University of Chicago Press, 2013); Mark Solovey, *Shaky Foundations: The Politics–Patronage–Social Science Nexus in Cold War America* (New Brunswick, NJ: Rutgers University Press, 2013); Sarah Bridger, *Scientists at War: The Ethics of Cold War Weapons Research* (Cambridge, MA: Harvard University Press, 2015); Rebecca Lemov, *Database of Dreams: The Lost Quest to Catalog Humanity* (New Haven, CT: Yale University Press, 2015); and the essays in Solovey and Hamilton Cravens, eds., *Cold War Social Science: Knowledge Production, Liberal Democracy, and Human Nature* (New York: Palgrave Macmillan, 2012). An important overview is found in Ron Robin, *The Making of the Cold War Enemy: Culture and Politics in the Military-Intellectual Complex* (Princeton, NJ: Princeton University Press, 2001).

6. Megan Barnhart Sethi, "Information, Education, and Indoctrination: The Federation of American Scientists and Public Communication Strategies in the Atomic Age," *Historical Studies in the Natural Sciences* 42, no. 1 (February 2012): 1–29; Kaiser, "The Atomic Secret in Red Hands?"

7. George Gaylord Simpson, *The Meaning of Evolution* (New Haven, CT: Yale University Press, 1949), 345, 315.

8. Ashley Montagu, *On Being Human* (New York: H. Schuman, 1950), 54, 109, 80, 30, 92. Segregationists decried the egalitarianism of Montagu and other postwar social scientists: John P. Jackson Jr., *Science for Segregation: Race, Law, and the Case against Brown v. Board of Education* (New York: New York University Press, 2005).

9. Priscilla Robertson, "On Getting Values out of Science," *Humanist* 16, no. 4 (July–August 1956): 174; Edmund W. Sinnott, "The Biological Basis of Democracy," *Yale Review* 35, no. 1 (September 1945): 62–63; Sinnott, *The Biology of the Spirit* (New York: Viking, 1955), 155; Sinnott, *Two Roads to Truth: A Basis for Unity under the Great Tradition* (New York: Viking, 1953), 224–227. On such postwar expressions of what Maurizio Meloni calls "democratic biology," see Andrew Jewett, "Naturalizing Liberalism in the 1950s," in *Professors and Their Politics*, ed. Neil Gross and Solon J. Simmons (Baltimore: Johns Hopkins University Press, 2014), 191–216; and Meloni, *Political Biology: Science and Social Values in Human Heredity from Eugenics to Epigenetics* (New York: Palgrave Macmillan, 2016).

10. Nils Gilman, *Mandarins of the Future: Modernization Theory in Cold War America* (Baltimore: Johns Hopkins University Press, 2003); Robert Adcock, "Interpreting Behavioralism," in *Modern Political Science: Anglo-American Exchanges since 1880*, ed. Adcock, Mark Bevir, and Shannon C. Stimson (Princeton, NJ: Princeton University Press, 2007), 180–208.

11. Solovey, *Shaky Foundations,* esp. 20–55. Even the significant minority of postwar social scientists who claimed that values figured directly in scientific research—that strict value-neutrality was impossible for even the most rigorously empirical scholar—often couched their claims in technical language and wrote for specialist audiences, without making explicitly normative claims in their publications or seeking wider audiences for their arguments. For example, value theory became an established subfield of philosophy in this period. Detailed methodological reflections by scientific practitioners multiplied as well. But these discussants tended to address the value question in an abstract manner, analyzing the epistemological and methodological roles of values in general rather than advocating specific norms or moral tenets. Ethel M. Albert and Clyde Kluckhohn cataloged the postwar scholarship in *A Selected Bibliography on Values, Ethics, and Esthetics* (Glencoe, IL: Free Press, 1959).

12. Andrew Jewett, *Science, Democracy, and the American University: From the Civil War to the Cold War* (New York: Cambridge University Press, 2012).

13. Douglas Sloan, *Faith and Knowledge: Mainline Protestantism and American Higher Education* (Louisville, KY: Westminster/John Knox, 1994), 37, 45–46; Jewett, *Science, Democracy, and the American University.* Although many behavioral scientists described this stance as value-neutral, others presented it as an antidote to detached, sterile scholarship. Gabriel Almond, a fixture of the behavioral science establishment, adopted the latter view when he embraced the Christian political theorist John Hallowell's critique of "scientificism" in social inquiry (Almond, "Politics, Science, and Ethics," *American Political Science Review* 40, no. 2 [April 1946]: 285).

14. Gilman helpfully summarizes this centrist liberalism and the associated sense of its inevitability in *Mandarins of the Future.*

15. Theodor W. Adorno et al., *The Authoritarian Personality* (New York: Harper, 1950); W. Lloyd Warner, Marchia Meeker, and Kenneth Eells, *Social Class in America: A Manual of Procedure for the Measurement of Social Status* (Chicago: Science Research Associates, 1949), 5. At the same time, these psychologists identified "tolerance toward ambiguity" as a hallmark of mental health, at a time when many Americans sought moral clarity above all. Else Frenkel-Brunswik, "Intolerance of Ambiguity as an Emotional and Perceptual Personality Variable," *Journal of Personality* 18, no. 1 (September 1949): 108–136; Jamie Cohen-Cole, *The Open Mind: Cold War Politics and the Sciences of Human Nature* (Chicago: University of Chicago Press, 2014). For other uses of psychological models, see Harry A. Overstreet, *The Mature Mind* (New York: W. W. Norton, 1949); and Erik H. Erikson, *Childhood and Society* (New York: W. W. Norton, 1950); for social-structural approaches, see Stuart A. Queen, *The American Social System: Social Control, Personal Choice, and Public Decision* (Boston: Houghton Mifflin, 1956); Joseph A. Kahl, *The American Class Structure* (New York: Rinehart, 1957); and Bernard Barber, *Social Stratification: A Comparative Analysis of Structure and Process* (New York: Harcourt, Brace, 1957).

16. Gilman, *Mandarins of the Future*; David C. Engerman et al., eds., *Staging Growth: Modernization, Development, and the Global Cold War* (Amherst: University of Massachusetts Press, 2003).

17. Harvard University Committee on the Objectives of a General Education in a Free Society, *General Education in a Free Society* (Cambridge, MA: Harvard University Press, 1945), 57; Philip E. Jacob, *Changing Values in College: An Exploratory Study of the Impact of College Teaching* (New York: Harper, 1957); Vera M. Moss, "Evaluating 'Values,'" *Journal of Higher Education* 31 (1960): 155. When Jacob argued that colleges and universities were not actually making students more liberal, numerous respondents deemed that conclusion simply wrong; American higher education had a clear "liberalizing impact" (Alex S. Edelstein, "Since Bennington: Evidence of Change in Student Political Behavior," *Public Opinion Quarterly* 26 [1962]: 573). On this controversy, see Jewett, "Naturalizing Liberalism in the 1950s."

18. Peter J. Westwick, *The National Labs: Science in an American System, 1947–1974* (Cambridge, MA: Harvard University Press, 2003); Joy Rohde, *Armed with Expertise: The Militarization of American Social Research during the Cold War* (Ithaca, NY: Cornell University Press, 2013); Hunter Crowther-Heyck, *Herbert A. Simon: The Bounds of Reason in Modern America* (Baltimore: Johns Hopkins University Press, 2005); Paul Erickson et al., *How Reason Almost Lost Its Mind: The Strange Career of Cold War Rationality* (Chicago: University of Chicago Press, 2013); Robert M. Collins, *More: The Politics of Economic Growth in Postwar America* (New York: Oxford University Press, 2000).

19. On the fear of rule by distant experts, see David A. Horowitz, *America's Political Class under Fire: The Twentieth Century's Great Culture War* (New York: Routledge, 2003). Timothy Melley discusses more amorphous concerns about a loss of individual control—what he calls agency panic—in *Empire of Conspiracy: The Culture of Paranoia in Postwar America* (Ithaca, NY: Cornell University Press, 2000).

20. Milton Birnbaum, *Aldous Huxley's Quest for Values* (Knoxville: University of Tennessee Press, 1971), 140–141, 145 (quoted), 150 (quoted). Huxley responded warmly to the general semantics movement and considered working it into a novel that would provide a utopian counterstatement to *Brave New World* (Allen Walker Read, "Is General Semantics Compatible with Utopianism?," *General Semantics Bulletin* 52 [1985]: 32). A recent study of the brothers is R. S. Deese, *We Are Amphibians: Julian and Aldous Huxley on the Future of Our Species* (Oakland: University of California Press, 2015).

21. B. F. Skinner, *Walden Two* (New York: Macmillan, 1948); "Boxes for Babies," *Life*, November 3, 1947, 73–74; Skinner, *Beyond Freedom and Dignity* (New York: Knopf, 1971). Daniel C. Williams compared *Walden Two* to *Brave New World* in "The Social Scientist as Philosopher and King," *Philosophical Review* 58, no. 4 (July 1949): 345. On Skinner's reception, see also Daniel W. Bjork, *B. F. Skinner: A Life* (New York: Basic Books, 1993); and Alexandra Rutherford, *Beyond the Box: B. F. Skinner's Technology of Behavior from Laboratory to Life, 1950s–1970s* (Toronto: University of Toronto Press, 2009).

22. Bergen Evans, "Sunny View of Utopian Way of Life," *Chicago Daily Tribune*, June 13, 1948, B8; Gerald Roscoe, "Even Boston Is Better," *Boston Globe*, August 12, 1948, 15; Williams, "The Social Scientist as Philosopher and King," 345, 349–350, 353–354; Harold A. Larrabee, review of *Walden Two*, by B. F. Skinner, *Journal of Philosophy* 46, no. 20 (September 29, 1948): 654–655; Paul B. Foreman, review of *Walden Two*, by Skinner, *Southwestern Social Science Quarterly* 29, no. 2 (September 1948): 164–166; John K. Jessup, "Utopia Bulletin," *Fortune* (October 1948): 191–196.

23. Paul Boyer noted the postwar uptick in *By the Bomb's Early Light: American Thought and Culture at the Dawn of the Atomic Age* (New York: Pantheon, 1985), 166–172. For examples of the social engineering terminology, see Goodwin Watson, "How Social Engineers Came to Be," *Journal of Social Psychology* 21 (1945): 135–141; Ralph W. Burhoe, "On the Need for Social Engineering," *Science* 104, no. 2690 (July 19, 1946): 62; Ashley Montagu, "Anthropology and Social Engineering," *American Anthropologist* 48, no. 4 (October–December 1946): 666–667; Clifford T. Morgan, "Human Engineering," in *Current Trends in Psychology*, ed. Wayne Dennis (Pittsburgh: University of Pittsburgh Press, 1947), 169–195; Sidney Post Simpson and Ruth Field, "Social Engineering through Law: The Need for a School of Applied Jurisprudence," *New York University Law Quarterly Review* 22 (1947): 145; John Amherst Sexson, "School Administrator as a Social Engineer," *School Executive* 67 (June 1948): 21–23; Leonard C. Mead, "A Program of Human Engineering," *Personnel Psychology* 1, no. 3 (September 1, 1948): 303–317; Albert C. Ettinger, "Social Engineering in Race Relations," *Persona* 1 (1949): 17–35; Lawrence Edwin Abt, "The Human Engineering Seminar at New York University," *Science* 109, no. 2819 (January 7, 1949): 19; Philip M. Hauser, "Social Science and Social Engineering," *Philosophy of Science* 16, no. 3 (July 1, 1949): 209–218; David H. Jenkins, "Social Engineering in Educational Change: Outline of Method," *Progressive Education* 26 (May 1949): 193–197; and Eleanor Crook, "Human Engineering, for Better or for Worse," *Survey* 86

(July 1950): 347–350. Skinner's expansive vision of social engineering appears in B. F. Skinner, "The Control of Human Behavior," *Transactions of the New York Academy of Sciences* 17 (1955): 547–551; Skinner, "Freedom and the Control of Men," *American Scholar* 25, no. 1 (December 1955): 47–65; and Carl R. Rogers and Skinner, "Some Issues Concerning the Control of Human Behavior," *Science* 124 (November 30, 1956): 1057–1066.

24. Alan C. Petigny, *The Permissive Society: America, 1941–1965* (New York: Cambridge University Press, 2009), 22, 24; Elizabeth Lunbeck, *The Psychiatric Persuasion: Knowledge, Gender, and Power in Modern America* (Princeton, NJ: Princeton University Press, 1994); Ellen Herman, *The Romance of American Psychology: Political Culture in the Age of Experts* (Berkeley: University of California Press, 1995); James H. Capshew, *Psychologists on the March: Science, Practice, and Professional Identity in America, 1929–1969* (New York: Cambridge University Press, 1999).

25. Sarah E. Igo, *The Averaged American: Surveys, Citizens, and the Making of a Mass Public* (Cambridge, MA: Harvard University Press, 2007), esp. 185. For scientists in popular culture, see esp. Marcel C. LaFollette, *Science on the Air: Popularizers and Personalities on Radio and Early Television* (Chicago: University of Chicago Press, 2008), and LaFollette, *Science on American Television: A History* (Chicago: University of Chicago Press, 2013).

26. Petigny detailed these postwar shifts in *The Permissive Society*.

27. Petigny, *The Permissive Society*, 37–41, 22 (quoted); Julia Grant, *Raising Baby by the Book: The Education of American Mothers* (New Haven, CT: Yale University Press, 1998), 234. For additional criticism, see Ann Hulbert, *Raising America: Experts, Parents, and a Century of Advice about Children* (New York: Knopf, 2003). On changes in parenting patterns, see also Paula Fass, "The Child-Centered Family? New Rules in Postwar America," in *Reinventing Childhood After World War II*, ed. Fass and Michael Grossberg (Philadelphia: University of Pennsylvania Press, 2011), 1–18; and Fass, *The End of American Childhood: A History of Parenting from Life on the Frontier to the Managed Child* (Princeton, NJ: Princeton University Press, 2016).

28. Alfred C. Kinsey, *Sexual Behavior in the Human Male* (Philadelphia: W. B. Saunders, 1948); Institute for Sex Research, *Sexual Behavior in the Human Female* (Philadelphia: W. B. Saunders, 1953); W. Norman Pittenger, *The Christian View of Sexual Behavior: A Reaction to the Kinsey Report* (Greenwich, CT: Seabury Press, 1954), 16; Bruce V. Lewenstein, "'Public Understanding of Science' in America, 1945–1965" (PhD diss., University of Pennsylvania, 1987), 217; Igo, *The Averaged American*, 237–238, 260 (quoted); Gallup Poll [January 1948], USGALLUP.022148.RK11A, (EA) and Gallup Poll [January 1948], USGALLUP.022148.RK11A, (EA) both in Gallup Organization, Cornell University, Ithaca, NY, Roper Center for Public Opinion Research, iPOLL; quoted in Robert Cecil Johnson, "Kinsey, Christianity, and Sex: A Critical Study of Reactions in American Christianity to the Kinsey Reports on Human Sexual Behavior" (PhD diss., University of Wisconsin–Madison, 1973), 224. See also James H. Jones, *Alfred C. Kinsey: A Life* (New York: W. W. Norton, 2004). Kinsey's effort to gain veto power over publications discussing his work merely

reinforced the fear that an arrogant cadre of experts was imposing its morally relativistic practices on the masses (Lewenstein, "'Public Understanding of Science' in America, 1945–1965," 217).

29. Frank Wilson to the *Atlantic Monthly,* n.d., Jacques Barzun Papers, Columbia University, Rare Book and Manuscript Library, box 43, folder "Science corr. Jan. 1960–Dec. 1964"; William G. Mackenzie to Jacques Barzun, February 19, 1961, Barzun Papers, box 51, folder "The House of Intellect, Jan. 1961–Dec. 1964"; quoted in Diane Ravitch, *Left Back: A Century of Battles over School Reforms* (New York: Simon and Schuster, 2001), 333, 349; John Melling to *Food for Thought,* November 17, 1960, Barzun Papers, box 51, folder "The House of Intellect corr. Jan.–Dec. 1960"; Isabelle Holland to Joseph Wood Krutch, July 23, 1951, Joseph Wood Krutch Papers, Library of Congress, box 4, folder H. For summaries of the life adjustment model and its critics, see Ravitch, *Left Back,* 327–348; and Andrew Hartman, *Education and the Cold War: The Battle for the American School* (New York: Palgrave Macmillan, 2008), esp. 55–72, 91–136. On educational critiques from the right, see also Adam Laats, *The Other School Reformers: Conservative Activism in American Education* (Cambridge, MA: Harvard University Press, 2015).

30. David Riesman with Nathan Glazer and Reuel Denney, *The Lonely Crowd: A Study of the Changing American Character* (New Haven, CT: Yale University Press, 1950); Herbert J. Gans, "Best-Sellers by Sociologists: An Exploratory Study," *Contemporary Sociology* 26, no. 2 (March 1997): 134. Despite the era's image as a golden age of public sociology, science writers in 1954 considered the social sciences the least interesting of the sciences to the public (Lewenstein, "'Public Understanding of Science' in America," 224). On Riesman, see Daniel Horowitz, *Consuming Pleasures: Intellectuals and Popular Culture in the Postwar World* (Philadelphia: University of Pennsylvania Press, 2012), esp. 121–136; and Daniel Geary, "Children of the Lonely Crowd: David Riesman, the Young Radicals, and the Splitting of Liberalism in the 1960s," *Modern Intellectual History* 10, no. 3 (November 2013): 603–633.

31. William H. Whyte, *The Organization Man* (New York: Simon and Schuster, 1956), 6–7, 30, 26, 35; Vance Packard, *The Hidden Persuaders* (New York: David McKay, 1957), 5, 255, 265; quoted in Daniel Horowitz, *Vance Packard and American Social Criticism* (Chapel Hill: University of North Carolina Press, 1994), 160. "The organization man is not in the grip of vast social forces about which it is impossible for him to do anything," Whyte concluded his book. "The options are there, and with wisdom and foresight he can turn the future away from the dehumanized collective that so haunts our thoughts" (*The Organization Man,* 447–448). As he put it elsewhere in the book, "a status quo cannot long endure without an ideology to sustain it" (249). On Whyte's critique of sociology and the capacity of Packard's books to undermine sociological authority, see David Paul Haney, *The Americanization of Social Science: Intellectuals and Public Responsibility in the Postwar United States* (Philadelphia: Temple University Press, 2008), 175–177, 205–207.

32. Gallup Poll [April 1949], USGALLUP.052849.R11B; Gallup Poll [April 1953], USGALLUP.052353.RK14B; Gallup Poll (AIPO) [September 1949], USGAL

LUP.447K.QK16; Gallup Poll (AIPO) [September 1949], USGALLUP.447T. QT16; Gallup Poll [September 1947], USGALLUP.101047.RK13C; Gallup Poll (AIPO) [January 1947], USGALLUP.47-388.QK13C; all in Gallup Organization, Cornell University, Ithaca, NY, Roper Center for Public Opinion Research, iPOLL; Susan Tyler Hitchcock, *Frankenstein: A Cultural History* (New York: W. W. Norton, 2007); Johann Wolfgang von Goethe, *Faust* (Baltimore: Penguin, 1949); Goethe, *The Urfaust: Goethe's Faust in the Original Form*, trans. Douglas M. Scott (Woodbury, NY: Barron's Educational Series, 1958); Tom Moylan, *Scraps of the Untainted Sky: Science Fiction, Utopia, Dystopia* (Boulder, CO: Westview Press, 2000), 170; Margaret Mead and Rhoda Metraux, "Image of the Scientist among High-School Students," *Science* 126 (August 30, 1957): 387 (emphasis removed), 384.

5. MODERNITY AND SCIENTISM

1. All quoted in Paul S. Boyer, *By the Bomb's Early Light: American Thought and Culture at the Dawn of the Atomic Age* (Chapel Hill: University of North Carolina Press, 1985), 271, 231.

2. Buell G. Gallagher, "False Alternatives in Education," *Christian Century* 63, no. 21 (May 22, 1946): 653–654.

3. Ronald L. Numbers, *The Creationists: From Scientific Creationism to Intelligent Design,* expanded ed. (Cambridge, MA: Harvard University Press, 2006).

4. Carl F. H. Henry, *Remaking the Modern Mind* (Grand Rapids, MI: W. B. Eerdmans, 1948), 22.

5. Quoted in Seth Jacobs, "'Our System Demands the Supreme Being': The U.S. Religious Revival and the 'Diem Experiment,' 1954–55," *Diplomatic History* 25, no. 4 (Fall 2001): 589, 593–594; Educational Policies Commission, *Moral and Spiritual Values in the Public Schools* (Washington, DC: National Education Association, 1951); John Foster Dulles, "The Secretary of State on Faith of Our Fathers," Department of State Publication 5300, General Foreign Policy Series 84 (Washington, DC: Government Printing Office, 1954), 10; quoted in Mark Solovey, *Shaky Foundations: The Politics-Patronage-Social Science Nexus in Cold War America* (New Brunswick, NJ: Rutgers University Press, 2013), 176–177. Other recent studies of the postwar climate include Jacobs, *America's Miracle Man in Vietnam: Ngo Dinh Diem, Religion, Race, and U.S. Intervention in Southeast Asia, 1950–1957* (Durham, NC: Duke University Press, 2004); William Inboden III, *Religion and American Foreign Policy, 1945–1960: The Soul of Containment* (Cambridge, MA: Cambridge University Press, 2008); T. Jeremy Gunn, *Spiritual Weapons: The Cold War and the Forging of an American National Religion* (Westport, CT: Praeger, 2009); and Jonathan P. Herzog, *The Spiritual-Industrial Complex: America's Religious Battle against Communism in the Early Cold War* (New York: Oxford University Press, 2011).

6. Quoted in "Truman's Simple Truths," *America* 83, no. 18 (August 5, 1950): 460; quoted in Inboden, *Religion and American Foreign Policy,* 107–109, 1,

122–123, 17. *America*'s editors traced Truman's moral clarity to his avoidance of "the pseudo-sophistication of what goes by the name of university training" ("Truman's Simple Truths," 460). Before Truman took office, Franklin D. Roosevelt's 1939 State of the Union address identified religion as not just the first of the "three institutions indispensable to Americans" but also "the source of the other two—democracy and international good faith" (quoted in H. W. Brands, *Traitor to His Class: The Privileged Life and Radical Presidency of Franklin Delano Roosevelt* [New York: Doubleday, 2008], 512).

7. Quoted in Jacobs, "'Our System Demands the Supreme Being,'" 596, 590; quoted in Inboden, *Religion and American Foreign Policy,* 232, 267, 259, 261, 270.

8. K. Healan Gaston, *Imagining Judeo-Christian America: Religion, Secularism, and the Cold War Redefinition of Democracy* (Chicago: University of Chicago Press, 2019).

9. Will Herberg, *Protestant-Catholic-Jew: An Essay in American Religious Sociology* (Garden City, NY: Doubleday, 1955), 270. On Herberg's book, which sounded all of the themes described in this paragraph, see K. Healan Gaston, "The Cold War Romance of Religious Authenticity: Will Herberg, William F. Buckley Jr., and the Rise of the New Right," *Journal of American History* 99 (March 2013): 1133–1158.

10. Carl F. H. Henry, *The Uneasy Conscience of Modern Fundamentalism* (Grand Rapids, MI: W. B. Eerdmans, 1947), 57–58; Henry, *Remaking the Modern Mind* (Grand Rapids, MI: W. B. Eerdmans, 1948), 23; Henry, *The Protestant Dilemma: An Analysis of the Current Impasse in Theology* (Grand Rapids, MI: W. B. Eerdmans, 1948), 37. On Henry, see esp. David R. Swartz, *Moral Minority: The Evangelical Left in an Age of Conservatism* (Philadelphia: University of Pennsylvania Press, 2012) and Owen Strachan, *Awakening the Evangelical Mind: An Intellectual History of the Neo-Evangelical Movement* (Grand Rapids, MI: Zondervan, 2015).

11. Thurston Davis, "Drama in the Laboratory," *America* 91, no. 5 (May 1, 1954): 131; Edward A. Marciniak, review of *The Proper Study of Mankind,* by Stuart Chase, *American Catholic Sociological Review* 9, no. 4 (December 1948): 287; Thomas I. Cook, "The Scientific Attitude in Human Relations," *Review of Politics* 11, no. 2 (April 1949): 240; Frederick Wilhelmsen, "The Alienated Professor," *Commonweal* 64, no. 1 (April 6, 1956): 12; "Religion and the Social Sciences," *America* 92, no. 2 (October 9, 1954): 34; J. G. Demaray, "The Kremlin's Hidden Ally," *Catholic Educational Review* 49, no. 1 (January 1951), 667–674.

12. Gaston, *Imagining Judeo-Christian America* (Cogley cited on 163); Herberg, *Protestant-Catholic-Jew,* 273; quoted in Nathan Glazer, "Daniel P. Moynihan on Ethnicity," in *Daniel Patrick Moynihan: The Intellectual in Public Life,* ed. Robert A. Katzmann (Washington, DC: Woodrow Wilson Center Press, 1998), 19.

13. Catholic Church and Joseph Clifford Fenton, *Humani generis: Encyclical Letter of Pope Pius XII concerning Some False Opinions which Threaten to Undermine the Foundations of Catholic Doctrine* (1950).

14. Quoted in Patrick Allitt, *Catholic Intellectuals and Conservative Politics in America, 1950–1985* (Ithaca, NY: Cornell University Press, 1993); John LaFarge, "Let's Build the Spiritual Front," *America* 83, no. 20 (August 19, 1950): 509; quoted in Boyer, *By the Bomb's Early Light,* 231.

15. John Courtney Murray, "On the Necessity for Not Believing: A Roman Catholic Interpretation," *Yale Scientific Magazine* 23, no. 5 (February 1949): 22.

16. Jacques Maritain, "Christian Humanism: Life with Meaning and Discretion," *Fortune* 25, no. 4 (April 1942): 107, 164. On Maritain in the wartime context, see Alan Jacobs, *The Year of Our Lord 1943: Christian Humanism in an Age of Crisis* (New York: Oxford University Press, 2018).

17. Fulton J. Sheen, *Communism and the Conscience of the West* (Indianapolis, IN: Bobbs-Merrill, 1948), 52, 7–8; Mark S. Massa, *Catholics and American Culture: Fulton Sheen, Dorothy Day, and the Notre Dame Football Team* (New York: Crossroad, 1999).

18. Clare Boothe Luce, *The Twilight of God* (Chicago: Regnery, 1949), 22, 28; quoted in Mark Silk, *Spiritual Politics: Religion and America Since World War II* (New York: Simon and Schuster, 1989), 37.

19. Henry Luce, "A Speculation about 1980," in *The Ideas of Henry Luce,* ed. John K. Jessup (New York: Atheneum, 1969), 314.

20. "Letters: Science v. Theology," *Time* 76, no. 8 (August 22, 1960).

21. Douglas Sloan, *Faith and Knowledge: Mainline Protestantism and American Higher Education* (Louisville, KY: Westminster/John Knox, 1994), 47; Reinhold Niebuhr, *The Irony of American History* (New York: Scribner's, 1952); Niebuhr, *Faith and History: A Comparison of Christian and Modern Views of History* (New York: Scribner's, 1949), 12. Niebuhr was among those who dismissed the popular religious revival of the era as a theologically empty and implicitly secular phenomenon, aligned with modern scientism in its guise as consumer capitalism.

22. Sloan, *Faith and Knowledge,* 51, 43–47, 41. Lowry's book is *The Mind's Adventure: Religion and Higher Education* (Philadelphia: Westminster Press, 1950).

23. James A. Pike, *Doing the Truth: A Summary of Christian Ethics* (Garden City, NY: Doubleday, 1955), 23; Pike, "Report of the Chaplain of the University for the Academic Year Ending June 30, 1950," Office of the President, Central Files, University Archives, Columbia University, box 414, folder 14; Pike, "Religion in Higher Education and the Problem of Pluralism," *Christianity and Crisis* 10, no. 23 (January 8, 1951): 179; Robert E. Fitch, "Christian Faith and the Inquiring Mind," *Christian Scholar* 38, no. 4 (December 1955): 253–254. Niebuhr likewise argued that "the various sciences, pretending to be presuppositionless, insinuate metaphysical presuppositions into their analysis" (*Faith and History,* 53).

24. Robert S. Ellwood, *The Politics of Myth: A Study of C. G. Jung, Mircea Eliade, and Joseph Campbell* (Albany: State University of New York Press, 1999); Joseph Campbell, *The Hero with a Thousand Faces* (New York: Pantheon, 1949); Thomas Merton, *The Seven-Storey Mountain* (New York: Harcourt, Brace, 1948), 138; Merton, "The Contemplative Life in the Modern World,"

in *Thomas Merton: Selected Essays*, ed. Patrick F. O'Connell (Maryknoll, NY: Orbis, 2013).

25. James Luther Adams, "What Kind of Religion Has a Place in Higher Education?," *Journal of Bible and Religion* 13 (November 1945): 191.

26. Gaston, *Imagining Judeo-Christian America*. For Frank's views, see his *The Rediscovery of Man: A Memoir and a Methodology of Modern Life* (New York: G. Braziller, 1958).

27. Gaston, "The Cold War Romance of Religious Authenticity"; Will Herberg, *Judaism and Modern Man: An Interpretation of Jewish Religion* (Philadelphia: Jewish Publication Society of America, 1951), 27, 11–12, 34. Herberg summarized his view of the modern predicament in "Prophetic Faith in an Age of Crisis," *Judaism* 1, no. 3 (July 1952): 195–202.

28. Herberg, *Judaism and Modern Man*, 7, 27.

29. Robert Gordis, *A Faith for Moderns* (New York: Bloch, 1960), 18–20, 15–16, 22–24.

6. THE HUMANISTIC OPPOSITION

1. "Report by the Commission on Trends in Education," *PMLA* 77, no. 2 (May 1962): 87. On the humanities during this period, see esp. David A. Hollinger, "Literary Culture," in *The United States in the Twentieth Century: An Encyclopedia*, vol. 4, ed. Stanley Kutler, Robert Dallek, Hollinger, and Thomas McCraw (New York: Scribner's, 1996), 1435–1459; Hollinger, ed., *The Humanities and the Dynamics of Inclusion since World War II* (Baltimore: Johns Hopkins University Press, 2006); and Thomas Bender and Carl E. Schorske, eds., *American Academic Culture in Transformation: Fifty Years, Four Disciplines* (Princeton, NJ: Princeton University Press, 1998).

2. Stephen Schryer explores the portrayal of the postwar humanities as the antidote to bureaucratization and collectivization in his *Fantasies of the New Class: Ideologies of Professionalism in Post–World War II American Fiction* (New York: Columbia University Press, 2011). On the political context for the humanities revival, see also Charles M. Dorn, *American Education, Democracy, and the Second World War* (New York: Palgrave Macmillan, 2007). An early announcement of the shift is Patricia Beesley, *The Revival of the Humanities in American Education* (New York: Columbia University Press, 1940).

3. Bernard Phillips, "The Humanities and the Idea of Man," *Journal of General Education* 2, no. 2 (January 1948): 131–132, 136, 129. The literary critic Richard Chase mused that "if one had read and understood Melville one would not vote for Henry Wallace," the left-liberal presidential candidate of 1948, because Melville "presents his reader with a vision of life so complexly true that it exposes the ideas of Henry Wallace as hopelessly childish and superficial." Chase, "Art, Nature, Politics," *Kenyon Review* 12, no. 4 (Autumn 1950): 580–594. "If only all Marxists, all capitalist Chambers of Commerce, and all Gallup Poll predictors would read Dostoevski's *Notes from the Underground*," the conservative poet and historian Peter Viereck similarly lamented. Viereck, *Conservatism Revisited* (1949; repr., New York: Free Press, 1962), 44.

4. Lionel Trilling, "The Kinsey Report" (1948), in *The Moral Obligation to Be Intelligent: Selected Essays,* ed. Leon Wieseltier (New York: Macmillan, 2000), 123. The social sciences, Trilling continued, "no longer pretend that they can merely describe what people do; they now have the clear consciousness of their ability to manipulate and adjust"—and, he added, gave potent aid to governments as well as industry (Trilling, "The Kinsey Report," 123).

5. Max Lerner, "The Frontiers of the Human Condition," *American Scholar* 27, no. 1 (Winter 1957–1958): 17. See also Lerner, *Education and a Radical Humanism: Notes toward a Theory of the Educational Crisis* (Columbus: Ohio State University Press, 1962).

6. Howard Mumford Jones, *One Great Society: Humane Learning in the United States* (New York: Harcourt, Brace, 1959), 164–167; Arthur M. Schlesinger Jr., "The Statistical Soldier," *Partisan Review* 16, no. 8 (August 1949): 855. See also Schlesinger, "The Humanist Looks at Empirical Social Research," *American Sociological Review* 27, no. 6 (December 1962): 768–771 (and note that journal's decision to publish it). The "Atheists for Niebuhr" epithet appeared in Morton White, "Religion, Politics, and the Higher Learning," *Confluence* 3, no. 4 (1954): 404.

7. Hyatt Howe Waggoner, *The Heel of Elohim: Science and Values in Modern American Poetry* (Norman: University of Oklahoma Press, 1950), 90, 10–11; Hayward Keniston, "Literature as a Barometer of Modern European Society," in Walter R. Agard et al., *The Humanities for Our Time* (Lawrence: University of Kansas Press, 1949), 44. Jones occasionally worried about specialization in the universities, but he found that shortcoming in all branches of learning: e.g., Jones, *Education and World Tragedy* (Cambridge, MA: Harvard University Press, 1946), 48–51. Joan Shelley Rubin discusses Jones's postwar work in "The Scholar and the World: Academic Humanists and General Readers in Postwar America," in Hollinger, *The Humanities and the Dynamics of Inclusion since World War II,* 73–103; see also Peter Brier, *Howard Mumford Jones and the Dynamics of Liberal Humanism* (Columbia: University of Missouri Press, 1994).

8. C. P. Snow, *The Two Cultures and the Scientific Revolution* (New York: Cambridge University Press, 1959).

9. Snow was best known for his "Strangers and Brothers" novels on politics and power in the mid-twentieth century. The most recent installments were *The New Men* (New York: Scribner, 1954), featuring postwar nuclear physicists; *Homecoming* (New York: Scribner, 1956); and *The Conscience of the Rich* (New York: Scribner, 1958).

10. Snow, *The Two Cultures and the Scientific Revolution,* 10–11, 7.

11. Snow, *The Two Cultures and the Scientific Revolution.*

12. F. R. Leavis, *Two Cultures? The Significance of C. P. Snow* (London: Chatto and Windus, 1962); James Feron, "Leavis' Attack on C. P. Snow Stuns British Literary World," *New York Times,* March 10, 1962, 23; Feron, "C. P. Snow Backed in Leavis Attack," *New York Times,* March 16, 1962, 33; Kenneth Campbell, "Snow in the Wings," *New York Times,* September 16, 1962, 129; Seth S. King, "C. P. Snow Bids Rich Lands Show Magnanimity," *New York Times,* April 14, 1962, 3; "Topics," *New York Times,* April 1, 1962, 182; Feron,

"C. P. Snow Hopeful of Culture Unity," *New York Times,* October 25, 1963, 33; H. Margolis, "Intellectual Life in England: Leavis Views C. P. Snow; Boothby Views Leavis," *Science* 135 (March 30, 1962): 1114–1115; Mary M. Simpson, "The Snow Affair," *Bulletin of the Atomic Scientists* 19, no. 4 (April 1963): 28–32; "The Corridors of Power," *Time,* May 16, 1960, 105–106; Mary Ann Callan, "Are We in Danger of Losing Human Race?," *Los Angeles Times,* December 4, 1960, M1, M11; "Television Program Highlights for Tuesday," *Washington Post, Times Herald,* May 16, 1961, A26; Jean White, "ICA Calls Big Figures in the Arts," *Washington Post, Times Herald,* September 29, 1960, B8; C. P. Snow, "Three Dragons to Slay," *The Michigan Alumnus* 69, no. 10 (July 1963): 316–317; C. P. Snow et al., "The Moral Un-Neutrality of Science," *Science,* n.s., 133, no. 3448 (January 27, 1961): 255–262; "Sir Charles and Lady Snow," *Vogue,* March 1, 1961, 141; Sigmund Koch, "Psychological Science versus the Science-Humanism Antinomy: Intimations of a Significant Science of Man," *American Psychologist* 16, no. 10 (October 1961): 629; Claudio Giorgio Segre, "C. P. Snow and the 'Two Cultures'" (MA thesis, Stanford University, 1961); David K. Cornelius and Edwin St. Vincent, eds., *Cultures in Conflict: Perspectives on the Snow-Leavis Controversy* (Chicago: Scott, Foresman, 1964). Meanwhile, a team of social scientists set out, in good American fashion, to assess Snow's thesis by administering surveys to undergraduates at four colleges. Comparing Snow's assertions to students' perceptions, they left no detail unexamined. "If one were to study [the scientist's] recreational habits," the authors wrote, "one would find him most frequently at chess, rarely playing bridge, and never playing poker." Poker was presumably the domain of the engineer, whom the students deemed "a more 'regular guy' than the scientist . . . the engineer is believed more likely to have a pretty wife" (David C. Beardslee and Donald D. O'Dowd, "The College-Student Image of the Scientist," *Science,* n.s., 133 [March 31, 1961]: 1000–1001, 998).

13. On the debate's British contexts, see Guy Ortolano, *The Two Cultures Controversy: Science, Literature and Cultural Politics in Postwar Britain* (New York: Cambridge University Press, 2009), and Peter Mandler, "The Two Cultures Revisited: The Humanities in British Universities since 1945," *20th Century British History* 26, no. 3 (September 2015): 400–423.

14. James B. Conant, *Modern Science and Modern Man* (New York: Columbia University Press, 1952); Conant, *The Citadel of Learning* (New Haven, CT: Yale University Press, 1956); J. Robert Oppenheimer, *Atom and Void: Essays on Science and Community* (Princeton, NJ: Princeton University Press, 1989); Warren Weaver, "The Imperfections of Science," *Proceedings of the American Philosophical Society* 104, no. 5 (October 17, 1960): 427–428; Jones, *One Great Society: Humane Learning in the United States* (New York: Harcourt, Brace, 1959), 164–167. For more on this understanding of science, see Andrew Jewett, *Science, Democracy, and the American University: From the Civil War to the Cold War* (New York: Cambridge University Press, 2012), 314–320. Jamie Cohen-Cole explores the postwar vogue of creativity in *The Open Mind: Cold War Politics and the Sciences of Human Nature* (Chicago: University of Chicago Press, 2014).

15. Michael Wolff, "The Uses of Context: Aspects of the 1860's," *Victorian Studies* 9, supp. (September 1965): 51.

16. Bentley Glass, "The Academic Scientist, 1940–1960," *Science,* n.s., 132, no. 3427 (September 2, 1960): 603.

17. James B. Conant, "Science and Society in the Post-War World," *Vital Speeches of the Day* 9, no. 13 (April 15, 1943): 394–395; Conant, *Our Fighting Faith: Five Addresses to College Students* (Cambridge, MA: Harvard University Press, 1942), 12, 77. On Conant, see also Joel Isaac, *Working Knowledge: Making the Human Sciences from Parsons to Kuhn* (Cambridge, MA: Harvard University Press, 2012); and George A. Reisch, *The Politics of Paradigms: Thomas S. Kuhn, James B. Conant, and the Cold War "Struggle for Men's Minds"* (Albany: State University of New York Press, 2019).

18. Theodore H. Von Laue, "Modern Science and the Old Adam," *Bulletin of the Atomic Scientists* 19, no. 1 (January 1963): 2–5.

19. M. Brewster Smith, "'Mental Health' Reconsidered: A Special Case of the Problem of Values in Psychology," *American Psychologist* 16 (1961): 206; Lloyd Fallers, "C. P. Snow and the Third Culture," *Bulletin of the Atomic Scientists* 17, no. 8 (October 1961): 307–308; Harry S. Kantor, review of Snow, *The Two Cultures and the Scientific Revolution, Monthly Labor Review* 83, no. 10 (October 1960): 1100.

20. C. P. Snow, *The Two Cultures: And a Second Look* (New York: Cambridge University Press, 1963), 84.

21. Robert M. MacIver, "Science as a Social Phenomenon," *Proceedings of the American Philosophical Society* 105, no. 5 (October 1961): 500–505; James F. Davidson, "Political Science and Political Fiction," *American Political Science Review* 55, no. 4 (December 1961): 855; James P. Scanlan, "The Philosophy of Social Science," *Review of Politics* 23, no. 1 (January 1961): 107, 111; Robert Lekachman, review of *Science and Government,* by C. P. Snow, *Political Science Quarterly* 76, no. 3 (September, 1961): 465.

22. Jewett, *Science, Democracy, and the American University.*

23. David Riesman et al., *The Lonely Crowd: A Study of the Changing American Character* (New Haven, CT: Yale University Press, 1950); Riesman, *Individualism Reconsidered and Other Essays* (Glencoe, IL: Free Press, 1954); Barrington Moore Jr., *Political Power and Social Theory* (Cambridge, MA: Harvard University Press, 1958); Nathan Glazer, "'The American Soldier' as Science: Can Sociology Fulfill Its Ambitions?," *Commentary* 8, no. 5 (November 1949): 487–496. Meanwhile, William H. Whyte argued that American children should receive "a rigorously fundamental schooling" rooted in the humanities (*The Organization Man* [New York: Simon and Schuster, 1956], 447).

24. Crane Brinton et al., "The Application of Scientific Method to the Study of Human Behavior," *American Scholar* 21, no. 2 (Spring 1952): 212–213, 219. The other two participants were B. F. Skinner and Joseph Wood Krutch, whose sharp criticism of scientism is discussed in Chapter 8. *American Scholar* editor Hiram Haydn had hoped for a vigorous debate between Krutch and Skinner, but Skinner ignored Krutch's attempts to bait him and even agreed with Krutch that social scientists were manipulative and frightening. Whereas Skinner's critics viewed him as the consummate social scientist, he considered himself a natural scientist.

25. Cushing Strout, "The Twentieth-Century Enlightenment," *American Political Science Review* 49, no. 2 (June 1955): 324, 339, 333, 328.

26. Ted V. McAllister, *Revolt against Modernity: Leo Strauss, Eric Voegelin, and the Search for a Postliberal Order* (Lawrence: University Press of Kansas, 1996), 79–80; Eric Voegelin, "The Origins of Scientism," *Social Research* 15, no. 4 (December 1948): 464, 494, 490–491, 488. Voegelin's less pessimistic argument appears in *The New Science of Politics: An Introduction* (Chicago: University of Chicago Press, 1952). Strauss summarized his analysis in "Social Science and Humanism" (1956), in *The Rebirth of Classical Political Rationalism: An Introduction to the Thought of Leo Strauss,* ed. Thomas L. Pangle (Chicago: University of Chicago Press, 1989), 3–12. For a less conservative émigré perspective, see Reinhard Bendix, *Social Science and the Distrust of Reason* (Berkeley: University of California Press, 1951).

27. Hans Morgenthau, *Scientific Man vs. Power Politics* (Chicago: University of Chicago Press, 1946), 5–6.

28. William Barrett, *Irrational Man: A Study in Existential Philosophy* (Garden City, NY: Doubleday, 1958). On American responses to existentialism, see esp. George Cotkin, *Existential America* (Baltimore: Johns Hopkins University Press, 2003).

29. Barrett, *Irrational Man,* 24, 21–22, 30, 239, 244, 247.

30. David Paul Haney, *The Americanization of Social Science: Intellectuals and Public Responsibility in the Postwar United States* (Philadelphia: Temple University Press, 2008), 147–149, 150–151, 156–157; Albert H. Hobbs, *The Claims of Sociology: A Critique of Textbooks* (Harrisburg, PA: Stackpole, 1951); Hobbs, *Social Problems and Scientism* (Harrisburg, PA: Stackpole, 1953). Many of Hobbs's admirers later welcomed Mills's book (Haney, *The Americanization of Social Science,* 167–169). More recently, the philosopher of science Susan Haack opened an article with a quote from Hobbs: Haack, "Six Signs of Scientism: Part I," *Skeptical Inquirer* 37, no. 6 (November–December 2013): 40.

31. Pitirim A. Sorokin, *Fads and Foibles in Modern Sociology and Related Sciences* (Chicago: Regnery, 1956), 314, 304, 301. For a more direct statement of Sorokin's social criticism, see Sorokin, *The Reconstruction of Humanity* (Boston: Beacon Press, 1948). Haney analyzes Sorokin's critique in *The Americanization of Social Science,* 124–137; see also Barry V. Johnston, *Pitirim A. Sorokin: An Intellectual Biography* (Lawrence: University Press of Kansas, 1995).

32. Katherine Pandora, *Rebels within the Ranks: Psychologists' Critiques of Scientific Authority and Democratic Realities in New Deal America* (New York: Cambridge University Press, 1997); quoted in Jessica Grogan, *Encountering America: Humanistic Psychology, Sixties Culture, and the Shaping of the Modern Self* (New York: HarperCollins, 2012), x; quoted in Ellen Herman, *The Romance of American Psychology: Political Culture in the Age of Experts* (Berkeley: University of California Press, 1995), 267.

33. Herman, *The Romance of American Psychology,* 264–275 (quoted at 269); Carl Rogers and B. F. Skinner, "Some Issues Concerning the Control of Human Behavior," *Science* 124 (November 30, 1956): 1057–1066; Grogan, *Encountering America,* 15, 86–87.

34. Margaret Mead, "Anthropology and an Education for the Future," in *The Teaching of Anthropology*, Memoir 94, ed. David G. Mandelbaum, Gabriel W. Lasker, and Ethel M. Albert (Arlington, VA: American Anthropological Association, 1963), 379.

7. A NEW RIGHT

1. E. Case, review of *The Conscience of the Rich*, by C. P. Snow, *National Review* 5, no. 10 (March 8, 1958): 238; Joan Didion, "Inadequate Mirrors," *National Review* 8, no. 27 (July 2, 1960): 430–431.
2. Frederick Wilhelmsen, "The Sorcerer's Apprentices," *National Review* 8, no. 8 (February 27, 1960): 144–145.
3. William F. Buckley Jr., "Operation Snow Removal," *National Review* 12, no. 12 (March 27, 1962): 194; Buckley, "C. P. Snow, R I P," *National Review* 32, no. 15 (July 25, 1980): 879–880.
4. Many of these figures also shared an intellectual hero in the anti-Enlightenment thinker Edmund Burke, who was frequently cited by postwar critics of rationalism and scientism from across the political spectrum. A helpful overview of the postwar movement is found in Patrick Allitt, *The Conservatives: Ideas and Personalities throughout American History* (New Haven, CT: Yale University Press, 2009), 158–190. For a fuller account, see George H. Nash, *The Conservative Intellectual Movement in America Since 1945*, 30th anniversary ed. (Wilmington, DE: ISI Books, 2006).
5. William F. Buckley Jr., "Publisher's Statement," *National Review* 1 (November 19, 1955): 5; Albert H. Hobbs to Henry Regnery, August 31, 1949, Henry Regnery Papers, Hoover Institution Archives (hereafter Regnery Papers), box 30, folder 14. On the tendency to read liberalism into nature, see Andrew Jewett, "Naturalizing Liberalism in the 1950s," in *Professors and Their Politics*, ed. Neil Gross and Solon J. Simmons (Baltimore: Johns Hopkins University Press, 2014), 191–216, and Jamie Cohen-Cole, *The Open Mind: Cold War Politics and the Sciences of Human Nature* (Chicago: University of Chicago Press, 2014).
6. Frank S. Meyer, "The Scholarly Journals: Confusion in the Court," *National Review* 1, no. 8 (January 11, 1956): 23.
7. Richard M. Weaver, "On Social Science," *National Review* 1, no. 25 (May 9, 1956): 20; Frank S. Meyer, "Retreat from Relativism," *National Review* 1, no. 4 (December 14, 1955): 24–25.
8. Frank S. Meyer, "Politics and Responsibility," *National Review* 1, no. 20 (April 4, 1956): 21; Meyer, "Principles and Heresies: A Conscious Conservatism," *National Review* 2, no. 9 (August 18, 1956): 16; Meyer, "The Scholarly Journals," 23.
9. Buckley, "Publisher's Statement," 5; Richard M. Weaver, "From Poetry to Bitter Fruit," *National Review* 1, no. 10 (January 25, 1956): 26.
10. Hobbs to Regnery, August 31, 1949; E. Merrill Root, *Collectivism on the Campus: The Battle for the Mind in American Colleges* (New York: Devin-Adair,

1955), 322–335; Raymond Moley, "Scientism," *Newsweek,* June 28, 1954, 88; *Tax-Exempt Foundations: Hearings before the Special Committee to Investigate Tax-Exempt Foundations and Comparable Organizations, Part 1* (Washington, DC: Government Printing Office, 1954), 73; Mark Solovey, *Shaky Foundations: The Politics–Patronage–Social Science Nexus in Cold War America* (New Brunswick, NJ: Rutgers University Press, 2013), 122–127, 140–145; Charles W. Lowry to Monroe Bush, September 26, 1956, and Lowry to William Y. Elliott, May 3, 1957, William Y. Elliott Papers, Hoover Institution Archives, box 6, folder "FRASCO (Committee on Communist Education in Schools)." On the Reece Committee, see also David A. Horowitz, *America's Political Class under Fire: The Twentieth Century's Great Culture War* (New York: Routledge, 2003), 142–147. On FRASCO, see William Inboden III, *Religion and American Foreign Policy, 1945–1960: The Soul of Containment* (New York: Cambridge University Press, 2008); K. Healan Gaston, "The Cold War Romance of Religious Authenticity: Will Herberg, William F. Buckley Jr., and the Rise of the New Right," *Journal of American History* 99, no. 4 (March 2013): 1133–1158; and Gaston, *Imagining Judeo-Christian America: Religion, Secularism, and the Cold War Redefinition of Democracy* (Chicago: University of Chicago Press, 2019), 180–183, 197–198.

11. F. A. Hayek, "Planning, Science and Freedom," *Nature* 148 (November 15, 1941): 580–584. For Mises's critique of mechanistic social science, see his "Social Science and Natural Science," *Journal of Social Philosophy and Jurisprudence* 7, no. 3 (April 1942): 240–253.

12. F. A. Hayek, "Scientism and the Study of Society. Part I," *Economica* 9, no. 35 (August 1942): 279; Hayek, "Scientism and the Study of Society. Part III," *Economica* 11, no. 49 (February 1944): 31; Hayek, *The Counter-Revolution of Science: Studies on the Abuse of Reason* (Glencoe, IL: Free Press, 1952).

13. Allitt, *The Conservatives,* 165–172.

14. Gaston, *Imagining Judeo-Christian America;* Richard M. Weaver, *Ideas Have Consequences* (Chicago: University of Chicago Press, 1948), 2. On Catholicism and conservatism, see Patrick Allitt, *Catholic Intellectuals and Conservative Politics in America: 1950–1985* (Ithaca, NY: Cornell University Press, 1993). By contrast, traditionalists stood at a great distance, both socially and intellectually, from fundamentalists and other conservative evangelicals.

15. Jennifer Burns, "Godless Capitalism: Ayn Rand and the Conservative Movement," *Modern Intellectual History* 1, no. 3 (November 2004): 359–385.

16. Willmoore Kendall, "The Liberal Line," *National Review* 1, no 15 (February 29, 1956): 8; Buckley, "Publisher's Statement," 5; William F. Buckley Jr., *God and Man at Yale: The Superstitions of "Academic Freedom"* (Chicago: Regnery, 1951), 25.

17. Hans J. Morgenthau, *Scientific Man vs. Power Politics* (Chicago: University of Chicago Press, 1946), 9; Eric Voegelin, *The New Science of Politics: An Introduction* (Chicago: University of Chicago Press, 1952); Leo Strauss, *Natural Right and History* (Chicago: University of Chicago Press, 1953); Ted V. McAllister, *Revolt against Modernity: Leo Strauss, Eric Voegelin, and the Search for a Postliberal Order* (Lawrence: University Press of Kansas, 1996), 79; Peter

Viereck, *Conservatism Revisited* (1949; repr., New York: Free Press, 1962), 39, 43–45; Viereck, "The Revolution in Values: Roots of the European Catastrophe, 1870–1952," *Political Science Quarterly* 67, no. 3 (September 1952): 343–344, 348–349, 339, 356.

18. Russell Kirk, "Prospects for a Conservative Bent in the Human Sciences," *Social Research* 35, no. 4 (Winter 1968): 580–581.

19. Edward Case, "Big Brotherly Reference Book," *National Review* 1, no. 23 (April 25, 1956): 20–21. For another postwar cultural campaign, see Alan Filreis, *Counter-Revolution of the Word: The Conservative Attack on Modern Poetry, 1945–1960* (Chapel Hill: University of North Carolina Press, 2008). Kevin Mattson briefly mentions the cultural side of the conservative analysis in *Rebels All! A Short History of the Conservative Mind in Postwar America* (New Brunswick, NJ: Rutgers University Press, 2008), 121–122.

20. The phrase is from Arthur M. Schlesinger Jr., *The Vital Center: The Politics of Freedom* (Boston: Houghton Mifflin, 1949). Key examples include Theodor W. Adorno, Else Frenkel-Brunswik, Daniel J. Levinson, and R. Nevitt Sanford, *The Authoritarian Personality* (New York: Harper, 1950); and Daniel Bell, ed., *The New American Right* (New York: Criterion, 1955). In one of the droller responses to this phenomenon, the conservative biologist Mark Graubard proposed titling his autobiography "Pilgrims in the Murky Swamp and Their Dull Pedestrian Kin." Graubard to Henry Regnery, n.d. [1961 or 1962], Regnery Papers, box 26, folder 4.

21. Albert H. Hobbs to Raymond Moley, January 6, 1954, Raymond Moley Papers, Hoover Institution Archives, box 23, folder 51.

22. Cohen-Cole, *The Open Mind.*

23. Henry Regnery to Donald Meiklejohn, January 19, 1956, Regnery Papers, box 50, folder 13; Regnery to Robert M. Hutchins, March 10, 1954, Regnery Papers, box 33, folder 1. An array of such cases, large and small, appear in Russell Kirk, *Academic Freedom: An Essay in Definition* (Chicago: Regnery, 1955), 59–71, and Root, *Collectivism on the Campus,* 302–322. Such episodes hit particularly close to home for some of the conservative movement's leading architects. M. Stanton Evans, who wrote the 1961 book *Revolt on the Campus* and the 1962 "Sharon Statement" of Young Americans for Freedom, saw his father Medford B. Evans, a John Birch Society member and an ardent segregationist, dismissed by Louisiana's Northwestern State College in 1959. Other conservatives watched as beloved teachers were ignored, harassed, or ousted, as did Buckley when his mentor Willmoore Kendall was ostracized and eventually shown the door at Yale.

24. Kirk, *Academic Freedom;* Root, *Collectivism on the Campus.*

25. Russell Kirk to editor of *The Reporter,* April 6, 1955, Regnery Papers, box 39, folder 9; Buckley, *God and Man at Yale.* For another expression of faith in the university's openness, see William T. Couch to Henry Regnery, October 5, 1958, Regnery Papers, box 82, folder 1.

26. "Remarks of Benjamin H. Namm," January 25, 1951, Hugh Gibson Papers, Hoover Institution Archives, box 86, folder "*The Educational Reviewer,* New York, 1950–1951." Namm and Buckley were among the many conservatives

who decried the widespread use of Paul Samuelson's Keynesian economics textbook.

27. Kirk, *Academic Freedom*, 7.

28. Russell Kirk to Henry Regnery, December 2, 1954, Regnery Papers, box 39, folder 9. A number of important edited volumes featured conservative writers alongside other critics of scientism: e.g., Felix Morley, ed., *Essays on Individuality* (Philadelphia: University of Pennsylvania Press, 1958), and Helmut Schoeck and James W. Wiggins, eds., *Scientism and Values* (Princeton, NJ: Van Nostrand, 1960).

29. John U. Nef, "Monthly Meeting of Instructional Staff Held October 9 [1951]," Friedrich Hayek Papers, Hoover Institution Archives (hereafter Hayek Papers), box 63, folder 10; Robert M. Hutchins to Henry Regnery, January 27, 1954, Regnery Papers, box 33, folder 1; Russell Kirk to Regnery, January 25, 1954, Regnery Papers, box 39, folder 9.

30. Pitirim A. Sorokin to Henry Regnery, April 2, 1956, Regnery Papers, box 70, folder 9; Regnery to Russell Kirk, July 31, 1953, Regnery Papers, box 39, folder 9; Regnery to Sorokin, n.d., attached to Sorokin to Regnery, May 3, 1957, Regnery Papers, box 70, folder 9; Pitirim A. Sorokin to Henry Regnery, April 9, 1956, Regnery Papers, box 70, folder 9. Regnery also sought to build bridges by finding readers for Kirk's books beyond the usual circle of conservatives, with some success. Among the admirers of *The Conservative Mind*, according to Kirk, was the liberal historian Eric Goldman, who was "*giving* copies of it to friends in Vienna" (Kirk to Regnery, September 4, 1954, Regnery Papers, box 39, folder 9). Regnery later sent Kirk's *Academic Freedom* to Sidney Hook, writing, "I think you would be more in agreement with him than not." Regnery to Hook, February 7, 1955, Regnery Papers, box 30, folder 27. Regnery had long sought to cultivate Hook, whose pragmatically tinged Marxism gave way after World War II to an equally pugnacious anticommunism.

31. Russell Kirk, *Decadence and Renewal in the Higher Learning: An Episodic History of American University and College Since 1953* (South Bend, IN: Gateway, 1978), 3–13 (quote on 5); Kirk, *Academic Freedom*, 105; Kirk, *The Conservative Mind: From Burke to Santayana* (Chicago: Regnery, 1953). The account in *Academic Freedom* does not identify Kirk as the protagonist but reflects his own experience.

32. Henry Regnery to Russell Kirk, October 8, 1953, Regnery to Kirk, October 12, 1953, Kirk to Regnery, October 27, 1953, Kirk to Henry Regnery, November 1, 1953, Kirk to Regnery, November 9, 1953, Kirk to Regnery, March 18, 1956, Kirk to Regnery, May 11, 1954, and Regnery to Kirk, December 15, 1955, Regnery Papers, box 39, folder 9.

33. Henry Regnery to Russell Kirk, October 12, 1953, Kirk to Regnery, July 31, 1956, and Kirk to Regnery, November 15, 1955, Regnery Papers, box 39, folder 9; Regnery to Kirk, July 28, 1958, Regnery Papers, box 40, folder 1. Colegrove, a political scientist, also sought to remake the social sciences. He urged Regnery to solicit and publish a series of social-scientific works that would counter the "anti-conservative" bias of those disciplines and the liberal political

culture that they shaped. Colegrove identified a particular need for books on "The Do-Gooder in Politics," "The Rich Liberal in Politics," "The Pathology of the Civil Service," and, above all, the life of Republican champion Robert A. Taft (Colegrove, "Fields Needing the Publication of Conservative Books: Proposals for Books in These Fields," reprinted as part of Regnery, "A Proposal to Establish a New Series of Books on Education: International Relations, Government, Economics and Society," Regnery Papers, box 82, folder 1).

34. Henry Regnery to Friedrich Hayek, March 26, 1984, Hayek Papers, box 24, folder 19; "Report of a Meeting Held December 5, 1953," Regnery Papers, box 16, folder 5; Russell Kirk to Regnery, February 17, 1954, Regnery to various, October 23, 1953, Regnery to Kirk, April 27, 1953, and Regnery to Kirk, May 26, 1953, Regnery Papers, box 39, folder 9; Kirk to Regnery, November 23, 1957, Regnery Papers, box 40, folder 1.

35. Russell Kirk to Henry Regnery, October 27, 1953, and Kirk to Regnery, November 9, 1953, Regnery Papers, box 39, folder 9; Kirk to Regnery, December 11, 1957, Regnery Papers, box 40, folder 1; Regnery to Kirk, July 31, 1953, and Regnery to Kirk, September 1, 1954, Regnery Papers, box 39, folder 9; Frank Barnett to Regnery, April 17, 1953, Regnery Papers, box 6, folder 2; Albert H. Hobbs to Regnery, January 15, 1951, Regnery Papers, box 30, folder 14. Regnery did not publish Hobbs's book, but Pew and Hobbs may have made the same arrangement with its eventual publisher, Stackpole. On corporate support for the conservative movement in general, see esp. Kim Phillips-Fein, *Invisible Hands: The Making of the Conservative Movement from the New Deal to Reagan* (New York: Norton, 2009).

36. "Luncheon for Russell Kirk," February 19, 1959, Russell Kirk to Henry Regnery, May 7, 1959, and Kirk to Regnery, January 20, 1957, Regnery Papers, box 40, folder 1.

37. For Regnery's revisionism, see his exchanges with Barnes in Regnery Papers, box 6, folder 1, and his letter to Russell Kirk, February 21, 1958, Regnery Papers, box 40, folder 1.

8. CROSS-FERTILIZATION

1. Russell Kirk, "Is Social Science Scientific?," *New York Times*, June 25, 1961, 16, 18.

2. Merton's response is "Now the Case for Sociology," *New York Times*, July 16, 1961, 14. On this exchange, see David Paul Haney, *The Americanization of Social Science: Intellectuals and Public Responsibility in the Postwar United States* (Philadelphia: Temple University Press, 2008), 208–221.

3. Kirk, "Is Social Science Scientific?," 11, 14, 17. Haney discusses Kirk and other conservatives' appreciation of Mills in *The Americanization of Social Science,* 167–169. For his part, William F. Buckley Jr. drew on the writings of Protestant educators, theologians, and social scientists in his classic *God and Man at Yale,* citing Reinhold Niebuhr, Gordon Allport, and even Margaret

Mead. Buckley, *God and Man at Yale: The Superstitions of "Academic Freedom"* (Chicago: Regnery, 1951), 8, 18, 23, 200–203.

4. Jessica Grogan, *Encountering America: Humanistic Psychology, Sixties Culture, and the Shaping of the Modern Self* (New York: HarperCollins, 2012), 86–87; William H. Whyte, *The Organization Man* (New York: Simon and Schuster, 1956), 26; Joseph Wood Krutch, *The Measure of Man: On Freedom, Human Values, Survival, and the Modern Temper* (Indianapolis, IN: Bobbs-Merrill, 1954).

5. John W. Boyer, *The University of Chicago: A History* (Chicago: University of Chicago Press, 2015).

6. Friedrich Hayek, *The Counter-Revolution of Science: Studies on the Abuse of Reason* (Glencoe, IL: Free Press, 1952); Hayek, *The Sensory Order: An Inquiry into the Foundations of Theoretical Psychology* (Chicago: University of Chicago Press, 1952); Hayek to Robert M. Hutchins, July 13, 1950, Friedrich Hayek Papers, Hoover Institution Archives (hereafter Hayek Papers), box 55, folder 1. The Chicago economists largely returned this distrust; figures such as Milton Friedman believed that Hayek had strayed too far from the fold and abandoned serious work in economics. Hayek had indeed turned from questions of economic organization to political theory during the war years and then taken up deeper concerns in the philosophy of knowledge and psychology, resulting in his 1952 books.

7. Robert S. Thomas, "Enlightenment and Authority: The Committee on Social Thought and the Ideology of Postwar Conservatism (1927–1950)" (PhD diss., Columbia University, 2010), 190–194, 243–244, 468, 472, 604; John U. Nef, "Articles of C. F. von Weizsacker and Henry Jacoby," March 2, 1951, Henry Regnery Papers, Hoover Institution Archives (hereafter Regnery Papers), box 50, folder 11. Elsewhere at Chicago, figures such as Hans Morgenthau in the political science department helped reinforce the university's largely critical stance toward scientism (John Gunnell, *The Descent of Political Theory: The Genealogy of an American Vocation* [Chicago: University of Chicago Press, 1993], 195–196).

8. Philip Gleason, *Contending with Modernity: Catholic Higher Education in the Twentieth Century* (New York: Oxford University Press, 1995); Peter M. Rutkoff and William B. Scott, *New School: A History of the New School for Social Research* (New York: Free Press, 1986); Claus-Dieter Krohn, *Intellectuals in Exile: Refugee Scholars and the New School for Social Research* (Amherst: University of Massachusetts Press, 1993); Abram Leon Sachar, *Brandeis University: A Host at Last,* rev. ed. (Hanover, NH: University Press of New England for Brandeis University Press, 1995); Stephen J. Whitfield, "Brandeis University," in *Encyclopedia Judaica,* 2nd ed., ed. Fred Skolnik (Detroit: Macmillan Reference, 2007); Martin Duberman, *Black Mountain: An Experiment in Community* (New York: E. P. Dutton, 1972), 193, 251–252, 292–293, 473n74; Reamer Kline, *Education for the Common Good: A History of Bard College—The First 100 Years, 1860–1960* (Annandale-on-Hudson, NY: Bard College, 1982); Thomas P. Brockway, *Bennington College, in the Beginning* (Bennington, VT: Bennington College Press, 1981); Thomas Boardman Greenslade, *Kenyon Col-*

lege: Its Third Half Century (Gambier, OH: Kenyon College, 1975); Ronald Schwartz, "Riverside Days: Recollection of Robert Nisbet as a Teacher," *American Sociologist* 45, no. 1 (March 2014): 34–49, esp. 34–35.

9. Gunnell, *The Descent of Political Theory*; Hannah Arendt, *The Origins of Totalitarianism* (New York: Harcourt, Brace, 1951); John Hallowell, *Main Currents in Modern Political Thought* (New York: Holt, 1950), 636; Hallowell, *The Moral Foundation of Democracy* (Chicago: University of Chicago Press, 1954), 83.

10. H. A. W. Myrin, "The Myrin Institute, Inc.," *Proceedings of the Myrin Institute* (Fall 1954): 4–5. Waldorf schooling is usually considered a form of progressive education today, but its postwar proponents were deeply skeptical of scientific approaches to the study of the human world.

11. Russell Kirk, "The American College: A Proposal for Reform," *Proceedings of the Myrin Institute* (Spring 1957): 5–20; "Guest Speakers at the Institute Seminar During 1953–54," *Proceedings of the Myrin Institute* (Fall 1954): 8–9; Leonard E. Read, "Let the Method Fit the Objective," *Proceedings of the Myrin Institute* (Winter 1960–1961): 19–31; "Report on Activities during 1956–57," *Proceedings of the Myrin Institute* (Fall 1957): 30–31; "On the Campus of Adelphi College," *Proceedings of the Myrin Institute* (Spring 1958): 23; "A Report on Current Activities of the Myrin Institute, Inc.," *Proceedings of the Myrin Institute* (Spring 1955): 33–36; Sylvester M. Morey, "By Way of Introduction," *Proceedings of the Myrin Institute* (Spring 1963): 2–3; Myrin, "The Myrin Institute, Inc.," 4–5. Myrin's president, the physician Franz E. Winkler, spelled out his views in *Man: A Bridge between Two Worlds* (New York: Harper, 1960).

12. W. C. Mullendore to Henry Regnery, n.d., Regnery Papers, box 82, folder 1; James Sloan Allen, *The Romance of Commerce and Culture: Capitalism, Modernism, and the Chicago-Aspen Crusade for Cultural Reform* (Boulder, CO: University Press of Colorado, 2002).

13. Elizabeth Flower to Sidney Hook, July 9, 1953, W. Rex Crawford to Hook, September 29, 1954, and Crawford to Hook, November 1, 1954, Sidney Hook Papers, Hoover Institution Archives, box 23, folder 47. "If you don't mind," program head Crawford cautioned Hook, "don't do anything about publicity for the course."

14. Joseph Wood Krutch, unpublished essay ("Science Is Wonderful . . ."), n.d., Joseph Wood Krutch Papers, Library of Congress (hereafter Krutch Papers), box 15, folder "Untitled, Unpublished." His works of the 1950s included Krutch et al., *Is the Common Man Too Common? An Informal Survey of Our Cultural Resources and What We Are Doing about Them* (Norman: University of Oklahoma Press, 1954); *The Measure of Man: On Freedom, Human Values, Survival, and the Modern Temper* (Indianapolis: Bobbs-Merrill, 1954); and *Human Nature and the Human Condition* (New York: Random House, 1959); as well as the essays collected in *If You Don't Mind My Saying So . . . Essays on Man and Nature* (New York: W. Sloane, 1964), and *And Even If You Do: Essays on Man, Manners & Machines* (New York: Morrow, 1967).

15. Joseph Wood Krutch, *More Lives than One* (New York: W. Sloane, 1962); Peter Gregg Slater, "The Negative Secularism of *The Modern Temper:* Joseph Wood

Krutch," *American Quarterly* 33, no. 2 (July 1981): 186–188, 200 (quoted at 186); "Masses' Failure to Pray Laid to Material Life," *Los Angeles Times,* July 18, 1938, A18.

16. Krutch, *More Lives than One;* Joseph Wood Krutch, "A Liberal Examines His Conscience" (unpublished), 1941, Krutch Papers, box 14, folder "Unpublished Articles, 1941–1969"; Samuel C. Florman, "Mr. Krutch and the Scientific View," *American Scholar* 29, no. 2 (Spring 1960): 270. Florman ascribed to Krutch a "neo-orthodox anthropocentrism" that flew in the face of science's findings (275).

17. James I. McClintock, *Nature's Kindred Spirits: Aldo Leopold, Joseph Wood Krutch, Edward Abbey, Annie Dillard, and Gary Snyder* (Madison: University of Wisconsin Press, 1994).

18. Krutch, *The Measure of Man.*

19. Krutch, untitled, undated fragment ("From My Own Childhood . . ."), Krutch Papers, box 15, folder "Fragments"; Krutch, *The Measure of Man,* 250; Krutch, unpublished essay ("That Nineteenth-Century Stand-by . . ."), Krutch Papers, box 15, folder "Untitled, Unpublished." Outside of literature, Krutch identified Reinhold Niebuhr, Eric Voegelin, and the German philosopher Nicolai Hartmann as three of the leading critics of modern determinism but feared their influence was quite limited (*The Measure of Man,* 95).

20. Krutch, *More Lives than One;* Joseph Wood Krutch to Arthur Pack, October 22, 1962, Krutch Papers, box 7, folder "A—uncertain"; Krutch, unpublished essay ("On My Radio Is Inscribed . . ."), Krutch Papers, box 15, folder "Untitled, Unpublished."

21. Paul N. Pavich to Joseph Wood Krutch, "Easter Sunday," Krutch Papers, box 5, folder P; Alfred C. Ames, "On 'Helplessness' of Mankind," *Chicago Daily Tribune,* April 4, 1954, C4; Ames, "A Hopeful Look at Man," *Chicago Daily Tribune,* August 11, 1957, B8; John LaFarge, "Return to Realities," *America,* May 1, 1954; John T. Owens to Krutch, March 11, 1966, Krutch Papers, box 5, folder N–O; Ammon Hennacy to Krutch, January 31, 1957, Krutch Papers, box 4, folder H; Harold Debrest to Krutch, April 10, 1954, Krutch Papers, box 4, folder D; Emery Neff to Marcelle and Joseph Wood Krutch, July 29, 1956, Krutch Papers, box 5, folder "Emery Neff"; John R. Kirk, review of *The Measure of Man,* by Joseph Wood Krutch, *Humanist* 15 (January 1955): 39; William M. Weber to Krutch, October 12, 1954, Krutch Papers, box 7, folder W; unknown to Krutch, July 14, 1967, Krutch Papers, box 7, folder "Fragments." Catholic respondents proved particularly enthusiastic: e.g., Avery Dulles, "The Contemporary Flight from Ideas," *Loyola Law Review* 8 (1955–1956): 41; Eleanor O'Byrne to Krutch, December 20, 1954, Krutch Papers, box 5, folder N–O).

22. Joseph Wood Krutch, "Commencement Address, June 1, 1960," Krutch Papers, box 14, folder "Unpublished articles, 1941–1969." A leading Tucson businessman raved about the speech to the university's president, and reprint requests poured in (Richard A. Harvill to Krutch, June 13, 1960, Krutch Papers, box 4, folder H).

23. Henry Hazlitt to Joseph Wood Krutch, July 5, 1967, Krutch Papers, box 4, folder H; Hazlitt, *The Free Man's Library: A Descriptive and Critical Bibliog-*

raphy (Princeton, NJ: Van Nostrand, 1956); John Chamberlain, "The Dignity of Man," Krutch Papers, box 18, folder "Reviews: *The Measure of Man*"; "Sterling North Reviews the Books," *New York World-Telegram,* April 5, 1954; Norman R. Phillips, "The Conservative Implications of Skepticism," *Journal of Politics* 18, no. 1 (February 1956): 35–36, 29, 38, 28; Arthur Kemp to Krutch, Krutch Papers, box 4, folder K; Gladys Adams to Krutch, January 25, 1954, Krutch Papers, box 3, folder A; Peggy R. Walker, "Fears of Pupils' Parents," *Los Angeles Times,* July 4, 1954, B4.

24. Edmund A. Opitz to Joseph Wood Krutch, Krutch Papers, box 5, folder N–O; Krutch to Frank S. Meyer, December 3, 1962, Krutch Papers, box 7, folder "A—uncertain"; Felix Morley, ed., *Essays on Individuality* (Philadelphia: University of Pennsylvania Press, 1958); Mark Van Doren to Joseph and Marcelle Krutch, February 25, 1963, Krutch Papers, box 6, folder "Van Doren family, 1960–1963"; Krutch, *The Measure of Man,* 95; Bernard Iddings Bell to Krutch, March 14, 1953, Krutch Papers, box 3, folder B; Max Eastman to Krutch, February 20, 1953, Krutch Papers, box 4, folder E; F. A. Voigt to Miss Norton, October 14, 1958, Krutch Papers, box 6, folder V; Krutch, "Let's Be Prejudiced," *Freeman* 4, no. 20 (June 28, 1954): 703–704; Jameson G. Campaigne Jr. to Krutch, July 4, 1963, Krutch Papers, box 3, folder C; Eliseo Vivas to Krutch, May 25, 1968, Krutch Papers, box 6, folder V. A reader also drew the comparison to Bell, as did another author (Phyllis G. Cochrane to Krutch, July 22, 1951, Krutch Papers, box 3, folder C; Harllee Branch Jr., "The Crowd and the Commonplace," *Georgia Review* 9, no. 3 [October 1955]: 253). Bell approved of Krutch's educational arguments and sent a copy of his *Crowd Culture* (Bell to Krutch, March 2, 1953, Krutch Papers, box 3, folder B). For business leaders, see Richard H. Andrews to Krutch, September 18, 1957, Krutch Papers, box 3, folder A; Bruno R. Neumann to Krutch, Krutch Papers, box 5, folder N–O; untitled speech ("Fourteen Years Ago . . ."), 1958, Krutch Papers, box 9, folder "Speeches, 1953–1958"; "Agenda, Chamber of Commerce Forum Meeting, February 23, 1966," Krutch Papers, box 9, folder "Speeches, 1966–1970"; "Man the Enemy" (reprint from *The American Scholar*), Krutch Papers, box 7, folder W; Peter Kyropoulos to Krutch, February 4, 1965, Krutch Papers, box 4, folder K; Wallace E. Pratt to Krutch, May 7, 1967, Krutch Papers, box 5, folder P–Q; Paul J. Barringer Jr. to Krutch, June 2, 1970, Krutch Papers, box 7, folder A–E; "Biology and Humanism" reprint for Industrial Indemnity Company, n.d., Krutch Papers, folder "Speeches, 1958–1962"; "Biology and Humanism" draft, n.d., Krutch Papers, box 15, folder "Untitled, Unpublished"; "Modern Literature and the Image of Man," reprint for Industrial Indemnity Company, n.d., Krutch Papers, box 9, folder "Speeches, 1953–1958"; W. A. Haluk to Krutch, February 3, 1958, Krutch Papers, box 4, folder H; "Joseph Wood Krutch, San Francisco, 1958," reprint for Industrial Indemnity Company, Krutch Papers, box 9, folder "Speeches, 1953–1958"; MacGregor Folsom to Krutch, July 8, 1968, Krutch Papers, box 4, folder F; "Merely a Humanist–Discussion," reprint for Industrial Indemnity Company, 1968, Krutch Papers, box 9, folder "Speeches, 1966–1970"; Krutch to Marcelle Krutch, September 3, 1958, Krutch Papers, box 3, folder "wife + mother";

Kenneth Bechtel to Krutch, November 18, 1960, Krutch Papers, box 3, folder B; "April Conception Bay Trip," March 25, 1963, Krutch Papers; Mark Van Doren to Joseph and Marcelle Krutch, June 2, 1959, Krutch Papers, box 6, folder "Van Doren family, 1958–1959"; Carl Henry to Krutch, April 3, 1965, Krutch Papers, box 4, folder H; unpublished fragment ("Every now and then . . ."), in Krutch Papers, box 15, folder "Untitled, unpublished."

25. Bernard J. Fried to Joseph Wood Krutch, November 30, 1955, Krutch Papers, box 4, folder F; Irving Polk to Krutch, June 5, 1962, Krutch Papers, box 5, folder P; Irving P. Crawford to Krutch, October 20, 1958, Krutch Papers, box 7, folder "Fragments"; Thomas Van Osdall to Krutch, July 21, 1969, Krutch Papers, box 6, folder V; R. E. Riederer to Krutch, January 20, 1966, Krutch Papers, box 6, folder R; Robert L. Sinsheimer to Krutch, May 6, 1968 and June 20, 1968, Krutch Papers, box 6, folder S.

26. Loren Eiseley to Krutch, October 13, 1969, Krutch Papers, box 4, folder E; Marston Bates to Krutch, October 14, 1959, Krutch Papers, box 3, folder B; Bates to Krutch, October 20, 1966, Krutch Papers, box 3, folder B; Paul B. Sears to Krutch, November 19, 1965, Krutch Papers, box 6, folder S; Sears to Krutch, October 18, 1968, Krutch Papers, box 6, folder B; Willis D. Nutting to Krutch, February 7, 1958, Krutch Papers, box 5, folder N–O; Irving J. Selikoff to Krutch, November 26, 1965, Krutch Papers, box 6, folder S; Arnold B. Grobman to Krutch, December 12, 1960, Krutch Papers, box 4, folder G; John A. Moore to Krutch, October 23, 1959, Krutch Papers, box 5, folder M; Roy M. Fisher to Krutch, February 17, 1965, Krutch Papers, box 4, folder F.

27. Kurt H. Wolff to Joseph Wood Krutch, July 10, 1957, Krutch Papers, box 7, folder W; Norman Jacobson to Krutch, June 29, 1959, and June 30, 1959, Krutch Papers, box 4, folder I–J; Gordon Atkins to Krutch, November 19, 1958, Krutch Papers, box 3, folder A; Arthur P. Noyes to Krutch, March 8, 1955, Krutch Papers, box 5, folder N–O; N. H. Pronko to Krutch, October 17, 1959, Krutch Papers, box 5, folder P–Q; Karl Menninger to Krutch, May 2, 1955, Krutch Papers, box 5, folder M.

28. Ruth Nanda Anshen to George La Piana, September 16, 1952, La Piana Papers, Andover-Harvard Theological Library, Harvard Divinity School (hereafter La Piana Papers), box 30, folder 4. Anshen's proposed journal would have opposed nationalism and realism in the name of human unity. Only "social and economic reforms," she argued, rather than "spheres of influence . . . a despotic bureaucracy or a balance of power system," would enable the world to realize the immense material and spiritual promise of modern knowledge. To that end, her journal would view international affairs "in relation to those principles and ideals which alone justify man's distinction from the beast." In short, "Reason, tempered by Good Will" and embodied in a global political and legal framework, could solve the world's problems (Anshen, untitled prospectus for journal, April 1944, La Piana Papers, box 30, folder 4).

29. Mark Teich, "Editing Einstein," *Omni* 10, no 10 (July 1988): 24, 110; I. I. Rabi, *Science: The Center of Culture* (New York: World, 1970); Erich Fromm, *The Art of Loving* (New York: Harper, 1956); Fromm, *Beyond the Chains of Illusion: My Encounter with Marx and Freud* (New York: Simon and Schuster,

1962); Fromm, *The Heart of Man: Its Genius for Good and Evil* (New York: Harper & Row, 1964); Sinnott quoted in "Science Is Not Enough," *Time* 50, no. 17 (October 27, 1947), 97. *Moral Principles of Action,* the Science of Culture volume closest to Anshen's heart, featured essays by Robert M. MacIver, Karl Jaspers, Jacques Maritain, H. Richard Niebuhr, Erich Fromm, Martin Buber, Jean Piaget, Martin D'Arcy, Muhammad Zafrullah Khan, D. T. Suzuki, Swami Nikhilananda, Paul Tillich, and Albert Schweitzer, among others.

30. A pair of iconic books by French authors helped to buttress the perception that science authorized a lofty, integrated view of reality: *Human Destiny* (New York: Longmans, Green, 1947), by the biophysicist-philosopher Pierre Lecomte du Noüy, and *The Phenomenon of Man* (New York: Harper, 1959), by the scientifically trained Jesuit Pierre Teilhard de Chardin.

31. James Gilbert, *Redeeming Culture: American Religion in an Age of Science* (Chicago: University of Chicago Press, 1997); Barry V. Johnston, *Pitirim A. Sorokin: An Intellectual Biography* (Lawrence: University Press of Kansas, 1995); Paul Tillich, *Can Religion Survive?* (Cambridge, MA: Sci-Art Publishers, 1962). On such ambitions, see also Mark Greif, *The Age of the Crisis of Man: Thought and Fiction in America, 1933–1973* (Princeton, NJ: Princeton University Press, 2015).

32. Jennifer Ratner-Rosenhagen, "The Longing for Wisdom in Twentieth-Century US Thought," in *The Worlds of American Intellectual History,* ed. Joel Isaac, James T. Kloppenberg, Michael O'Brien, and Ratner-Rosenhagen (New York: Oxford University Press, 2017), 182–201. This phenomenon partially overlapped with the revival of the concept of natural law outside Catholic circles. Behind the scenes, Anshen herself longed to convert to Catholicism and considered Maritain (whose Columbia University salary she paid), Hunter College president George N. Shuster, and George La Piana of Harvard Divinity School her spiritual mentors. But she needed her husband's blessing or a teaching position to provide financial independence, and neither was forthcoming. Undeterred, Anshen carried on with her editing and writing (each of the Science and Culture volumes featured one or more framing essays from her pen) and moved forward with the research institute concept. Ruth Nanda Anshen, *Biography of an Idea* (Mt. Kisco, NY: Moyer Bell, 1986), 4–5, 83; Jacques Maritain to George Shuster, March 31, 1949, Jacques Maritain Papers, Jacques Maritain Center, University of Notre Dame, box 30, folder 25; John U. Nef, Memorandum to Committee on Social Thought, March 30, 1951, Hayek Papers, box 63, folder 10.

33. "Confidential Memorandum: The Institute for the Study of Man," September 24, 1957, and "Report of Meeting, December 27, 1957," in Jacques Barzun Papers, Columbia University, Rare Book and Manuscript Library (hereafter Barzun Papers), box 74, folder "Humanistic Institute."

34. Ruth Nanda Anshen to Jacques Barzun, June 28, 1957; Barzun to Anshen, July 1, 1957; and "The Institute for the Study of Man," précis, n.d.; all in Barzun Papers, box 74, folder "Humanistic Institute."

35. "Brief Resumé of the Origin and Development of the Plans for the Institute for the Study of Man"; "Specifications for the Institute for the Study of Man"; and

"Confidential Memorandum"; all in Barzun Papers, box 74, folder "Humanistic Institute."

36. Ruth Nanda Anshen to Jacques Barzun, June 28, 1957; "Report of Meeting, December 27, 1957"; and "The Institute for the Study of Man," précis, n.d.; all in Barzun Papers, box 74, folder "Humanistic Institute"; Chester Barnard to Richard Courant, Jacques Barzun, and Telford Taylor, May 24, 1958; Barzun to Ruth Nanda Anshen, July 16, 1958; Anshen to Barzun, May 6, 1958; and Anshen to Barzun, July 18, 1958; all in Barzun Papers, box 74, folder "Humanistic Institute." Almost thirty-five years later, Barzun would finally tame Anshen's overgrown prose by convincing her to rewrite an entire book as a series of aphorisms (Anshen, *Morals Equals Manners* [Mount Kisco, NY: Moyer Bell, 1992], xi).

37. L. Richardson Preyer to Telford Taylor, March 26, 1958; Ruth Nanda Anshen to Jacques Barzun, June 28, 1957; "Report of Meeting, December 27, 1957"; and "The Institute for the Study of Man" précis, n.d.; all in Barzun Papers, box 74, folder "Humanistic Institute."

38. Lincoln Reis, review of *The Measure of Man*, by Joseph Wood Krutch, *Commentary* 17 (January 1954): 581. Reviewers of Krutch's grant applications agreed that his argument was familiar, although the philosopher Siegfried Kracauer found it hackneyed and repetitive, whereas another thought that Krutch had a fresh take (Elinore Marvel, "Report on Joseph Wood Krutch," and E. B. Jr., "Memorandum for File," Bollingen Foundation Records, Library of Congress, series 1, box 21, folder "Krutch, Joseph Wood").

9. A NEW LEFT

1. Irwin Abrams, "What's Missing on the Campus?," *Phi Delta Kappan* 39, no. 7 (April 1958): 313.

2. Arthur M. Schlesinger Jr., *The Vital Center: The Politics of Freedom* (Boston: Houghton Mifflin, 1949), 243–244, 248.

3. Quoted in Margot A. Henriksen, *Dr. Strangelove's America: Society and Culture in the Atomic Age* (Berkeley: University of California Press, 1997), 174.

4. Paul Potter, quoted in Michael Kazin, *The Populist Persuasion: An American History*, rev. ed. (Ithaca, NY: Cornell University Press, 1998), 190.

5. Quoted in Stuart W. Leslie, *The Cold War and American Science: The Military-Industrial-Academic Complex at MIT and Stanford* (New York: Columbia University Press, 1993), 238. On conceptions of knowledge and politics in the military-industrial complex, see esp. Joy Rohde, *Armed with Expertise: The Militarization of American Social Research during the Cold War* (Ithaca, NY: Cornell University Press, 2013); and Sarah Bridger, *Scientists at War: The Ethics of Cold War Weapons Research* (Cambridge, MA: Harvard University Press, 2015).

6. Nils Gilman, *Mandarins of the Future: Modernization Theory in Cold War America* (Baltimore: Johns Hopkins University Press, 1993); Deborah Shapley, *Promise and Power: The Life and Times of Robert McNamara* (Boston: Little,

Brown, 1993); Don K. Price, *The Scientific Estate* (Cambridge, MA: Belknap Press of Harvard University Press, 1965); Spencer Klaw, *The New Brahmins: Scientific Life in America* (New York: Morrow, 1968); Ralph E. Lapp, *The New Priesthood: The Scientific Elite and the Uses of Power* (New York: Harper & Row, 1965). Cf. Donald W. Cox, *America's New Policy Makers: The Scientists' Rise to Power* (Philadelphia: Chilton Books, 1964).

7. Michael E. Latham, *Modernization as Ideology: American Social Science and "Nation Building" in the Kennedy Era* (Chapel Hill: University of North Carolina Press, 2000); Bridger, *Scientists at War.*

8. Alice O'Connor, *Poverty Knowledge: Social Science, Social Policy, and the Poor in Twentieth-Century U.S. History* (Princeton, NJ: Princeton University Press, 2001); Mical Raz, *What's Wrong with the Poor? Psychiatry, Race, and the War on Poverty* (Chapel Hill: University of North Carolina Press, 2013); Daniel Geary, *Beyond Civil Rights: The Moynihan Report and Its Legacy* (Philadelphia: University of Pennsylvania Press, 2015).

9. Gilman, *Mandarins of the Future;* Jamie Cohen-Cole, *The Open Mind: Cold War Politics and the Sciences of Human Nature* (Chicago: University of Chicago Press, 2014).

10. Ellen Herman, *The Romance of American Psychology: Political Culture in the Age of Experts* (Berkeley: University of California Press, 1995); Jessica Grogan, *Encountering America: Humanistic Psychology, Sixties Culture, and the Shaping of the Modern Self* (New York: HarperCollins, 2012); Doug Rossinow, *The Politics of Authenticity: Liberalism, Christianity, and the New Left in America* (New York: Columbia University Press, 1998); Barry V. Johnston, *Pitirim A. Sorokin: An Intellectual Biography* (Lawrence: University Press of Kansas, 1995); Floyd W. Matson, *The Broken Image: Man, Science, and Society* (New York: G. Braziller, 1964). Matson assumed that his readers had also digested Joseph Wood Krutch's *The Measure of Man* (79).

11. K. Healan Gaston, "The Cold War Romance of Religious Authenticity: Will Herberg, William F. Buckley, Jr., and the Rise of the New Conservatism," *Journal of American History* 99, no. 4 (March 2013): 1137; Kathleen L. Riley, *Fulton J. Sheen: An American Catholic Response to the Twentieth Century* (Staten Island, NY: St. Pauls/Alba House, 2004), 74–76; Thomas C. Reeves, *America's Bishop: The Life and Times of Fulton J. Sheen* (San Francisco: Encounter, 2001), 172. On conversion to Catholicism generally, see Patrick Allitt, *Catholic Converts: British and American Intellectuals Turn to Rome* (Ithaca, NY: Cornell University Press, 1997).

12. Dwight Macdonald, *The Root Is Man: Two Essays in Politics* (Alhambra, CA: Cunningham, 1953), 54, 17–18.

13. Macdonald, *The Root Is Man,* 18, 27, 39, 45; Paul Goodman, *Growing up Absurd: Problems of Youth in the Organized System* (New York: Random House, 1960), 200.

14. C. Wright Mills, *The Sociological Imagination* (New York: Oxford University Press, 1959). Similarly, a reviewer of Jacques Barzun's critical study of *The House of Intellect* (New York: Harper, 1959) argued that *The Sociological Imagination* fleshed out one aspect of the broader phenomenon that Barzun

identified (Charles Rolo, "Reader's Choice: The Decay of Intellect," *Atlantic* 203, no. 6 [June 1959]: 82). As Daniel Geary has noted, the articles leading up to Mills's book identified a significant subset of morally engaged practitioners, but Mills excised that group from his portrait of the profession in the book itself (Geary, *Radical Ambition: C. Wright Mills, the Left, and American Social Thought* [Berkeley: University of California Press, 2009], 173).

15. Herbert Marcuse, *One-Dimensional Man* (Boston: Beacon, 1964).

16. Mills, *The Sociological Imagination,* 61, 169, 173–175, 187, 166; Geary, *Radical Ambition,* 143–178; David Paul Haney, *The Americanization of Social Science: Intellectuals and Public Responsibility in the Postwar United States* (Philadelphia: Temple University Press, 2008), 137–171. Although Mills died in 1962, another figure of his generation, Alvin W. Gouldner, sounded similar themes in *The Coming Crisis of Western Sociology* (New York: Basic Books, 1970); see also Gouldner, "Anti-Minotaur: The Myth of a Value-Free Sociology," *Social Problems* 9, no. 3 (1962): 199–213.

17. Erich Fromm, *The Sane Society* (New York: Rinehart, 1955), 282, 120, 87.

18. Mills, *The Sociological Imagination,* 18.

19. Clark Kerr, *The Uses of the University* (Cambridge, MA: Harvard University Press, 1963).

20. Kerr, *The Uses of the University,* 50, 90–94, 123–125. On these tendencies in higher education, see Ethan Schrum, *The Instrumental University: Education in the Service of the National Agenda after World War II* (Ithaca, NY: Cornell University Press, 2019).

21. Thomas D. Snyder, "Higher Education," in *120 Years of American Education: A Statistical Portrait,* ed. Thomas D. Snyder (Washington, DC: National Center for Education Statistics, 1993), 65–66. In general, liberals insisted that postwar America should be compared to the societies of other times and places, not what one bluntly called "a non-existent ideal." The Catholic socialist Michael Harrington retorted, "My standard of comparison is not how much worse things used to be. It is how much better they could be if only we were stirred" (W. David Maxwell, "Some Dimensions of Relevance," *AAUP Bulletin* 55 [1969]: 338); Harrington, *The Other America: Poverty in the United States* [New York: Macmillan, 1962], 18).

22. Doug Rossinow, "Mario Savio and the Politics of Authenticity," in *The Free Speech Movement: Reflections on Berkeley in the 1960s,* ed. Robert Cohen and Reginald E. Zelnik (Berkeley: University of California Press, 2002), 541; Free Speech Movement, "We Want a University," in *The Berkeley Student Revolt: Facts and Interpretations,* ed. Seymour Martin Lipset and Sheldon S. Wolin (Garden City, NY: Anchor Books, 1965), 208; Greg Calvert, "In White America: Radical Consciousness and Social Change," in *The New Left: A Documentary History,* ed. Massimo Teodori (London: J. Cape, 1970), 412–418; Paul Goodman, "Thoughts on Berkeley," *New York Review of Books* 3 (1965): 5.

23. Free Speech Movement, "We Want a University," 210, 212; Robert Cohen and Reginald E. Zelnik, eds., *The Free Speech Movement: Reflections on Berkeley in the 1960s* (Berkeley: University of California Press, 2002).

24. Kerr, *The Uses of the University,* 164.

25. Tom Hayden, *The Port Huron Statement* (1964; repr., New York: Thunder's Mouth Press, 2005), 166–167.

26. A helpful overview is Howard Brick, *Age of Contradiction: American Thought and Culture in the 1960s* (New York: Twayne, 1998).

27. Harland G. Bloland and Sue M. Bloland, *American Learned Societies in Transition: The Impact of Dissent and Recession* (New York: McGraw-Hill, 1974); Andrew Jewett, "The Politics of Knowledge in 1960s America," *Social Science History* 36, no. 4 (Winter 2012): 551–581; Julie A. Reuben, "Challenging Neutrality: Sixties Activism and Debates over Political Advocacy in the American University," in *Professors and Their Politics,* ed. Neil Gross and Solon J. Simmons (Baltimore: Johns Hopkins University Press, 2014), 217–239; Ellen Schrecker, "The Disciplines and the Left: The Radical Caucus Movement," *Radical Teacher* 114 (Summer 2019): 8–11; Fabio Rojas, *From Black Power to Black Studies: How a Radical Social Movement Became an Academic Discipline* (Baltimore: Johns Hopkins University Press, 2007); Ibram X. Kendi, *The Black Campus Movement: Black Students and the Racial Reconstitution of Higher Education, 1965–1972* (New York: Palgrave Macmillan, 2012); Marilyn J. Boxer, *When Women Ask the Questions: Creating Women's Studies in America* (Baltimore: Johns Hopkins University Press, 1998); Evelyn Hu-DeHart, "The History, Development, and Future of Ethnic Studies," *Phi Delta Kappan* 75, no. 1 (1993): 50–54; Alice E. Ginsberg, *The Evolution of American Women's Studies: Reflections on Triumphs, Controversies, and Change* (New York: Palgrave Macmillan, 2008). H. L. Nieburg, *In the Name of Science* (Chicago: Quadrangle Books, 1966), offered an early expression of this 1960s sensibility, which also had predecessors among advocates of the "social responsibility" of scientists in the 1930s, the nuclear activists of the late 1940s, and other groups.

28. Hayden, *The Port Huron Statement,* 50–51; Eugene D. Genovese and Christopher Lasch, "The Education and the University We Need Now," *New York Review of Books* 13 (1969): 26; John H. Schaar and Sheldon S. Wolin, "Education and the Technological Society," *New York Review of Books* 13 (1969): 4.

29. Leslie, *The Cold War and American Science;* Kelly Moore, *Disrupting Science: Social Movements, American Scientists, and the Politics of the Military, 1945–1975* (Princeton, NJ: Princeton University Press, 2008); Noam Chomsky, "Responsibility," in *March 4: Scientists, Students, and Society,* ed. Jonathan Allen (Cambridge, MA: MIT Press, 1970), 13; Joel Feigenbaum, "Students and Society," in Allen, *March 4,* 5.

30. Union of Concerned Scientists, "Faculty Statement," in Allen, *March 4,* xxii; Bill Zimmerman et al., "Towards a Science for the People" (unpublished manuscript, 1972), https://www.ocf.berkeley.edu/~schwrtz/SftP/Towards.html. Cf. Frank Von Hippel and Joel Primack, "Public Interest Science," *Science* 177 (September 29, 1972): 1166–1171. For parallel debates among engineers, see Matthew H. Wisnioski, *Engineers for Change: Competing Visions of Technology in 1960s America* (Cambridge, MA: MIT Press, 2012).

31. Noam Chomsky, "The Responsibility of Intellectuals," in *The Dissenting Academy,* ed. Theodore Roszak (New York: Pantheon, 1968), 256.

32. Chomsky, "The Responsibility of Intellectuals," 271.

33. On the new class idea, see esp. B. Bruce-Briggs, *The New Class?* (New Brunswick, NJ: Transaction, 1979); Iván Szelenyi and Bill Martin, "The Three Waves of New Class Theories," *Theory and Society* 17, no. 5 (1988): 645–667; and Lawrence P. King and Szelenyi, *Theories of the New Class: Intellectuals and Power* (Minneapolis: University of Minnesota Press, 2004).

34. Joel Feigenbaum, "Students and Society," in Allen, *March 4, 5*; quoted in Moore, *Disrupting Science,* 42.

35. Paul A. Baran and Paul M. Sweezy, *Monopoly Capital: An Essay on the American Economic and Social Order* (New York: Monthly Review Press, 1966), 353, 357.

36. Martin Jay, *The Dialectical Imagination: A History of the Frankfurt School and the Institute of Social Research, 1923–1950* (Berkeley: University of California Press, 1996); Thomas Wheatland, *The Frankfurt School in Exile* (Minneapolis: University of Minnesota Press, 2009); Max Horkheimer and Theodor W. Adorno, *Dialectic of Enlightenment: Philosophical Fragments,* trans. Edmund Jephcott (Stanford, CA: Stanford University Press, 2002). Widespread frustration with mechanistic models and claims of value-neutrality also shaped the enthusiastic uptake of Thomas S. Kuhn's *The Structure of Scientific Revolutions* by radical students, despite Kuhn's desire to uphold the authority of science; see esp. George A. Reisch, *The Politics of Paradigms: Thomas S. Kuhn, James B. Conant, and the Cold War "Struggle for Men's Minds"* (Albany: State University of New York Press, 2019).

37. Herbert Marcuse, *One-Dimensional Man: Studies in the Ideology of Advanced Industrial Society* (Boston: Beacon Press, 1964), xvi, 22, 1.

38. Marcuse, *One-Dimensional Man,* x, 16; Wheatland, *The Frankfurt School in Exile.*

39. Genovese and Lasch, "The Education and the University We Need Now," 22–24, 26–27.

40. Thomas Szasz, *The Myth of Mental Illness: Foundations of a Theory of Personal Conduct* (New York: Harper and Row, 1961), 7, 43–44, 1, 218, 4; Szasz, "Toward the Therapeutic State," *New Republic* (December 11, 1965): 26–29; Michael E. Staub, *Madness Is Civilization: When the Diagnosis Was Social, 1948–1980* (Chicago: University of Chicago Press, 2011).

41. Paulo Freire, *Pedagogy of the Oppressed* (New York: Continuum, 1970); Ivan Illich, *Deschooling Society* (New York: Harper and Row, 1971), 74, 108, 112. On the American side, the influential education theorist John Holt lamented that modern societies "all worship the same gods: science, bigness, efficiency, growth, progress" and considered writing a book titled "Progress: The Road to Fascism" (quoted in Ron Miller, *Free Schools, Free People: Education and Democracy after the 1960s* [Albany: State University of New York Press, 2002], 85).

42. Robert Hunter, *The Storming of the Mind* (Garden City, NY: Doubleday, 1972). For overviews, see Peter Braunstein and Michael William Doyle, eds., *Imagine Nation: The American Counterculture of the 1960s and '70s* (New York: Routledge, 2002); and James R. Lewis and J. Gordon Melton, eds.,

Perspectives on the New Age (Albany: State University of New York Press, 1992).

43. Charles A. Reich, *The Greening of America* (New York: Random House, 1970), 256–257, 352, 41. "Wisdom demands a new orientation of science and technology towards the organic, the gentle, the non-violent, the elegant and beautiful," the German-British economist E. F. Schumacher wrote a few years later in his iconic book *Small Is Beautiful: Economics as If People Mattered* (New York: Harper & Row, 1973). A recent convert to Catholicism, Schumacher turned his attention squarely to materialism and scientism in *A Guide for the Perplexed* (New York: Harper & Row, 1977).

44. Theodore Roszak, *The Making of a Counter Culture: Reflections on the Technocratic Society and Its Youthful Opposition* (Garden City, NY: Doubleday, 1969), 7, 208, 220, 216, 229, 233, 31–33, 50–51.

45. Grogan, *Encountering America*; Herman, *The Romance of American Psychology*, esp. 273–274; Jeffrey J. Kripal, *Esalen: America and the Religion of No Religion* (Chicago: University of Chicago Press, 2007); Nadine Weidman, "Between the Counterculture and the Corporation: Abraham Maslow and Humanistic Psychology in the 1960s," in *Groovy Science: Knowledge, Innovation, and American Counterculture*, ed. David Kaiser and Patrick McCray (Chicago: University of Chicago Press, 2016), 109–141; Wendy Kline, "The Little Manual That Started a Revolution: How Hippie Midwifery Became Mainstream," in Kaiser and McCray, *Groovy Science*, 172–204; Donald Worster, *Nature's Economy: A History of Ecological Ideas*, 2nd ed. (New York: Cambridge University Press, 1994); Don Lattin, *The Harvard Psychedelic Club: How Timothy Leary, Ram Dass, Huston Smith, and Andrew Weil Killed the Fifties and Ushered in a New Age for America* (New York: HarperOne, 2010).

46. Lewis Mumford, *The Myth of the Machine: Technics and Human Development* (New York: Harcourt, Brace Jovanovich, 1967); Jacques Ellul, *The Technological Society*, trans. John Wilkinson (New York: Knopf, 1964), 135; Fred Turner, *From Counterculture to Cyberculture: Stewart Brand, the Whole Earth Network, and the Rise of Digital Utopianism* (Chicago: University of Chicago Press, 2006). Other critics likewise vested their hopes in computers and other forms of liberatory technology: e.g., Illich, *Deschooling Society* and Murray Bookchin (writing as Louis Herber), *Post-Scarcity Anarchism* (Berkeley, CA: Ramparts Press, 1971).

47. Kaiser and McCray, *Groovy Science*; David Kaiser, *How the Hippies Saved Physics: Science, Counterculture, and the Quantum Revival* (New York: Norton, 2011); Fritjof Capra, *The Tao of Physics: An Exploration of the Parallels between Modern Physics and Eastern Mysticism* (Berkeley, CA: Shambhala, 1975); Margot A. Henriksen, *Dr. Strangelove's America: Society and Culture in the Atomic Age* (Berkeley: University of California Press, 1997).

48. Peter L. Berger, *A Rumor of Angels: Modern Society and the Rediscovery of the Supernatural* (Garden City, NY: Doubleday, 1969); Andrew M. Greeley, "After Secularity: The Neo-Gemeinschaft Society: A Post-Christian Postscript," *Sociological Analysis* 27, no. 3 (Autumn 1966): 119–127.

10. SKEPTICISM INSTANTIATED

1. "Second Thoughts about Man," *Time* 101, no. 14 (April 2, 1973): 88–89; "Reaching beyond the Rational," *Time* 101, no. 17 (April 23, 1973): 95–100.
2. "Second Thoughts about Man," 89. The other articles in the series are "The Rediscovery of Human Nature," *Time* 101, no. 14 (April 2, 1973): 89–93; "Searching Again for the Sacred," *Time* 101, no. 15 (April 9, 1973): 92–97; and "What the Schools Cannot Do," *Time* 101, no. 16 (April 16, 1973): 78–84.
3. Mark Solovey, "Project Camelot and the 1960s Epistemological Revolution: Rethinking the Politics–Patronage–Social Science Nexus," *Social Studies of Science* 31, no. 2 (April 2001): 171–206; Joy Rohde, *Armed with Expertise: The Militarization of American Social Research during the Cold War* (Ithaca, NY: Cornell University Press, 2013); Stephen Kinzer, *Poisoner in Chief: Sidney Gottlieb and the CIA Search for Mind Control* (New York: Holt, 2019); Mark H. Lytle, *The Gentle Subversive: Rachel Carson, Silent Spring, and the Rise of the Environmental Movement* (New York: Oxford University Press, 2007); Barry Commoner, *Science and Survival* (New York: Viking, 1966); Michael Egan, *Barry Commoner and the Science of Survival: The Remaking of American Environmentalism* (Cambridge, MA: MIT Press, 2007); Kelly Moore, *Disrupting Science: Social Movements, American Scientists, and the Politics of the Military, 1945–1975* (Princeton, NJ: Princeton University Press, 2008); Thomas Blass, *The Man Who Shocked the World: The Life and Legacy of Stanley Milgram* (New York: Basic Books, 2004); Gina Perry, *Behind the Shock Machine: The Untold Story of the Notorious Milgram Psychology Experiments*, rev. ed. (New York: New Press, 2013); Philip G. Zimbardo, Christina Maslach, and Craig Haney, "Reflections on the Stanford Prison Experiment: Genesis, Transformations, Consequences," in *Obedience to Authority: Current Perspectives on the Milgram Paradigm*, ed. Thomas Blass (Mahwah, NJ: Lawrence Erlbaum, 1999), 193–238; Harriet A. Washington, *Medical Apartheid: The Dark History of Medical Experimentation on Black Americans from Colonial Times to the Present* (New York: Doubleday, 2006); Susan M. Reverby, *Examining Tuskegee: The Infamous Syphilis Study and Its Legacy* (Chapel Hill: University of North Carolina Press, 2009); Ralph V. Katz and Rueben C. Warren, eds., *The Search for the Legacy of the USPHS Syphilis Study at Tuskegee* (Lanham, MD: Lexington Books, 2011); quoted in Carole Gallagher, *American Ground Zero: The Secret Nuclear War* (New York: Random House, 1993), xxiii.
4. Amy Sue Bix, *Inventing Ourselves out of Jobs? America's Debate over Technological Unemployment, 1929–1981* (Baltimore: Johns Hopkins University Press, 2000); Bernardo Bátiz-Lazo, *Cash and Dash: How ATMs and Computers Changed Banking* (New York: Oxford University Press, 2018); Erik M. Conway, *High-Speed Dreams: NASA and the Technopolitics of Supersonic Transportation, 1945–1999* (Baltimore: Johns Hopkins University Press, 2005); Christopher P. Toumey, *Conjuring Science: Scientific Symbols and Cultural Meanings in American Life* (New Brunswick, NJ: Rutgers University Press,

1996); R. Allan Freeze and Jay H. Lehr, *The Fluoride Wars: How a Modest Public Health Measure Became America's Longest-Running Political Melodrama* (Hoboken, NJ: Wiley, 2009); quoted in Jon Turney, *Frankenstein's Footsteps: Science, Genetics and Popular Culture* (New Haven, CT: Yale University Press, 1998), 147, 157; Sheldon Krimsky, *Genetic Alchemy: A Social History of the Recombinant DNA* Controversy (Cambridge, MA: MIT Press, 1982). Taylor's book is *The Biological Time Bomb* (New York: World, 1968).

5. Hubert Bloch, "The Problem Defined," in *Civilization & Science: In Conflict or Collaboration? A Ciba Foundation Symposium* (New York: Associated Scientific Publishers, 1972), 1; Frank Trippett, "Science: No Longer a Sacred Cow," *Time* 109, no. 10 (March 7, 1977): 72–73. Stephen Toulmin dated the skepticism back "thirty or forty years" (Toulmin, "The Historical Background to the Anti-Science Movement," in *Civilization & Science,* 27).

6. Helpful overviews include Maurice Isserman and Michael Kazin, *America Divided: The Civil War of the 1960s,* 6th ed. (New York: Oxford University Press, 2020), and Edward D. Berkowitz, *Something Happened: A Political and Cultural Overview of the Seventies* (New York: Columbia University Press, 2006).

7. Jefferson Cowie and Nick Salvatore, "The Long Exception: Rethinking the Place of the New Deal in American History," *International Labor and Working-Class History* 74, no. 1 (Fall 2008): 3–32; Cowie, *The Great Exception: The New Deal and the Limits of American Politics* (Princeton, NJ: Princeton University Press, 2016).

8. John A. Andrew, *The Other Side of the Sixties: Young Americans for Freedom and the Rise of Conservative Politics* (New Brunswick, NJ: Rutgers University Press, 1997); Rebecca E. Klatch, *A Generation Divided: The New Left, the New Right, and the 1960s* (Berkeley: University of California Press, 1999); Jennifer Burns, "Godless Capitalism: Ayn Rand and the Conservative Movement," *Modern Intellectual History* 1, no. 3 (November 2004): 359–385.

9. "The Sharon Statement," *National Review* 9, no. 12 (September 24, 1960): 173.

10. Barry Goldwater, *The Conscience of a Conservative* (Shepherdsville, KY: Victor, 1960), 73, 10, 84.

11. Phyllis Schlafly, *A Choice Not an Echo* (Alton, IL: Pere Marquette, 1964). On the fusion of religious and economic conservatism in the emerging Republican stronghold of Southern California, see Lisa McGirr, *Suburban Warriors: The Origins of the New American Right* (Princeton, NJ: Princeton University Press, 2001); and Darren Dochuk, *From Bible Belt to Sunbelt: Plain-Folk Religion, Grassroots Politics, and the Rise of Evangelical Conservatism* (New York: Norton, 2011).

12. Francis A. Schaeffer and C. Everett Koop, *Whatever Happened to the Human Race?* (Old Tappan, NJ: F. H. Revell, 1979), 20–21, 24; Jerry Falwell, *Listen, America!* (Garden City, NY: Doubleday, 1980), 66; Christopher Toumey, *God's Own Scientists: Creationists in a Secular World* (New Brunswick, NJ: Rutgers University Press, 1994), 80–81.

13. Francis Schaeffer, *Escape from Reason* (London: InterVarsity, 1968); Schaeffer, *How Should We Then Live? The Rise and Decline of Western Thought and*

Culture (Old Tappan, NJ: Revell, 1976), 146–147; Tim LaHaye, *The Battle for the Mind* (Old Tappan, NJ: Revell, 1980), 101–103, 67, 97, 122, 112.

14. Ann Hulbert, *Raising America: Experts, Parents, and a Century of Advice about Children* (New York: Knopf, 2003), 260; Thomas Maier, *Dr. Spock: An American Life* (New York: Harcourt Brace, 1998); James Dobson, *Dare to Discipline* (Wheaton, IL: Tyndale House, 1971); Jamie Cohen-Cole, *The Open Mind: Cold War Politics and the Sciences of Human Nature* (Chicago: University of Chicago Press, 2014). For another educational flashpoint, see Christopher J. Phillips, *The New Math: A Political History* (Chicago: University of Chicago Press, 2015). "Permissive parenting" appeared in Diana Baumrind, "Effects of Authoritative Parental Control on Child Behavior," *Child Development* 37 (1966): 887–907; and Baumrind, "Child Care Practices Anteceding Three Patterns of Preschool Behavior," *Genetic Psychology Monographs* 75 (1967): 43–88.

15. On the continuities with earlier critiques, see Angus Burgin, *The Great Persuasion: Reinventing Free Markets since the Depression* (Cambridge, MA: Harvard University Press, 2012).

16. Milton Friedman and Rose D. Friedman, *Free to Choose: A Personal Statement* (New York: Harcourt Brace Jovanovich, 1980), 2–3.

17. Sheila Jasanoff, *The Fifth Branch: Science Advisers as Policymakers* (Cambridge, MA: Harvard University Press, 1990); Friedman and Friedman, *Free to Choose*, 191; Falwell, *Listen, America!*, 166. "The air is in general far cleaner and the water safer today than one hundred years ago," the Friedmans insisted, and the industrialized nations had the best air and water (*Free to Choose*, 218).

18. Thomas Medvetz, *Think Tanks in America* (Chicago: University of Chicago Press, 2012); Jason M. Stahl, *Right Moves: The Conservative Think Tank in American Political Culture since 1945* (Chapel Hill: University of North Carolina Press, 2016).

19. Lewis F. Powell Jr., "Attack on American Free Enterprise System" (1971), available at https://scholarlycommons.law.wlu.edu/powellmemo.

20. Helpful overviews include Mark Gerson, *The Neoconservative Vision: From the Cold War to the Culture Wars* (Lanham, MD: Madison Books, 1997); Murray Friedman, *The Neoconservative Revolution: Jewish Intellectuals and the Shaping of Public Policy* (New York: Cambridge University Press, 2005); and Justin Vaïsse, *Neoconservatism: The Biography of a Movement* (Cambridge, MA: Belknap Press of Harvard University Press, 2010).

21. Irving Kristol, "American Conservatism: 1945–1995" (1995), reprinted in *The Neoconservative Persuasion: Selected Essays, 1942–2009* (New York: Basic Books, 2011), 175; Daniel Patrick Moynihan, "Liberalism and Knowledge," in *Coping: Essays on the Practice of Government* (New York: Random House, 1973), 267.

22. "The dilemma of social science is exquisite and unavoidable," Moynihan wrote. "To assert that nothing can be done overnight is to be accused of supporting those who want nothing done ever. To assert the contrary is to debase the science" ("Liberalism and Knowledge," 263).

23. Moynihan, "Liberalism and Knowledge," 253–254, 256; Daniel Patrick Moynihan, *Maximum Feasible Misunderstanding: Community Action in the War on Poverty* (New York: Free Press, 1969), 171–172. Moynihan's famed study of the culture of poverty is *The Negro Family: The Case for National Action* (Washington, DC: United States Department of Labor, Office of Policy Planning and Research, 1965).

24. Irving Kristol, "Utopianism, Ancient and Modern" (1973), reprinted in *Two Cheers for Capitalism* (New York: Basic Books, 1978), 160–166; Kristol, "About Equality" (1972), reprinted in *Neoconservatism: The Autobiography of an Idea* (New York: Free Press, 1995), 168; Kristol, "On Conservatism and Capitalism," *Wall Street Journal,* September 11, 1975, 20; Kristol, "'When Virtue Loses All Her Loveliness': Some Reflections on Capitalism and 'the Free Society,'" in *Capitalism Today,* ed. Daniel Bell and Irving Kristol (New York: Basic Books, 1971). Such passages reflected, in part, Kristol's engagement with the work of Leo Strauss, a fierce critic of scientism in political thought.

25. Kristol, "Utopianism, Ancient and Modern," 1978, 162; quoted in Gary Dorrien, *The Neoconservative Mind: Politics, Culture, and the War of Ideology* (Philadelphia: Temple University Press, 1993), 103; Kristol, "On Corporate Capitalism in America" (1975), reprinted in *Neoconservatism: The Autobiography of an Idea,* 221–222. Key writings from 1979 included B. Bruce-Briggs, *The New Class?* (New Brunswick, NJ: Transaction Books, 1979), and Alvin Ward Gouldner, *The Future of Intellectuals and the Rise of the New Class* (New York: Seabury Press, 1979). An important earlier work is David T. Bazelon, *Power in America: The Politics of the New Class* (New York: New American Library, 1967). For a cogent summary of the new class discourse and its critics, see David A. Horowitz, *America's Political Class under Fire: The Twentieth Century's Great Culture War* (New York: Routledge, 2003), 3–7; a fuller analysis appears in Lawrence P. King and Iván Szelenyi, *Theories of the New Class: Intellectuals and Power* (Minneapolis: University of Minnesota Press, 2004).

26. Peter L. Berger, "The Worldview of the New Class: Secularity and Its Discontents," in Bruce-Briggs, *The New Class?,* 51–54. Berger's classic study of secularization is *The Sacred Canopy: Elements of a Sociological Theory of Religion* (Garden City, NY: Doubleday, 1967).

27. Michael Novak, "Needing Niebuhr Again," *Commentary* 54, no. 3 (September 1972): 52–61; Novak, *The Rise of the Unmeltable Ethnics: Politics and Culture in the Seventies* (New York: Macmillan, 1972), 32, 263.

28. Christopher Lasch, *The Culture of Narcissism: American Life in an Age of Diminishing Expectations* (New York: Norton, 1979), 176, 182–184. The earlier book is *Haven in a Heartless World: The Family Besieged* (New York: Basic Books, 1977); see also Eric Miller, *Hope in a Scattering Time: A Life of Christopher Lasch* (Grand Rapids, MI: William B. Eerdmans, 2010).

29. Kate Millett, *Sexual Politics* (Garden City, NY: Doubleday, 1970), 93, 177–178.

30. Peter Schrag and Diane Divoky, *The Myth of the Hyperactive Child: And Other Means of Child Control* (New York: Pantheon, 1975), 27, 16, xvi, 20, xii.

31. Betty Friedan, *The Feminine Mystique* (New York: Norton, 1963), 103–104, 125; Ellen Herman, *The Romance of American Psychology: Political Culture in the Age of Experts* (Berkeley: University of California Press, 1995); Mari Jo Buhle, *Feminism and Its Discontents: A Century of Struggle with Psychoanalysis* (Cambridge, MA: Harvard University Press, 2000); Phyllis Chesler, *Women and Madness* (Garden City, NY: Doubleday, 1972); Millett, *Sexual Politics*, 178.

32. Daryl Michael Scott, *Contempt and Pity: Social Policy and the Image of the Damaged Black Psyche, 1880–1996* (Chapel Hill: University of North Carolina Press, 1997); Daniel Geary, *Beyond Civil Rights: The Moynihan Report and Its Legacy* (Philadelphia: University of Pennsylvania Press, 2015); Joseph White, "Toward a Black Psychology: White Theories Ignore Ghetto Life Styles," *Ebony* 25, no. 11 (September 1970): 45; quoted in Scott, *Contempt and Pity*, 180. See also Reginald L. Jones, ed., *Black Psychology* (New York: Harper & Row, 1972).

33. Joyce A. Ladner, ed., *The Death of White Sociology* (New York: Random House, 1973); Robert Staples, "The Myth of the Impotent Black Male," *Black Scholar* 2, no. 10 (1971): 2.

34. Friedan, *The Feminine Mystique*, 126–127; Millett, *Sexual Politics*, 220, 228, 233. For critiques from within the disciplines, see Michelle Zimbalist Rosaldo and Louise Lamphere, eds., *Woman, Culture, and Society* (Stanford, CA: Stanford University Press, 1974); Rayna Rapp Reiter, ed., *Toward an Anthropology of Women* (New York: Monthly Review Press, 1975); and Marcia Millman and Rosabeth Moss Kanter, eds., *Another Voice: Feminist Perspectives on Social Life and Social Science* (Garden City, NY: Anchor/Doubleday, 1975).

35. Edward O. Wilson, *Sociobiology: The New Synthesis* (Cambridge, MA: Belknap Press of Harvard University Press, 1975); Richard Dawkins, *The Selfish Gene* (New York: Oxford University Press, 1976); William H. Tucker, *The Science and Politics of Racial Research* (Urbana: University of Illinois Press, 1994); Neil Jumonville, "The Cultural Politics of the Sociobiological Debate," *Journal of the History of Biology* 35, no. 3 (Autumn 2002): 569–593; Erika Lorraine Milam, *Creatures of Cain: The Hunt for Human Nature in Cold War America* (Princeton, NJ: Princeton University Press, 2019).

36. Jumonville, "The Cultural Politics of the Sociobiological Debate"; Myrna Perez Sheldon, "The Public Life of Scientific Orthodoxy: Stephen Jay Gould, Evolutionary Biology and American Creationism, 1965–2002" (PhD diss., Harvard University, 2014); Milam, *Creatures of Cain*; Sarah Blaffer Hrdy, *The Langurs of Abu: Female and Male Strategies of Reproduction* (Cambridge, MA: Harvard University Press, 1977); Hrdy, *The Woman That Never Evolved* (Cambridge, MA: Harvard University Press, 1981); Dorothy Burnham, "Biology and Gender: False Theories about Women and Blacks," *Freedomways* 17, no. 1 (January 1977): 8–13; Burnham, "Biology and Gender," *Genes and Gender* 1 (1978): 51–59. The various styles of radical criticism appear in an important reader from the time: Rita Arditti, Pat Brennan, and Steve Cavrak, eds., *Science and Liberation* (Boston: South End Press, 1980).

37. Rita Arditti, "Women's Biology in a Man's World: Some Issues and Questions," *Science for the People* 5, no. 4 (July 1973): 39. An important collection is Ruth Hubbard, Mary Sue Henifin, and Barbara Fried, *Women Look at Biology Looking at Women: A Collection of Feminist Critiques* (Boston: G. K. Hall, 1979).

38. Evelyn Fox Keller, "Gender and Science," *Psychoanalysis and Contemporary Thought* 1, no. 3 (September 1978): 409–433.

39. Rosemary Radford Ruether, *New Woman, New Earth: Sexist Ideologies and Human Liberation* (New York: Seabury Press, 1975); Susan Griffin, *Woman and Nature: The Roaring inside Her* (New York: Harper & Row, 1978), xiii–xiv; Carolyn Merchant, *The Death of Nature: Women, Ecology, and the Scientific Revolution* (San Francisco: Harper & Row, 1980), 37.

40. Vine Deloria Jr., *Custer Died for Your Sins: An Indian Manifesto* (New York: Macmillan, 1969); Deloria, *God Is Red* (New York: Grosset and Dunlap, 1973); Steve Pavlik and Daniel R. Wildcat, eds., *Destroying Dogma: Vine Deloria, Jr. and His Influence on American Society* (Golden, CO: Fulcrum, 2006); David Martinez, *Life of the Indigenous Mind: Vine Deloria Jr. and the Birth of the Red Power Movement* (Lincoln: University of Nebraska Press, 2019).

41. Sheila Jasanoff, *The Ethics of Invention: Technology and the Human Future* (New York: Norton, 2016); Audra J. Wolfe, *Competing with the Soviets: Science, Technology, and the State in Cold War America* (Baltimore: Johns Hopkins University Press, 2013); Sarah Bridger, *Scientists at War: The Ethics of Cold War Weapons Research* (Cambridge, MA: Harvard University Press, 2015); Sheila Slaughter and Gary Rhoades, "The Emergence of a Competitiveness Research and Development Policy Coalition and the Commercialization of Academic Science and Technology," *Science, Technology & Human Values* 21, no. 3 (July 1996): 303–339.

11. SCIENCE AS CULTURE

1. Jeffrey Stout, *Democracy and Tradition* (Princeton, NJ: Princeton University Press, 2004), 3. On "possessive individualism," see C. B. Macpherson, *The Political Theory of Possessive Individualism: Hobbes to Locke* (Oxford: Clarendon Press, 1962).

2. Phyllis Chesler, *Women and Madness* (Garden City, NY: Doubleday, 1972); "Science and Black People" [editorial], *Black Scholar* 5, no. 6 (March 1974), n.p.; Shulamith Firestone, *The Dialectic of Sex: The Case for Feminist Revolution* (New York: Morrow, 1970).

3. Helpful overviews include Bruce J. Schulman, *The Seventies: The Great Shift in American Culture, Society, and Politics* (New York: Free Press, 2001); John D. Skrentny, *The Minority Rights Revolution* (Cambridge, MA: Belknap Press of Harvard University Press, 2002); and Edward Berkowitz, *Something Happened: A Political and Cultural Overview of the Seventies* (New York: Columbia University Press, 2006).

4. For details and criticism, see Victoria E. Bonnell and Lynn Hunt, eds., *Beyond the Cultural Turn: New Directions in the Study of Society and Culture* (Berkeley: University of California Press, 1999).

5. John Hartley, *A Short History of Cultural Studies* (Thousand Oaks, CA: Sage, 2003).

6. A useful overview is Jackson Lears, "The Concept of Cultural Hegemony: Problems and Possibilities," *American Historical Review* 90, no. 3 (June 1985): 567–593.

7. On this approach, see David R. Hiley, James Bohman, and Richard Shusterman, eds., *The Interpretive Turn: Philosophy, Science, Culture* (Ithaca, NY: Cornell University Press, 1991). Early statements include Charles Taylor, "Interpretation and the Sciences of Man," *Review of Metaphysics* 25, no. 1 (September 1971): 3–51; and Clifford Geertz, *The Interpretation of Cultures* (New York: Basic Books, 1973).

8. E.g., David M. Ricci, *The Tragedy of Political Science: Politics, Scholarship, and Democracy* (New Haven, CT: Yale University Press, 1984); Edward T. Silva and Sheila Slaughter, *Serving Power: The Making of the Academic Social Science Expert* (Westport, CT Greenwood Press, 1984).

9. Robert M. Young, "Malthus and the Evolutionists: The Common Context of Biological and Social Theory," *Past & Present* 43 (May 1969): 109–145; Barry Barnes, *Scientific Knowledge and Sociological Theory* (Boston: Routledge & Kegan Paul, 1974); David Bloor, *Knowledge and Social Imagery* (Boston: Routledge & Kegan Paul, 1976); Trevor Pinch, "The Construction of the Paranormal: Nothing Unscientific Is Happening," in *On the Margins of Science: The Social Construction of Rejected Knowledge*, ed. Roy Wallis (Keele, UK: University of Keele, 1979), 237–270. A polemical history of the field is John H. Zammito, *A Nice Derangement of Epistemes: Post-Positivism in the Study of Science from Quine to Latour* (Chicago: University of Chicago Press, 2004).

10. David F. Noble, *America by Design: Science, Technology, and the Rise of Corporate Capitalism* (New York: Knopf, 1977), 322; Stanley Aronowitz, *Science as Power: Discourse and Ideology in Modern Society* (Minneapolis: University of Minnesota Press, 1988), 27–28, 33–34, x, 18, 147.

11. Evelyn Fox Keller, *Reflections on Gender and Science* (New Haven, CT: Yale University Press, 1985); Anne Fausto-Sterling, *Myths of Gender: Biological Theories about Women and Men* (New York: Basic Books), 1985; Elizabeth Fee, "A Feminist Critique of Scientific Objectivity," *Science for the People* 14, no. 4 (July–August 1982): 5–8, 30–33; Alison M. Jaggar, *Feminist Politics and Human Nature* (Lanham, MD: Rowman and Littlefield, 1983); Sandra Harding, *The Science Question in Feminism* (Ithaca, NY: Cornell University Press, 1986); Sue V. Rosser, "Are There Feminist Methodologies Appropriate for the Natural Sciences and Do They Make a Difference?," *Women's Studies International Forum* 15, no. 5 (September–December 1992): 539–540.

12. Steven Shapin and Simon Schaffer, *Leviathan and the Air-Pump: Hobbes, Boyle, and the Experimental Life* (Princeton, NJ: Princeton University Press, 1985),

344; Bruno Latour, "The Impact of Science Studies on Political Philosophy," *Science, Technology & Human Values* 16, no. 1 (Winter 1991): 10, 12, 14.

13. Useful points of entry include François Cusset, *French Theory: How Foucault, Derrida, Deleuze, & Co. Transformed the Intellectual Life of the United States* (Minneapolis: University of Minnesota Press, 2008); and Alan D. Schrift, *Poststructuralism and Critical Theory's Second Generation* (New York: Routledge, 2014).

14. Linda J. Nicholson, ed., *Feminism/Postmodernism* (New York: Routledge, 1990); Judith Butler and Joan Wallach Scott, *Feminists Theorize the Political* (New York: Routledge, 1992); Patrick Williams and Laura Chrisman, eds., *Colonial Discourse and Post-Colonial Theory: A Reader* (New York: Columbia University Press, 1994); Leela Gandhi, *Postcolonial Theory: A Critical Introduction* (New York: Columbia University Press, 1998).

15. Michel Foucault, *Power/Knowledge: Selected Interviews and Other Writings, 1972–1977,* ed. Colin Gordon (New York: Pantheon, 1980), 114. Donna Haraway played a key role in bringing poststructuralist insights into STS: e.g., *Simians, Cyborgs, and Women: The Reinvention of Nature* (New York: Routledge, 1991).

16. Haraway's challenge to prevailing understandings of universalism and objectivity was particularly influential: "Situated Knowledges: The Science Question in Feminism and the Privilege of Partial Perspective," *Feminist Studies* 14, no. 3 (October 1988): 575–599. See also Andrew Ross, ed., *Universal Abandon? The Politics of Postmodernism* (Minneapolis: University of Minnesota Press, 1988).

17. Foucault, *Power/Knowledge;* Thomas Nagel, *The View from Nowhere* (New York: Oxford University Press, 1986); Haraway, "Situated Knowledges"; Zygmunt Bauman, *Intimations of Postmodernity* (New York: Routledge, 1992), 37.

18. On such constructions, see Keith Michael Baker and Peter Hanns Reill, eds., *What's Left of Enlightenment? A Postmodern Question* (Stanford, CA: Stanford University Press, 2001).

19. Zygmunt Bauman, *Modernity and the Holocaust* (Ithaca, NY: Cornell University Press, 1989), 72, 65, 68, 70, 108–110.

20. Bauman, *Modernity and the Holocaust,* 73, 154.

21. Early works by political theorists included James Der Derian and Michael J. Shapiro, eds., *International/Intertextual Relations: Postmodern Readings of World Politics* (Lexington, MA: Lexington Books, 1989); and William E. Connolly, *Identity \ Difference: Democratic Negotiations of Political Paradox* (Ithaca, NY: Cornell University Press, 1991); see also Nikolas Rose and Peter Miller, "Political Power beyond the State: Problematics of Government," *British Journal of Sociology* 43, no. 2 (June 1992): 173–205.

22. Rorty, *Consequences of Pragmatism: Essays 1972–1980* (Minneapolis: University of Minnesota Press, 1982), 150–151; Cusset, *French Theory.* Rorty developed his critique in *Philosophy and the Mirror of Nature* (Princeton, NJ: Princeton University Press, 1979); see also Rorty, *Contingency, Irony, and Solidarity* (New York: Cambridge University Press, 1989); and Rorty, *Objectivity, Relativism, and Truth* (Cambridge: Cambridge University Press, 1991).

23. Margaret Talbot, "Darwin in the Dock," *New Yorker* 81, no. 39 (December 5, 2005): 66–77; David Demeritt, "Science Studies, Climate Change, and the Prospects for Constructivist Critique," *Economy and Society* 35, no. 3 (2006): 453–479. Since 2000, the social construction metaphor has ceded ground to a new language of "co-production," which asserts that knowledge emerges in tandem with cultural formations and reflects the fact that the external world puts up some resistance in the laboratory and the field, constraining if not fully determining scientists' conclusions. See esp. Sheila Jasanoff, "Ordering Knowledge, Ordering Society," in *States of Knowledge: The Co-production of Science and the Social Order*, ed. Sheila Jasanoff (New York: Routledge, 2004), 13–45.

24. Shapin and Schaffer, *Leviathan and the Air-Pump*; Latour, "The Impact of Science Studies on Political Philosophy"; Andrew Ross, ed., *Science Wars* (Durham, NC: Duke University Press, 1996); Noretta Koertge, ed., *A House Built on Sand: Exposing Postmodernist Myths about Science* (New York: Oxford University Press, 1998).

25. Paul Gross and Norman Levitt, *Higher Superstition: The Academic Left and Its Quarrels with Science* (Baltimore: Johns Hopkins University Press, 1994); Alan D. Sokal and Jean Bricmont, *Fashionable Nonsense: Postmodern Intellectuals' Abuse of Science* (New York: Picador USA, 1998); Editors of Lingua Franca, *The Sokal Hoax: The Sham That Shook the Academy* (Lincoln: University of Nebraska Press, 2000); Marvin Harris, "Post-Modern Anti-Scientism," in *The Objectivity Crisis: Rethinking the Role of Science in Society*, ed. G. E. Brown (Washington, DC: Government Printing Office, 1993), 23.

26. The charge of conservatism recurs in Ross, *Science Wars*.

27. Andrew Hartman, *A War for the Soul of America: A History of the Culture Wars* (Chicago: University of Chicago Press, 2015); Allan Bloom, *The Closing of the American Mind* (New York: Simon and Schuster, 1987), 379.

28. Hartman, *A War for the Soul of America*; Ronald L. Numbers, *The Creationists: From Scientific Creationism to Intelligent Design*, exp. ed. (Cambridge, MA: Harvard University Press, 2006).

29. Sheila Slaughter and Gary Rhoades, "The Emergence of a Competitiveness Research and Development Policy Coalition and the Commercialization of Academic Science and Technology," *Science, Technology & Human Values* 21, no. 3 (July 1996): 303–339; Audra J. Wolfe, *Competing with the Soviets: Science, Technology, and the State in Cold War America* (Baltimore: Johns Hopkins University Press, 2013); Michael Riordan, Lillian Hoddeson, and Adrienne W. Kolb, *Tunnel Visions: The Rise and Fall of the Superconducting Super Collider* (Chicago: University of Chicago Press, 2015).

30. Fred Turner, *From Counterculture to Cyberculture: Stewart Brand, the Whole Earth Network, and the Rise of Digital Utopianism* (Chicago: University of Chicago Press, 2006); Clemens Apprich, *Technotopia: A Media Genealogy of Net Cultures*, trans. Aileen Derieg (New York: Rowman & Littlefield International, 2017); Anthony Elliott, *The Culture of AI: Everyday Life and the Digital Revolution* (New York: Routledge, 2019); Neil Postman, *Technopoly: The Surrender of Culture to Technology* (New York: Vintage, 1993). Theo-

dore Roszak also weighed in: *The Cult of Information: The Folklore of Computers and the True Art of Thinking* (New York: Pantheon, 1986).

31. Paul Ramsey, *Fabricated Man; the Ethics of Genetic Control* (New Haven, CT: Yale University Press, 1970); Jeremy Rifkin, *Algeny: A New Word—A New World* (New York: Viking, 1983); Rifkin, *The Biotech Century: Harnessing the Gene and Remaking the World* (New York: Jeremy P. Tarcher/Putnam, 1998); Leon Kass, "The Wisdom of Repugnance," *New Republic* 216, no. 22 (June 2, 1997): 17–26; Bill Joy, "Why the Future Doesn't Need Us," *Wired* 8, no. 4 (April 2000): 240–242; Elaine L. Graham, *Representations of the Post/Human: Monsters, Aliens and Others in Popular Culture* (Manchester: Manchester University Press, 2002); Jerome H. Barkow, Leda Cosmides, and John Tooby, eds., *The Adapted Mind: Evolutionary Psychology and the Generation of Culture* (New York: Oxford University Press, 1992); Russell Jacoby and Naomi Glauberman, eds., *The Bell Curve Debate: History, Documents, Opinions* (New York: Times Books, 1995); Sheila Slaughter and Larry L. Leslie, *Academic Capitalism: Politics, Policies, and the Entrepreneurial University* (Baltimore: Johns Hopkins University Press, 1997); Vandana Shiva, *Biopiracy: The Plunder of Nature and Knowledge* (Boston: South End Press, 1997); Paul Raeburn, *The Last Harvest: The Genetic Gamble That Threatens to Destroy American Agriculture* (New York: Simon & Schuster, 1995); Peter D. Kramer, Listening to Prozac (New York: Viking, 1993); Elizabeth Wurtzel, *Prozac Nation: Young and Depressed in America* (Boston: Houghton Mifflin, 1994). As in a number of other areas, some critics tied controversial technologies to both scientism and postmodernism: e.g., Richard John Neuhaus, "The Return of Eugenics," *Commentary* 85, no. 4 (April 1988): 15–26.

32. Frank S. Zelko, *Make It a Green Peace! The Rise of Countercultural Environmentalism* (New York: Oxford University Press, 2013); Carolyn Merchant, *Radical Ecology: The Search for a Livable World* (New York: Routledge, 1992); Spencer R. Weart, *The Discovery of Global Warming,* rev. and exp. ed. (Cambridge, MA: Harvard University Press, 2008); Joshua P. Howe, *Behind the Curve: Science and the Politics of Global Warming* (Seattle: University of Washington Press, 2014).

33. K. Healan Gaston, *Imagining Judeo-Christian America: Religion, Secularism, and the Redefinition of Democracy* (Chicago: University of Chicago Press, 2019), 276–277.

34. For dismissals of Niebuhr, see George M. Marsden, *The Soul of the American University: From Protestant Establishment to Established Nonbelief* (New York: Oxford University Press, 1994), 373–374, 397–398; and Douglas Sloan, *Faith and Knowledge: Mainline Protestantism and American Higher Education* (Louisville, KY: Westminster/John Knox Press, 1994), 112–121.

35. George M. Marsden, "The Soul of the American University: A Historical Overview," in *The Secularization of the Academy,* ed. George M. Marsden and Bradley J. Longfield (New York: Oxford University Press, 1992), 40. Such claims about science's presuppositions had been heard earlier: e.g., Niebuhr, "Religion in a Secular Age" (1937), in *The Essential Reinhold Niebuhr: Selected Essays and Addresses,* ed. Robert McAfee Brown (New Haven, CT: Yale University Press, 1986), 79–80.

36. The phrase "social construction" was launched by the 1960s writings of Peter L. Berger and Thomas Luckmann, émigré critics of scientism whose work on shared symbolic worlds overlapped their accounts of secularization: Berger and Luckmann, *The Social Construction of Reality: A Treatise in the Sociology of Knowledge* (Garden City, NY: Doubleday, 1966).

37. "Multipresuppositionalism" would be more precise, though even more awkward.

38. Alasdair C. MacIntyre, *After Virtue: A Study in Moral Theory* (Notre Dame, IN: University of Notre Dame Press, 1981); MacIntyre, *Whose Justice? Which Rationality?* (Notre Dame, IN: University of Notre Dame Press, 1988), 9. Note the echoes in Sandra G. Harding, *Whose Science? Whose Knowledge? Thinking from Women's Lives* (Ithaca, NY: Cornell University Press, 1991).

39. Alvin Plantinga, "Reason and Belief in God," in *Faith and Rationality: Reason and Belief in God,* ed. Alvin Plantinga and Nicholas Wolterstorff (Notre Dame, IN: University of Notre Dame Press, 1983), 16–93. See also Wolterstorff, *Reason within the Bounds of Religion* (Grand Rapids, MI: Eerdmans, 1976); Plantinga, *Warrant: The Current Debate* (New York: Oxford University Press, 1993); and Plantinga, *Warrant and Proper Function* (New York: Oxford University Press, 1993). Plantinga completed his argument in *Warranted Christian Belief* (New York: Oxford University Press, 2000).

40. Hartman, *A War for the Soul of America;* Richard John Neuhaus, *The Naked Public Square: Religion and Democracy in America* (Grand Rapids, MI: Eerdmans, 1984); Stephen L. Carter, *The Culture of Disbelief: How American Law and Politics Trivialize Religious Devotion* (New York: Basic Books, 1993).

41. For the earlier argument, see Francis A. Schaeffer and C. Everett Koop, *Whatever Happened to the Human Race?* (Old Tappan, NJ: F. H. Revell, 1979); and Jerry Falwell, *Listen, America!* (Garden City, NY: Doubleday, 1980).

42. Frank S. Ravitch, *Masters of Illusion: The Supreme Court and the Religion Clauses* (New York: University Press, 2007); John Rawls, *A Theory of Justice* (Cambridge, MA: Belknap Press of Harvard University Press, 1971); Charles Larmore, "Public Reason," in *The Cambridge Companion to Rawls,* ed. Samuel Freeman (New York: Cambridge University Press, 2002), 368–393. A sampling of criticism appears in Paul J. Weithman, *Religion and Contemporary Liberalism* (Notre Dame, IN: University of Notre Dame Press, 1997).

43. Michael Sandel, "The Procedural Republic and the Unencumbered Self," *Political Theory* 12, no. 1 (February 1984): 81–96; Sandel, "Religious Liberty—Freedom of Conscience or Freedom of Choice?," in *Articles of Faith, Articles of Peace: The Religious Liberty Clauses and the American Public Philosophy,* ed. James Davison Hunter and Os Guinness (Washington, DC: Brookings Institution, 1990); Charles Taylor, *Sources of the Self: The Making of the Modern Identity* (Cambridge, MA: Harvard University Press, 1989); Taylor, *The Explanation of Behaviour* (New York: Humanities Press, 1964); Taylor, "Interpretation and the Sciences of Man"; Taylor, *A Secular Age* (Cambridge, MA: Belknap Press of Harvard University Press, 2007). For additional criticism, see Neuhaus, *The Naked Public Square,* 257; and Stanley Hauerwas, *Truthfulness*

and *Tragedy: Further Investigations in Christian Ethics* (Notre Dame, IN: University of Notre Dame Press, 1977), 218.

44. Marsden, *The Soul of the American University,* 434, 442; Marsden, "The Soul of the American University," 36–37. Cf. Marsden, *The Outrageous Idea of Christian Scholarship* (New York: Oxford University Press, 1997).

45. Marsden, "The Soul of the American University," 39–40; Marsden, *The Soul of the American University,* 400, 389, 306, 375. Marsden argued that the surrounding society gave academia its basic shape, but some held that academic theories set the tone of the culture at large: e.g., Sloan, *Faith and Knowledge,* viii.

46. Of course, many poststructuralists and other theorists continued to see religion as an obstacle; see, e.g., Richard Rorty, "Religion as Conversation-Stopper," *Common Knowledge* 3, no. 1 (Spring 1994): 1–6, although many neopragmatists viewed religion more favorably. Poststructuralism influenced a diverse group of theologians, as in John Milbank, *Theology and Social Theory: Beyond Secular Reason* (Cambridge, MA: Basil Blackwell, 1990); and Carl A. Raschke, *The Next Reformation: Why Evangelicals Must Embrace Postmodernity* (Grand Rapids, MI: Baker Academic, 2004).

47. Theodore Roszak, *Where the Wasteland Ends; Politics and Transcendence in Postindustrial Society* (Garden City, NY: Doubleday, 1972); Morris Berman, *The Reenchantment of the World* (Ithaca, NY: Cornell University Press, 1981); Mary Daly, *Beyond God the Father: Toward a Philosophy of Women's Liberation* (Boston: Beacon Press, 1973); James H. Cone, *A Black Theology of Liberation* (Philadelphia: Lippincott, 1970); Emilie M. Townes, *Womanist Justice, Womanist Hope* (Atlanta: Scholars Press, 1993); Cornel West, *Prophesy Deliverance! An Afro-American Revolutionary Christianity* (Philadelphia: Westminster, 1982); Peter Gabel, "Creationism and the Spirit of Nature," *Tikkun* 2, no. 5 (November–December 1987): 59. For Lerner's views, see *The Politics of Meaning: Restoring Hope and Possibility in an Age of Cynicism* (Reading, MA: Addison-Wesley, 1996).

48. Talal Asad, *Genealogies of Religion: Discipline and Reasons of Power in Christianity and Islam* (Baltimore: Johns Hopkins University Press, 1993); Rajeev Bhargava, ed., *Secularism and Its Critics* (New York: Oxford University Press, 1998); Dipesh Chakrabarty, *Provincializing Europe: Postcolonial Thought and Historical Difference* (Princeton, NJ: Princeton University Press, 2000). After 2000, many postcolonial theorists redefined secularism as a regime enforcing the "privatization" of religion, an analysis that decentered science. However, the knowledge-politics link often persisted: e.g., Vincent William Lloyd and Ludger Viefhues-Bailey, "Introduction: Is the Postcolonial Postsecular?," *Critical Research on Religion* 3, no. 1 (April 2015): 18–19.

CONCLUSION

1. Leon Wieseltier, "'Perhaps Culture Is Now the Counterculture': A Defense of the Humanities," *New Republic* online, May 28, 2013, http://www.newrepublic
.com/article/113299/leon-wieseltier-commencement-speech-brandeis-university

-2013; Steven Pinker, "Science Is Not Your Enemy: A Plea for an Intellectual Truce," *New Republic* 244, no. 13 (August 19, 2013): 28–33.

2. Sohrab Ahmari, "They Blinded Us with Science: The History of a Delusion," *Commentary* 149, no. 5 (May 2020): 23–29; Devra Mandell, in "Letters to the Editor: Why America Tolerates a Lying, Hydroxychloroquine-Hawking President," *Los Angeles Times,* May 21, 2020, https://www.latimes.com/opinion /story/2020-05-21/why-america-tolerates-a-lying-hydroxychloroquine-popping -president; Steven Pinker on Twitter, May 21, 2020, https://twitter.com/sapinker /status/1263463995870269440.

3. It is worth noting that these categories emerged in Western contexts and do not always find exact parallels in other cultures and languages.

4. To be clear, these narratives are neither the only expressions nor the sole source of distrust toward scientific authority. But they have often corralled that sentiment into particular shapes. They have also directed frustration about other phenomena—technocratic modes of governance, say, or consumer capitalism—toward the scientific enterprise. These stories have created a stock reflex or habit, offering a ready interpretive frame and vocabulary that has been applied to virtually every societal phenomenon beyond traditional religion. Some religious critics have even identified biblical literalism as an outgrowth of science: e.g., Reinhold Niebuhr, "Religion in a Secular Age" [1937], in *The Essential Reinhold Niebuhr: Selected Essays and Addresses,* ed. Robert McAfee Brown (New Haven, CT: Yale University Press, 1986), 89.

5. Such explanations also omit many other factors, including those that modern social scientists tend to emphasize: economic interests, ethnoracial conflicts, status concerns, bureaucratic inertia, structural constraints, unforeseen side ef- fects, psychological rationalizations, and so forth.

6. Huston Smith, *Beyond the Post-Modern Mind* (New York: Crossroad, 1982), 68; Alasdair C. MacIntyre, *After Virtue: A Study in Moral Theory* (Notre Dame, IN: University of Notre Dame Press, 1981), 83. Of course, the evidence presented in this book strengthens the rebuttal that the modern era never fea- tured mindless conformity to scientific authority.

7. To take a few examples, it was not an illegitimate extrapolation from physics that produced the imperative to treat individuals equally under the law, as citi- zens rather than concrete persons. Nor did that philosophical error create a market-based mode of economic exchange that often caters to the baser de- sires, the vertical and horizontal integration of firms and the corresponding growth of complex management structures, or the extension of bureaucratic approaches to the administration of public policy. The champions of these practices sometimes compared them to science, insofar as they represented formally secular and "depersonalized" forms of behavior, but that pattern of legitimation hardly explains their emergence, spread, or appeal. Even at the philosophical level, moreover, there are many nonscientific alternatives to reli- gious forms of reasoning. Science and religion do not divide the entire world of thought between them.

8. Steven Shapin neatly captured this view of science in the title of his book *Never Pure: Historical Studies of Science as If It Was Produced by People with*

Bodies, Situated in Time, Space, Culture, and Society, and Struggling for Credibility and Authority (Baltimore: Johns Hopkins University Press, 2010).

9. Employing such heuristic distinctions does not require resolving the thorny and potentially insoluble "demarcation problem" of crafting a philosophical definition that categorically differentiates science from nonscience.

10. On the societal demand for quantification and related "technologies of trust," see Theodore M. Porter, *Trust in Numbers: The Pursuit of Objectivity in Science and Public Life* (Princeton, NJ: Princeton University Press, 1995).

ACKNOWLEDGMENTS

In many ways, I have been preparing to write this book for my whole life. As the child of a physics professor and a math teacher, I grew up in a sea of books, magazines, and documentaries about science. These came from many different periods, though I had little sense of that fact or its meaning; I marveled at Edwin Abbott's 1884 classic *Flatland* even as I devoured cutting-edge 1970s works such as *Omni* magazine, Carl Sagan's *The Dragons of Eden* and *Cosmos* series, and Douglas Hofstadter's *Gödel, Escher, Bach*. It may be a function of my own proclivities at the time, or the tastes of my parents, or even my selective memory since then, but I remember hearing (and caring) little about science's material contributions—new devices, higher standards of living, increased longevity—and a great deal about its intersections with other modes of creativity and human expression, including both art and religion. Whether the topic was particle physics, mathematics, or—most often—animal and human behavior, science appeared to me as an exercise in perspective-taking; it was a matter of altered perceptions, radical shifts in conceptual paradigms. This emphasis on new ways of seeing blended seamlessly with my commitment to animal rights: human beings needed to learn to value animals in and for themselves, rather than inflicting needless cruelty or simply using them for their own purposes.

Of course, I understood science in highly distinctive terms, though I was hardly alone in doing so. Arriving at Berkeley in 1988, I set out to study theoretical physics, astronomy, and mathematics, with a philosophy minor on the side. It never occurred to me that physics was largely a product of laboratories—often, laboratories tied to the military-industrial complex. After my third semester, I abandoned the physics major and turned to the social sciences and humanities, while involving myself with political activism on and off campus. To be sure, I had been learning about people and cultures all along, despite my primary focus on the natural sciences. Even science

fiction and fantasy novels, which I read with abandon through childhood and adolescence, had served as a kind of surrogate anthropology—a set of thought experiments about how human beings (or other sentient creatures) might behave under
very different, and often quite extreme, circumstances.

Although I was now studying individuals and societies directly in my coursework,
I continued to ruminate on how scientific sensibilities intersected with individual
personalities, on the one side, and social and political tendencies, on the other. Perhaps science was actually about control—psychological, interpersonal, social—
rather than creative leaps of imagination. That was clearly one of its main historical
effects. But did the story end there, or did science also bear more hopeful, even radical, possibilities?

Other kinds of historical questions also occupied me, especially regarding political
and social movements. Yet as I layered historical and social-scientific perspectives
onto my own experiences, I found myself returning again and again to the issue of
science's place in the modern world—including its complex relations to political and
social movements themselves. Back at Berkeley for a PhD after four years in publishing and music, I ended up exploring the shifting meanings of science in the
twentieth-century United States. I have since pursued that question far and wide, but
I remain ever grateful to my undergraduate and graduate teachers at Berkeley for
giving me the solid interdisciplinary foundation I needed to address it effectively as a
professional historian. David Hollinger and Cathryn Carson deserve special mention
for teaching an unorthodox student with many rough edges what it meant to be a
scholar, and not just someone with a bunch of ideas and opinions.

Of course, much of great import has happened since then. This book bears the
marks of innumerable people, places, and events, as well as changing historical
circumstances and intellectual currents. I began the project during my time at Harvard University, which helped fund a year of research leave, and completed it while
teaching at Boston College. I am deeply grateful to colleagues and administrators
at both institutions. During my leave year, the National Humanities Center also
provided invaluable financial support and technical assistance via the John G.
Medlin Jr. Fellowship. I want to thank the Center's library staff in particular for
their countless hours spent gathering books and articles from the local universities.
Although such printed sources anchor most of my discussion, I would also like to
thank archivists at the Library of Congress, the Hoover Institution Archives, Columbia University's Rare Book and Manuscript Library, the University of Notre
Dame's Jacques Maritain Center, and Harvard Divinity School's Andover-Harvard
Theological Library for helping me flesh out key portions of the story.

Teaching courses on science in the United States has also helped me immeasurably with this book. The lectures and discussions allowed me to clarify my thinking,
experiment with narrative formulations and explanatory strategies, and learn from
students—often one's best teachers. I benefited especially from conversations with
a remarkable collection of graduate teaching fellows for my science and religion
course at Harvard: Casey Bohlen, Colin Bossen, Collier Brown, Altin Gavranovic,
John Gee, Cristina Groeger, Terence Keel, Eva Payne, and Myrna Perez Sheldon.
Most of them are now professional colleagues as well. Several students also supported the book more directly. In the project's early phases, Casey Bohlen, Made-

line DeSantis, Katie Duggan, Robert Long, Elias Nelson, Julian Petri, Kip Richardson, and Fran Swanson helped me build up the research base. Later in the process, Will Holub-Moorman became a valued conversation partner as well as a superb research assistant.

Along the way, I also developed many insights through conversations with groups of scholars assembled for collaborative volumes. Most relevant topically were a special issue of *Social Science History,* overseen by Michael A. Bernstein; the *Professors and Their Politics* volume, organized by Neil Gross and Solon J. Simmons; the *Worlds of American Intellectual History* project, led by Joel Isaac, James Kloppenberg, Michael O'Brien, and Jennifer Ratner-Rosenhagen; the *American Labyrinth* volume, edited by Andrew Hartman and Raymond Haberski Jr.; and the forthcoming *Disenchantment's Enchantments* book, edited by Gaymon Bennett and J. Benjamin Hurlbut. In each case, I learned a great deal from the editors and my fellow contributors. The Knowledges and Contexts discussion group at the National Humanities Center and an Institute Vienna Circle workshop with Mark Brown and Heather Douglas also proved especially fruitful as the book came together. My deepest thanks to all involved.

I received helpful feedback on pieces of this project from organizers, panelists, and audiences in numerous contexts. These included a Cornell University conference on "Culture and Conflict"; the Production et Diffusion des Savoirs seminar at the École Normale Supérieure de Cachan; Princeton University's Shelby Cullom Davis Center for Historical Studies; the Contemporary History Colloquium at the Smithsonian Institution; the Department of Religious Studies at the University of North Carolina; the History Department Colloquium at Duke University; a York University conference on "Social Science, Ideology, and Public Policy in the United States, 1961 to the Present"; the Center for the Study of Work, Labor, and Democracy at the University of California, Santa Barbara; and a Columbia University event titled "The Engine of Modernity: Construing Science as the Driving Force of History in the First Half of the Twentieth Century"; as well as several forums at Harvard (the Mahindra Humanities Center; the History of Science Department; the Science, Religion, and Culture Seminar; the STS Circle; and the semicentennial event for C. P. Snow's *The Two Cultures*); and meetings of the Organization of American Historians; the American Historical Association; the History of Science Society; the U.S. Intellectual History Conference; the Conference on the History of Recent Economics; and the International Congress of History of Science, Technology, and Mathematics.

Individual colleagues, near and far, provided inspiration, assistance, and camaraderie as well. At Harvard, Maya Jasanoff, Heidi Voskuhl, and Kirsten Weld became especially close friends. In the Department of History, I also want to thank David Armitage, Sven Beckert, Ann Blair, Evelyn Brooks-Higginbotham, Vince Brown, Joyce Chaplin, Genevieve Clutario, Lizabeth Cohen, Nancy Cott, Emma Dench, Peter Gordon, James Kloppenberg, Mary Lewis, Jill Lepore, Lisa McGirr, Ian Miller, Daniel Smail, and Rachel St. John. Beyond the department, Danielle Allen, Anya Bernstein Bassett, Robin Bernstein, Janet Browne, David Hempton, David Holland, Sheila Jasanoff, Rebecca Lemov, Nonie Lesaux, Meira Levinson, Elizabeth Lunbeck, Dan McKanan, Everett Mendelsohn, Samuel Moyn, Ahmed Ragab, Julie Reuben, Sindhu Revuluri, Sophia Roosth, Moshik Temkin, Richard

Tuck, and Natasha Warikoo shone particularly brightly in the Harvard firmament. So, too, did the peerless community of Social Studies tutors. At Boston College, it was a rare treat to work with my co-conspirators Chris Kenaley and Jenna Tonn, and deeply informative to boot. Brian Gareau, Gregory Kalscheur, Kevin Kenny, David Quigley, and Sarah Ross supported my visiting position, and Julian Bourg went above and beyond.

Outside Boston, important friends and interlocutors have included Casey Blake, Howard Brick, Mark Brilliant, Thomas Broman, Mark Brown, Angus Burgin, John Carson, Jamie Cohen-Cole, George Cotkin, Henry Cowles, Heather Douglas, Lucas Fain, Lynn Festa, Philippe Fontaine, Mary Furner, Daniel Geary, Michael Gordin, Leah Gordon, David Greenberg, Neil Gross, Raymond Haberski Jr., Andrew Hartman, Jonathan Holloway, Daniel Horowitz, Sarah Igo, Daniel Immerwahr, Joel Isaac, Jason Josephson-Storm, David Kaiser, Amy Kittelstrom, Paul Kramer, Andrew Lakoff, Christopher Loss, Ajay Mehrotra, Erika Milam, Jill Morawski, Adam Nelson, Ronald Numbers, Michael Pettit, Jefferson Pooley, Jennifer Ratner-Rosenhagen, George Reisch, Joy Rohde, Gabriel Rosenberg, Dorothy Ross, Joan Shelley Rubin, John Rudolph, Ethan Schrum, Adam Shapiro, Jason Smith, Mark Solovey, Thomas Stapleford, Noël Sugimura, Jennifer Sutton, Marga Vicedo, Jessica Wang, Alex Wellerstein, and Audra Wolfe. I also want to take this chance to thank three scholars with whom I have rarely crossed paths but whose writings have influenced my work over the years: Steven Shapin, Theodore Porter, and the late Yaron Ezrahi.

I have been very fortunate to work with Harvard University Press on this project, which Joyce Seltzer, Jeff Dean, and Andrew Kinney shepherded through the various stages of its journey. At Westchester Publishing Services, Ingrid Burke and John Donohue prepared the manuscript for publication. Howard Brick and Ronald Numbers provided expert feedback on the original draft, while Daniel Geary read key portions of the final version.

On the home front, a host of friends and family members have kept me going. Among friends, let me give special thanks to Tim McLucas and Claire Rowberry, Craig Malkin, Prabal Chakrabarty and Vanessa Ruget, Norman and Cassie Fahrney, Kate Sonderegger, Jeremy Wallach, Chris Welbon, Danae Vu, and Eric Volkman. As for family, I could fill another volume with the wisdom and kindnesses of siblings, stepsiblings, aunts, uncles, cousins, nieces, nephews, and in-laws. It seems hopelessly inadequate to thank them en masse, but hope it will suffice to do so, while expressing particular gratitude to Joe and Kay Gaston, John and Lisa Jewett, and Mary and Howie Ditkof.

And then there is Healan Gaston, with whom I have enjoyed daily conversations of unparalleled richness for more than two decades. A formidable historian and an incisive thinker, she has made me and my work better in every way. Together, we have crafted a body of writing that I hope will inspire readers of all kinds to question conventional views of science, religion, and politics, just as the works I encountered as a child opened my horizons. At the same time, we learn constantly from the distinctive perspectives of our own children as well. Samuel Jewett completed our family as this project took shape and has happily joined his brother Joseph in the endless quest to remind us of what really matters. This book is for Healan, Joseph, and Samuel, with great love.

INDEX

Abbey, Edward, 164

Abortion, 2, 207, 209, 211–214, 251

Academic left, 251; cultural understandings of power, 230, 233–235, 239–242, 254; on the Enlightenment, 228, 238, 240–241; vs. essentialism, 239–240; feminists, 237–238, 240; vs. foundationalism, 230, 243; on hegemony, 234–236, 254, 256; on the Holocaust, 241–242; vs. humanism, 240; vs. individualism, 231, 238, 240–242; and Marxism, 230, 232, 234, 236–240; on natural sciences, 235–244, 247; overlap with conservatives, 238, 244, 255; on popular culture, 230, 233–234, 254; postcolonialism, 233–234, 238–239, 241, 249, 254–255, 327n46; postmodernism, 240–245; postsecularism, 254–255; poststructuralism, 238–245, 249–250, 327n46; quasi-universalism, 256–257; on the Scientific Revolution, 237–238; social constructionism, 228, 236–240, 243–244, 326n36; STS scholars, 236–238, 240, 243–244, 250, 254; vs. universalism, 230–234, 238–242, 252, 255. *See also* Radicals

Accommodationism, 86, 117, 252

Adams, James Luther, 116

Adelphi College, 161–162

Adler, Mortimer, 78, 159–160, 162, 274n19

Adorno, Theodor, 77, 182, 186, 195. *See also* Frankfurt School

Advertising, 39; as scientific phenomenon, 30, 40, 58, 86, 98–99, 102–103; scientists in, 99; Watson and, 40

Affirmative action, 217, 233

Agnew, Spiro, 213, 218

Ahmari, Sohrab, 259

Albers, Josef, 160

Allen, Frederick Lewis, 43

Allport, Gordon, 136, 303n3

America, 10, 110, 159, 166, 292n6

American Association for the Advancement of Science (AAAS), 126

American Catholic Philosophical Association, 61

American Civil Liberties Union (ACLU), 44

American Institute of Judaism, 81

American Jewish Committee, 13

American Psychiatric Association, 169

American Psychological Association, 125, 169

American Scholar, 159, 297n24

American Sociological Society, 70

Americans for Democratic Action, 153

Anglo-Catholics, 63–64